NIELS BOHR
COLLECTED WORKS
VOLUME 13

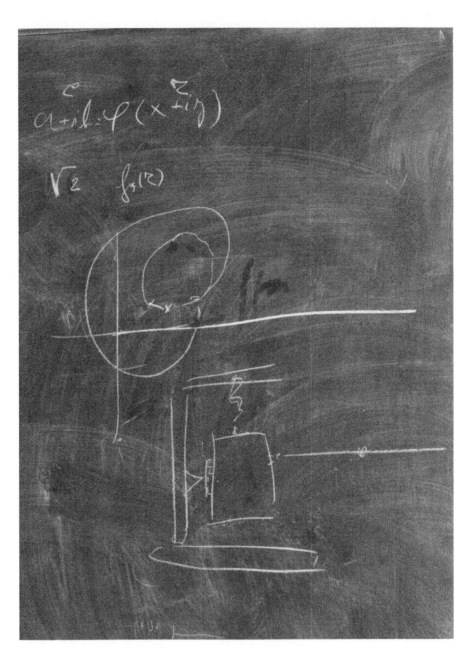

NIELS BOHR'S LAST BLACKBOARD, WITH A SKETCH OF TWO INTERWEAVING PLANES TO ILLUSTRATE THE AMBIGUITY OF LANGUAGE AND UNDERNEATH IT A DIAGRAM OF THE "EINSTEIN BOX".

NIELS BOHR

COLLECTED WORKS

GENERAL EDITOR

FINN AASERUD

THE NIELS BOHR ARCHIVE, COPENHAGEN

VOLUME 13
CUMULATIVE SUBJECT INDEX

EDITED BY

FINN AASERUD

2008

ELSEVIER

AMSTERDAM • BOSTON • HEIDELBERG • LONDON • NEW YORK • OXFORD
PARIS • SAN DIEGO • SAN FRANCISCO • SINGAPORE • SYDNEY • TOKYO

Elsevier
Radarweg 29, PO Box 211, 1000 AE Amsterdam, The Netherlands
Linacre House, Jordan Hill, Oxford OX2 8DP, UK

Special Limited Edition (300 sets): 2008
Copyright © 2008 Elsevier B.V. All rights reserved

Notice
No responsibility is assumed by the publisher for any injury and/or damage to persons
or property as a matter of products liability, negligence or otherwise, or from any use
or operation of any methods, products, instructions or ideas contained in the material
herein. Because of rapid advances in the medical sciences, in particular, independent
verification of diagnoses and drug dosages should be made

Library of Congress Cataloging-in-Publication Data
A catalog record for this book is available from the Library of Congress

British Library Cataloguing in Publication Data
A catalogue record for this book is available from the British Library
8
ISBN: 978-0-444-53286-2 (Limited edition set of 13 volumes)
      978-0-444-53283-1 (Vol. 1)
      978-0-444-53282-4 (Vol. 2)
      978-0-444-53287-9 (Vol. 3)
      978-0-444-53288-6 (Vol. 4)
      978-0-444-53279-4 (Vol. 5)
      978-0-444-53289-3 (Vol. 6)
      978-0-444-53290-9 (Vol. 7)
      978-0-444-53284-8 (Vol. 8)
      978-0-444-53277-0 (Vol. 9)
      978-0-444-53278-7 (Vol. 10)
      978-0-444-53280-0 (Vol. 11)
      978-0-444-53281-7 (Vol. 12)
      978-0-444-53291-6 (Vol. 13)
Set delivered in two boxes: Volumes 1–6 and Volumes 7–13.

For information on all Elsevier publications
visit our website at books.elsevier.com

Transferred to Digital Print 2008

Printed and bound by CPI Antony Rowe, Eastbourne

GENERAL PREFACE

TO THE NEW EDITION

The present volume is unique to this new complete edition of the Niels Bohr Collected Works, the first edition of which consisted only of the first twelve volumes. It seems therefore natural to place the General Preface to the new edition here.

Quantum physics, and more specifically quantum mechanics, may well be argued to have constituted the major scientific revolution in the twentieth century, with immense practical, social and philosophical implications. While the practical applications of quantum mechanics have literally changed our world, its many counterintuitive elements have defied simple understanding and remain a topic of debate among physicists and philosophers alike. Niels Bohr was the leader of the quantum revolution, in more than one sense of the word. As a physicist, he proposed his atomic model in 1913, subsequently perfecting it and showing its immense predictive power. As an entrepreneur, he established his Institute for Theoretical Physics in 1921, making it the Mecca for the younger generation of physicists from all over the world pursuing the implications of the quantum, under the guidance of their leader and teacher, Niels Bohr. As a philosopher and teacher, he was the principal person in formulating the "Copenhagen Interpretation" of quantum mechanics, incorporating the complementarity concept, which to Bohr had implications far beyond physics.

But Bohr's achievements went much further. In physics, he made crucial contributions not least to nuclear physics and the theory of collisions. From 1933 to 1940 he made his institute into a temporary haven for young physicists no longer welcome in Germany for reasons of race or politics. After his escape from Nazi-occupied Denmark in October 1943, he contributed to the development of the atomic bomb in America. At the same time, he pursued his own mission to convince the British Prime Minister Winston Churchill and the American President Franklin D. Roosevelt that they should inform the Soviet Union of the atomic bomb project in order to avoid a nuclear arms race after the war. After the war, Bohr continued his efforts for what

he called an "Open World", as evidenced, for example, in his Open Letter to the United Nations from 1950. While Bohr's orientation was thus genuinely international, he felt great obligations to Denmark, the land that he loved and never considered leaving in spite of many tempting offers from abroad, and in which, especially in the postwar years, he came to hold iconic status.

All of these aspects of Niels Bohr's varied life and work are documented in the Niels Bohr Collected Works, the first and last (Vol. 12) volumes of which were originally published in 1972 and 2006, respectively. During this long interval, many of the volumes have sold out, so that it has so far never been possible at any one time to obtain the entire set. The present complete edition of the Niels Bohr Collected Works has been produced to remedy this situation and to give individuals and libraries the opportunity to acquire all the volumes at once. As an extra bonus, it includes this new index volume (Vol. 13), allowing the reader to find his or her way around in the thousands of pages constituting the series.

The project to publish the Niels Bohr Collected Works was conceived by Bohr's close collaborator Léon Rosenfeld. In Vol. 1 Rosenfeld concluded the General Introduction with the following words:

> "It is hoped that this edition will come to include all Niels Bohr's writings: in the first place his great creative work in atomic and nuclear physics and his no less fundamental contributions to epistemology, which he so anxiously wished to be considered in the same scientific spirit in which they were conceived; but also his occasional writings on public affairs, which illustrate the width of his interests and the generosity and optimism with which he approached all human problems. Bringing together Niels Bohr's writings should not merely provide historians of science with a serviceable tool; it should above all give all those who value the spirit of science easy access to a life-work entirely devoted, with uncommon power and earnestness of purpose, to the rational analysis of the laws of nature and of the singular character of their meaning for us."

After all the years that have elapsed until the publication of the last volume in December 2006, it is remarkable how accurately these words describe the final product.

Nevertheless, it was inevitable that the Niels Bohr Collected Works should acquire its own history during these many years. As already noted, the project was conceived by Léon Rosenfeld (1904–1974), physicist, historian of

science and Bohr's close and longtime collaborator. Upon Rosenfeld's death, another of Bohr's colleagues, Jens Rud Nielsen (1894–1979), temporarily took responsibility for the publication. In 1977, Erik Rüdinger (1934–2007) was assigned Rosenfeld's combined tasks as leader of the Niels Bohr Archive and General Editor of the Niels Bohr Collected Works.

At the centennial of Bohr's birth in 1985, the Niels Bohr Archive, which previously had led an unofficial existence in offices provided by the Niels Bohr Institute, was established formally as an independent institution under the auspices of the Danish Ministry of Education on the basis of a deed of gift from Bohr's widow, Margrethe, who had died the year before. Rüdinger continued in his combined position, in the same quarters, until 1989, when he sought new challenges elsewhere. The position was then offered to me, and I have occupied it since.

The general organization of the material in this series is thematic rather than strictly chronological. This allows for the presentation of each paper (or group of papers) along with other relevant material – drafts, notes, letters and other items. Since themes sometimes overlap, and since Bohr often treated a number of topics in a single paper, it can be difficult to determine the volume in which a given article is to be found. In a few particularly complicated cases, it was even decided to present a paper or part of a paper in two volumes. To help the reader locate where particular publications can be found in the series, Vol. 12 includes a chronological bibliography of Bohr's publications with reference to where they appear in the Collected Works. Since this bibliography, like the general index, pertains to the entire series, it is also reproduced for the convenience of the reader in the present Vol. 13. While the Collected Works is complete with regard to Bohr's publications, both scientific and otherwise, the manuscripts and letters included are the result of a careful selection process. For the reader who wants to look further, an "Inventory of Relevant Manuscripts in the Niels Bohr Archive" is provided in the individual volumes.

With its natural emphasis on his scientific contributions, the Niels Bohr Collected Works document all aspects of Bohr's rich and eventful, yet remarkably unified, life and work. Each of the twelve volumes is introduced by its special editor – a physicist (some of whom knew Bohr personally) or scholar with particular knowledge of the subject in question. The general purpose of these introductions, allowing for the predilections of the individual editors, has been to make historical and conceptual connections between Bohr's writings, while otherwise letting each of these writings speak for itself. In order to help the reader place individual papers in the broader context of Bohr's

life and scientific career, a brief biography of Niels Bohr has been provided at the beginning of Vol. 1. Each volume is illustrated with rare photos and includes explanatory notes and a detailed subject index.

Most writings in German and all in Danish have been translated into English. The task of English translation presented major problems. Bohr's "notoriously difficult use of language", as Rudolf Peierls put it,[1] makes formidable demands on the translator. An attempt has always been made to strike a balance between idiomatic English and the preservation of Bohr's unmistakable personal style, which is an inseparable part of the tone and charm of his writings. Many of Bohr's papers were published in more than one language. Whenever an English version exists, it has been selected for reproduction. While references to other versions are given, the usually infrequent deviations from the English text are noted only when significant. When no English text exists, the paper is reproduced in its original version followed by an English translation. Unpublished texts have been edited and translated with a particular concern for accuracy. The editors have, however, used discretion and common sense in tacitly correcting misspellings and trivial grammatical errors.

It should be pointed out here that the volumes were not always published in the order of their numbering. Thus, as can be seen from the copyright years of the original volumes (included in the overview list of volumes on p. XI), Vol. 2 was published after Vols. 3 and 4, and Vol. 7 after Vols. 8 and 9.

Thanks are due here to Elsevier, represented in particular by Donna de Weerd–Wilson, Remco de Boer and Betsy Lightfoot. Donna's combination of enthusiasm and firmness with regard to getting the present complete edition of the Bohr Collected Works off the ground has been invaluable. When she went on maternity leave during the last preparatory stages, Remco's competent takeover assured a completely smooth transition. It has been a pleasure to resume the close collaboration with Betsy from the last few volumes. Betsy has led the technical work to produce the machine-generated index to the entire series from the indexes to the individual volumes. Christina Olausson has shown her usual competence and conscientiousness in the complex proofreading of the machine-generated index, as have Felicity Pors and Anne Lis Rasmussen, who also in other respects have shown the same dedication in the preparation of this edition as in the work with individual volumes in the original edition. When I was taken ill not long into the

---

[1] See the Introduction to Volume 9.

indexing process, Felicity willingly and efficiently took over the leadership of the work. For other acknowledgements pertaining to the Niels Bohr Collected Works I refer to the individual volumes.

Niels Bohr was a painstaking writer – both in his published works and in his private correspondence – who made every effort to ensure that his words accurately reflected his thoughts. In this sense, the present complete edition of the Niels Bohr Collected Works can facilitate a broader understanding of Bohr's accomplishments and a deeper appreciation of the intellectual developments which made them possible.

Finn Aaserud
The Niels Bohr Archive
April 2008

OVERVIEW OF VOLUMES 1–12 OF THE
NIELS BOHR COLLECTED WORKS

In addition to the Editor of each individual volume (see below), a General Editor has overseen the preparation of the entire series of the *Niels Bohr Collected Works*. The General Editors have been: Léon Rosenfeld (1904–1974) (Volumes 1 to 3), Erik Rüdinger (Volumes 5 to 9, Volume 7 jointly with Finn Aaserud) and Finn Aaserud (Volumes 10 to 12). All volumes are published by North-Holland/Elsevier.

Vol. 1, *Early Work (1905–1911)* (ed. J. Rud Nielsen), 1972.

Vol. 2, *Work on Atomic Physics (1912–1917)* (ed. Ulrich Hoyer), 1981.

Vol. 3, *The Correspondence Principle (1918–1923)* (ed. J. Rud Nielsen), 1976.

Vol. 4, *The Periodic System (1920–1923)* (ed. J. Rud Nielsen), 1977.

Vol. 5, *The Emergence of Quantum Mechanics (mainly 1924–1926)* (ed. Klaus Stolzenburg), 1984.

Vol. 6, *Foundations of Quantum Physics I (1926–1932)* (ed. Jørgen Kalckar), 1985.

Vol. 7, *Foundations of Quantum Physics II (1933–1958)* (ed. Jørgen Kalckar), 1996.

Vol. 8, *The Penetration of Charged Particles Through Matter (1912–1954)* (ed. Jens Thorsen), 1987.

Vol. 9, *Nuclear Physics (1929–1952)* (ed. Rudolf Peierls), 1986.

Vol. 10, *Complementarity Beyond Physics (1928–1962)* (ed. David Favrholdt), 1999.

Vol. 11, *The Political Arena (1934–1961)* (ed. Finn Aaserud), 2005.

Vol. 12, *Popularization and People (1911–1962)* (ed. Finn Aaserud), 2007.

CONTENTS

General Preface to the New Edition   V

Overview of Volumes 1–12   XI

Contents of Volume 13   XIII

Bibliography of Niels Bohr's Publications   1
   Reproduced in the Collected Works

Cumulative Index for the Collected Works   35

BIBLIOGRAPHY OF NIELS BOHR'S PUBLICATIONS REPRODUCED IN THE COLLECTED WORKS

INTRODUCTION

The contents of the Niels Bohr Collected Works have been organized in part by chronology, in part by topic, and it may not be obvious where a particular item can be found. In this last volume, it may therefore be helpful for the reader to have available a bibliography of material reproduced in the entire Collected Works. Because the series claims completeness only with regard to Bohr's publications, and because the several unpublished manuscripts and letters are included mainly to illustrate these, it has been deemed appropriate to limit the bibliography to publications proper, understood in a wide sense, so that Bohr's master's and doctor's theses at the University of Copenhagen as well as abstracts of lectures held at the Royal Academy of Sciences and Letters, published in the Academy's Proceedings, are included.

The list on the following pages is restricted to the versions of Bohr's publications reproduced in the Collected Works; for any other version the reader is referred to the information in the relevant volume itself. The number of the volume in which a publication is reproduced, as well as the page numbers it occupies there, are provided on the right-hand side of the page. The list is organized chronologically according to publication date, although there may be exceptions to this rule within each year.

(1) *Determination of the Surface-Tension of Water by the Method of Jet Vibration*, Phil. Trans. Roy. Soc. **209** (1909) 281–317   **1**, 25–65

(2) *An Account of the Application of the Electron Theory to Explain the Physical Properties of Metals*; *En Fremstilling af Elektrontheoriens Anvendelse til Forklaring af Metallernes fysiske Egenskaber*, Master's thesis, 1909
    Translation   **1**, 131–161

(3) *On the Determination of the Tension of a recently formed Water-Surface*, Proc. Roy. Soc. **A84** (1910) 395–403   **1**, 79–89

(4) *Studies on the Electron Theory of Metals*; *Studier over Metallernes Elektronteori*, Dissertation for the Degree of Doctor of Philosophy, V. Thaning & Appel, Copenhagen 1911   **1**, 163–290
    Translation   **1**, 291–395

(5) [Autobiography], "Festskrift udgivet af Kjøbenhavns Universitet i Anledning af Universitetets Aarsfest, November 1911", Schultz, Copenhagen 1911, pp. 76–77   **12**, 133–135
    Translation   **12**, 136

(6) *Note on the Electron Theory of Thermoelectric Phenomena*, Phil. Mag. **23** (1912) 984–988   **1**, 439–444

(7) *On the Theory of the Decrease of Velocity of Moving Electrified Particles on Passing through Matter*, Phil. Mag. **25** (1913) 10–31   **2**, 15–39; **8**, 47–71

(8) *On the Constitution of Atoms and Molecules*, Part I, Phil. Mag. **26** (1913), 1–25   **2**, 159–185

(9) *On the Constitution of Atoms and Molecules*, Part II: *Systems containing only a Single Nucleus*, Phil. Mag. **26** (1913) 476–502   **2**, 187–214

(10) *On the Constitution of Atoms and Molecules*, Part III: *Systems Containing Several Nuclei*, Phil. Mag. **26** (1913) 857–875   **2**, 215–233

(11) *The Spectra of Helium and Hydrogen*, Nature **92** (1913) 231–233   **2**, 273–276

(12) *On the Spectrum of Hydrogen*; *Om Brintspektret*, Fys. Tidsskr. **12** (1914) 97–114

English version in "The Theory of Spectra and Atomic Constitution", Cambridge University Press, 1922, pp. 1–19   **2**, 281–301

(13) *Atomic Models and X-Ray Spectra*, Nature **92** (1914) 553–554   **2**, 304

(14) *On the Effect of Electric and Magnetic Fields on Spectral Lines*, Phil. Mag. **27** (1914) 506–524   **2**, 347–368

(15) *On the Series Spectrum of Hydrogen and the Structure of the Atom*, Phil. Mag. **29** (1915) 332–335   **2**, 375–380

(16) *The Spectra of Hydrogen and Helium*, Nature **95** (1915) 6–7   **2**, 383–388

(17) *Modern Electrical Theory* [book review], Nature **95** (1915) 420–421   **12**, 19–20

(18) *On the Quantum Theory of Radiation and the Structure of the Atom*, Phil. Mag. **30** (1915) 394–415   **2**, 389–413

(19) *On the Decrease of Velocity of Swiftly Moving Electrified Particles in Passing Through Matter*, Phil. Mag. **30** (1915) 581–612   **2**, 57–90; **8**, 127–160

(20) *Henry Gwyn Jeffreys Moseley*, Phil. Mag. **31** (1916) 173–176   **12**, 283–288

(21) *On the Model of a Triatomic Hydrogen Molecule*, Medd. Kgl. Vet. Akad., Nobel Inst. **5**, No. 28 (1919)   **2**, 471–488

(22) *On the Quantum Theory of Line-Spectra*, Part I: *On the general theory*, Kgl. Dan. Vid. Selsk. Skr., 8. Række, IV.1 (1918–1922), pp. 5–36   **3**, 65–102

(23) *On the Quantum Theory of Line-Spectra*, Part II: *On the hydrogen spectrum*, Kgl. Dan. Vid. Selsk. Skr., 8. Række, IV.1 (1918–1922), pp. 37–100   **3**, 103–166

(24) *On the Quantum Theory of Line-Spectra*, Part III: *On the spectra of elements of higher atomic number*, Kgl. Dan. Vid. Selsk. Skr., 8. Række, IV.1 (1918–1922), pp. 101–111   **3**, 167–184

(25) *Professor Sir Ernest Rutherford and his Significance for the Recent Development of Physics*; *Professor Sir Ernest Rutherford og hans Betydning for Fysikens nyere Udvikling*, Politiken, 18 September 1920   **12**, 253–257
Translation   **12**, 258–261

(26) *Our Present Knowledge of Atoms*; *Unsere heutige Kenntnis vom Atom*, Die Umschau **25** (1921) 229–232
Translation   **4**, 84–89

(27) *On the Question of the Polarization of Radiation in the Quantum Theory*; *Zur Frage der Polarisation der Strahlung in der Quanten-theorie* Z. Phys. **6** (1921) 1–9   **3**, 339–349
Translation   **3**, 350–356

(28) *Geleitwort*, in N. Bohr, "Abhandlungen über Atombau aus den Jahren 1913–1916", Vieweg & Sohn, Brauschweig 1921   **3**, 303–324
Translation   **3**, 325–337

(29) *Atomic Structure*, Nature **107** (1921) 104–107   **4**, 71–82

(30) *Atomic Structure*, Nature **108** (1921) 208–209   **4**, 175–180

(31) *On the Result of Collisions between Atomic Systems and Free Electrical Particles*; *Om Virkningen af Sammenstød mellem Atomsystemer og fri elektriske Partikler*, Det nordiske H.C. Ørsted Møde i København 1920, H.C. Ørsted Komiteen, Copenhagen 1921, pp. 120–121   **8**, 195–198

Translation   **8**, 199–200

(32) *On the Explanation of the Periodic System*; *Om Forklaringen af det Periodiske System*; "Autoreferat av föredrag vid Andra Nordiska Fysikermötet i Upsala, den 24.–26. augusti", Edv. Berlings Boktryckeri A.B., Uppsala 1922, pp. 3–4; also Fys. Tidsskr. **20** (1922) 112–115

Translation   **4**, 421–424

(33) *On the Series Spectra of the Elements*, "The Theory of Spectra and Atomic Constitution", Cambridge University Press 1922, pp. 20–60   **3**, 241–282

(34) *The Structure of the Atom and the Physical and Chemical Properties of the Elements*; *Atomernes Bygning og Stoffernes fysiske og kemiske Egenskaber*, Jul. Gjellerup, Copenhagen 1922   **4**, 181–256

English version in "The Theory of Spectra and Atomic Constitution", Cambridge University Press, Cambridge 1924, iii–vii, x, 61–138.   **4**, 257–340
The English version includes an additional appendix. The pages in BCW also include the (untranslated) preface to the Danish book as well as *Preface* (May 1922) and *Preface to Second Edition* (May 1924) of the English book

(35) *The Difference between Series Spectra of Isotopes*, Nature **109** (1922) 745   **3**, 453–454

(36) *The Seventh Guthrie Lecture, Physical Society of London: The Effect of Electric and Magnetic Fields on Spectral Lines*, Fleetway Press, London 1922, pp. 275–302   **3**, 415–446

(37) *On the Selection Principle of the Quantum Theory*,   **3**, 447–452
Phil. Mag. **43** (1922) 1112–1116

(38) *On the application of the quantum theory to atomic
problems*; *L'application de la théorie des quanta aux
problèmes atomiques*, "Atomes et électrons", Rapports et discussions du Conseil de physique tenu à
Bruxelles du 1er au 6 avril 1921, Gauthier-Villars et
Cie, Paris 1923, pp. 228–247
Translation   **3**, 357–380

(39) [Speech at the banquet], "Les Prix Nobel en 1921–
1922", P.A. Norstedt, Stockholm 1923, pp. 102–104
Translation   **4**, 26–27

(40) *Niels Bohr, ibid.*, pp. 126–127   **12**, 143–145

(41) *The Structure of the Atom*; *Om Atomernes Bygning*,   **4**, 425–465
ibid., 1–37 (paginated independently)
English version in Nature **112** (1923) 29–44   **4**, 467–482

(42) *X-Ray Spectra and the Periodic System of the Ele-*   **4**, 483–518
ments; *Röntgenspektren und periodisches System der
Elemente* (with D. Coster), Z. Phys. **12** (1923) 342–
374
Translation   **4**, 519–548

(43) *Line Spectra and Atomic Structure*; *Linienspektren*   **4**, 549–610
und Atombau, Ann. d. Phys. **71** (1923) 228–288
Translation   **4**, 611–656

(44) *The Correspondence Principle*, Brit. Ass. Adv. Sci.,   **3**, 575–577
Report of the Annual Meeting, Liverpool, September
1923, London 1924, pp. 428–429

(45) *The Spectra of the Lighter Elements*, Nature **113**
(1924) 223–224
Contribution at the British Association Meeting at
Liverpool, September 1923
Transcription (from a slightly shorter manuscript)   **3**, 578–579

[8]

(46) *On the Application of the Quantum Theory to Atomic Structure*, Part I: *The Fundamental Postulates of the Quantum Theory*, Proc. Cambr. Phil. Soc. (Suppl.) 1924   **3**, 455–499

(47) *The Quantum Theory of Radiation*; *Über die Quantentheorie der Strahlung* (with H.A. Kramers and J.C. Slater) Z. Phys. **24** (1924) 69 (abstract only)   **5**, 97–98

     Translation   **5**, 98

(48) *The Quantum Theory of Radiation* (with H.A. Kramers and J.C. Slater), Phil. Mag. **47** (1924) 785–802   **5**, 99–118

(49) *The Foundations of Modern Atomic Research*; *Grundlaget for den Moderne Atomforskning*, Fys. Tidsskr. **23** (1925) 10–17   **5**, 125–135

     Translation   **5**, 136–142

(50) *On the Polarization of Fluorescent Light*; *Zur Polarisation des Fluorescenzlichtes*, Naturwiss. **12** (1924) 1115–1117   **5**, 143–147

     Translation   **5**, 148–154

(51) *On the Law of Conservation of Energy*; *Om Energisætningen*, Overs. Dan. Vidensk. Selsk. Forh. Juni 1924 – Maj 1925, p. 32   **5**, 173–174

     English version in Nature **116** (1925) 262   **5**, 174

(52) *On the Behaviour of Atoms in Collisions*; *Über die Wirkung von Atomen bei Stössen*, Z. Phys. **34** (1925) 142–157   **5**, 175–193

     Translation   **5**, 194–206

(53) *Atomic Theory and Mechanics*; *Atomteori og mekanik*, Mat. Tidsskr. B (1925) 104–107   **5**, 241–245

     Translation   **5**, 246–248

(54) *Atomic Theory and Mechanics*, Nature **116** (1925) 845–852   **5**, 269–280

[9]

(55) *Some Aspects of the Later Development of Atomic Theory*; *Nogle Træk fra Atomteoriens senere Udvikling*, Fys. Tidsskr. **24** (1926) 20–21   **5**, 281–283
Report by H.A. Kramers on Bohr's lecture at the Third Nordic Physicists' Meeting, Oslo, 24 August 1925
   Translation   **5**, 284–286

(56) *Sir Ernest Rutherford, O.M., P.R.S.*, Nature (Suppl.) **118** (18 Dec 1926) 51–52   **12**, 263–266

(57) *Atom*, Encyclopædia Britannica, 13th edition, Suppl., Vol. 1, London and New York 1926, pp. 262–267   **4**, 657–663

(58) *Spinning Electrons and the Structure of Spectra* (Letter to the Editor), Nature **117** (1926) 264–265   **5**, 287–289

(59) *Sir J.J. Thomson's Seventieth Birthday*, Nature **118** (1926) 879   **12**, 251–252

(60) *Atomic Theory and Wave Mechanics* (Abstract); *Atomteori og Bølgemekanik*, Overs. Dan. Vidensk. Selsk. Forh. Juni 1926 – Maj 1927, pp. 28–29   **6**, 55–56
English version in Nature **119** (1927) 262   **6**, 56

(61) *The Quantum Postulate and the Recent Development of Atomic Theory*; *Kvantepostulatet og Atomteoriens seneste Udvikling*, Overs. Dan. Vidensk. Selsk. Forh. Juni 1927 – Maj 1928, p. 27   **6**, 107–108
English version in Nature **121** (1928) 78   **6**, 108

(62) *The Quantum Postulate and the Recent Development of Atomic Theory*, Atti del Congresso Internazionale dei Fisici 11–20 Settembre 1927, Como–Pavia–Roma, Volume Secundo, Nicola Zanichelli, Bologna 1928, pp. 565–598   **6**, 109–146
Includes discussion remarks on Bohr's contribution (pp. 589–598)

[10]

(63) *The Quantum Postulate and the Recent Development*   **6**, 147–158
      of Atomic Theory, Nature (Suppl.) **121** (1928) 580–
      590

(64) *Discrepancies between Experiment and the Electro-*   **5**, 207–210
      magnetic Theory of Radiation; *Discordances entre*
      l'expérience et la théorie électromagnetique du ray-
      onnement, "Électrons et photons", Rapports et dis-
      cussions du cinquième Conseil de physique tenu
      à Bruxelles du 24 au 29 Octobre 1927, Gauthier-
      Villars, Paris 1928, pp. 91–92
      Discussion of A.H. Compton's contribution
      Translation   **5**, 211–212

(65) *General Discussion at the Fifth Solvay Conference*,
      ibid., pp. 253–256, 261–263, 264–265 (and unpub-
      lished manuscript)
      Translation   **6**, 99–106

(66) *Sommerfeld and the Theory of the Atom*; *Sommerfeld*   **12**, 343–344
      und die Atomtheorie, Naturwiss. **16** (1928) 1036
      Translation   **12**, 345–346

(67) *At Harald Høffding's 85th Birthday*; *Ved Harald*   **10**, 305–307
      Høffdings 85 Aars-Dag, Berlingske Tidende, 10
      March 1928
      Translation   **10**, 308–309

(68) [Autobiography], "Studenterne MCMIII: Personal-   **12**, 147–149
      historiske Oplysninger", Berlingske Bogtrykkeri,
      Copenhagen 1928, p. 275
      Translation   **12**, 150

(69) *Quantum Theory and Relativity* (Abstract); *Kvante-*   **6**, 199–200
      teori og Relativitet, Overs. Dan. Vidensk. Selsk.
      Forh. Juni 1928 – Maj 1929, p. 24
      English version in Nature **123** (1929) 434   **6**, 200

(70) *Professor Niels Bjerrum 50 Years*; *Professor Niels Bjerrum fylder 50 Aar*, Berlingske Tidende, 9 March 1929   **12**, 437–439

      Translation   **12**, 440–441

(71) *The Quantum of Action and the Description of Nature*; *Wirkungsquantum und Naturbeschreibung*, Naturwiss. **17** (1929) 483–486   **6**, 201–206

      English version in "Atomic Theory and the Description of Nature", Cambridge University Press, Cambridge 1934, pp. 92–101   **6**, 208–217

(72) *The Atomic Theory and the Fundamental Principles underlying the Description of Nature*; *Atomteorien og Grundprincipperne for Naturbeskrivelsen*, Beretning om det 18. skandinaviske Naturforskermøde i København 26.–31. August 1929, Frederiksberg Bogtrykkeri, Copenhagen 1929, pp. 71–83   **6**, 219–235

      English version in "Atomic Theory and the Description of Nature", Cambridge University Press, Cambridge 1934, pp. 102–119   **6**, 236–253

(73) *Introductory Survey* (with *Addendum* of 1931); *Indledende oversigt* (med *Tillæg* fra 1931), "Atomteori og Naturbeskrivelse", Festskrift udgivet af Københavns Universitet i Anledning af Universitetets Aarsfest 1929, Bianco Lunos Bogtrykkeri, Copenhagen 1929, pp. 5–19, and J.H. Schultz Forlag, Copenhagen 1958, pp. 23–25   **6**, 255–276

      English version in "Atomic Theory and the Description of Nature", Cambridge University Press, 1934, pp. 1–24 (includes additional unpaginated preface)   **6**, 277–302

(74) *Atom*, Encyclopædia Britannica, 14th edition, Vol. 2, London and New York 1929, pp. 642–648   **12**, 41–48

(75) *Maxwell and Modern Theoretical Physics*, Nature **128** (1931) 691–692   **6**, 357–360

(76) *The Magnetic Electron*; *L'électron magnetique*, "Le magnétisme", Rapports et discussions du sixième Conseil de physique tenu à Bruxelles du 20 au 25 octobre 1930, Gauthier-Villars, Paris 1932, pp. 276–280
Discussion of Wolfgang Pauli's lecture
Translation   **6**, 347–349

(77) *Philosophical Aspects of Atomic Theory* (Abstract),   **6**, 351–352
Nature **125** (1930) 958

(78) *The Use of the Concepts of Space and Time in*   **6**, 353–354
Atomic Theory (Abstract); *Om Benyttelsen af Begreb-*
erne Rum og Tid i Atomteorien, Overs. Dan. Vidensk.
Selsk. Forh. Juni 1930 – Maj 1931, p. 26
English version in Nature **127** (1931) 43   **6**, 354

(79) *Tribute to the Memory of Harald Høffding*; *Mindeord*   **10**, 311–318
over Harald Høffding, Overs. Dan. Vidensk. Selsk.
Virks. Juni 1931 – Maj 1932, pp. 131–136
Translation   **10**, 319–322

(80) *On Atomic Stability* (Abstract), Brit. Ass. Adv. Sci.,   **6**, 355–356
Report of the Centenary Meeting, London, 23–30
September 1931, London 1932, p. 333

(81) *Faraday Lecture: Chemistry and Quantum Theory of*   **6**, 371–408;
Atomic Constitution, J. Chem. Soc. (1932) 349–384   **9**, 91–97
Vol. 9 contains extract only

(82) *Atomic Stability and Conservation Laws*, Atti del   **9**, 99–114
Convegno di Fisica Nucleare della "Fondazione
Alessandro Volta", Ottobre 1931, Reale Accademia
d'Italia, Rome 1932, pp. 119–130

(83) *On the Properties of the Neutron* (Abstract); *Om*   **9**, 119–121
Neutronernes Egenskaber, Overs. Dan. Vidensk.
Selsk. Virks. Juni 1931 – Maj 1932, p. 52
Translation   **9**, 121

[13]

(84) *Light and Life*, Congress on Light Therapy in Copen-   **10**, 27–35
     hagen, 15 August 1932, Nature **131** (1933) 421–423,
     457–459

(85) *The Limited Measurability of Electromagnetic Fields*   **7**, 53–54
     of Force (Abstract); *Om den begrænsede Maalelighed*
     af elektromagtiske Kraftfelter (with L. Rosenfeld),
     Overs. Dan. Vidensk. Selsk. Virks. Juni 1932 – Maj
     1933, p. 35

     English version in Nature **132** (1933) 75   **7**, 54

(86) *On the Question of the Measurability of Electromag-*   **7**, 55–121
     netic Field Quantities; *Zur Frage der Messbarkeit*
     der elektromagnetischen Feldgrössen (with L. Rosen-
     feld), Mat.–Fys. Medd. Dan. Vidensk. Selsk. **12**,
     No. 8 (1933)

     English version in "Selected Papers of Léon   **7**, 123–166
     Rosenfeld" (eds. R.S. Cohen and J.J. Stachel),
     D. Reidel Publishing Company, Dordrecht 1979,
     pp. 357–400

(87) *On the Correspondence Method in Electron Theory*;   **7**, 167–182
     Sur la méthode de correspondance dans la théorie de
     l'électron, "Structure et propriétés des noyaux atom-
     iques", Rapports et discussions du septième Conseil
     de physique tenu à Bruxelles du 22 au 29 octobre
     1933, Gauthier-Villars, Paris 1934, pp. 216–228

        Translation   **7**, 183–191
        Translation of pp. 226–228   **9**, 129–132

(88) *Discussion Remarks*, *ibid.*, pp. 72, 175, 180, 214–215   **7**, 192–193;
     (in Vol. 7), 287–288, 327–328, 329–330, 331, 334   **9**, 133–141

        Translation   **7**, 193;
                                                           9, 135–141

(89) *Obituary for Christian Alfred Bohr: Born 25 Novem-*   **12**, 407–419
     ber 1916 – Died 2 July 1934; *Mindeord over Chris-*
     tian Alfred Bohr: født 25. November 1916 – død 2.
     Juli 1934, Private print, 1934

        Translation   **12**, 420–424

[14]

(90) *Zeeman Effect and Theory of Atomic Constitution*,   **10**, 335–340
"Zeeman Verhandelingen", Martinus Nijhoff, The
Hague 1935, pp. 131–134

(91) *Conversation with Niels Bohr*; *Samtale med Niels*   **12**, 151–162
Bohr, Berlingske Aftenavis, 2 October 1935
     Translation   **12**, 163–175

(92) *Quantum Mechanics and Physical Reality*, Nature   **7**, 289–290
136 (1935) 65

(93) *Friedrich Paschen on his Seventieth Birthday*;   **12**, 339–340
Friedrich Paschen zum siebzigsten Geburtstag,
Naturwiss. **23** (1935) 73
     Translation   **12**, 341–342

(94) *Can Quantum-Mechanical Description of Physical*   **7**, 291–298
Reality be Considered Complete?, Phys. Rev. **48**
(1935) 696–702

(95) *Properties and Constitution of Atomic Nuclei* (Ab-   **9**, 149–150
stract); *Om Atomkernernes Egenskaber og Opbyg-*
ning, Overs. Dan. Vidensk. Selsk. Virks. Juni 1935
– Maj 1936, p. 39
     English version in Nature **138** (1936) 695   **9**, 150

(96) *Neutron Capture and Nuclear Constitution (1)*,   **9**, 151–156
Nature **137** (1936) 344–348

(97) *Neutron Capture and Nuclear Constitution (2)*,   **9**, 157–158
Nature **137** (1936) 351

(98) *Conservation Laws in Quantum Theory*, Nature **138**   **5**, 215–216
(1936) 25–26

(99) *Properties of Atomic Nuclei*; *Atomkernernes Egen-*   **9**, 159–171
skaber, "Nordiska (19. skandinaviska) Naturforskar-
mötet i Helsingfors den 11.–15. augusti 1936",
Helsinki 1936, pp. 73–81
     Translation   **9**, 172–178

(100) [*Speech at the Meeting of Natural Scientists in Helsinki, 14 August 1936*]; [*Tale ved naturforskermødet i Helsingfors, 14. August 1936*], ibid., pp. 191–192   **11**,501–504

    Translation   **11**, 505–507

(101) *Causality and Complementarity*, Phil. Sci. **4** (1937) 289–298   **10**, 37–48

(102) *Transmutations of Atomic Nuclei*, Science **86** (1937) 161–165   **9**, 205–211

(103) *On the Transmutation of Atomic Nuclei*; *Om Spaltning af Atomkerner*, 5. nordiske Elektroteknikermøde, J.H. Schultz Bogtrykkeri, Copenhagen 1937, pp. 21–23   **9**, 213–217

    Translation   **9**, 218–221

(104) *On the Transmutation of Atomic Nuclei by Impact of Material Particles*, Part I: *General Theoretical Remarks* (with F. Kalckar), Mat.–Fys. Medd. Dan. Vidensk. Selsk. **14**, No. 10 (1937)   **9**, 223–264

(105) [Obituary for Rutherford], Nature **140** (1937) 752–753   **12**, 271–272

(106) [Obituary for Rutherford], Nature (Suppl.) **140** (1937) 1048–1049   **12**, 273–274

(107) *Nuclear Mechanics*; *Mécanique nucléaire*, "Actualités scientifiques et industrielles: Réunion internationale de physique–chimie–biologie, Congrès du Palais de la découverte, Paris, Octobre 1937, Vol. II: Physique nucléaire", Hermann et Cie, Paris 1938, pp. 81–82   **9**, 265–268

    Translation   **9**, 269

(108) *On Nuclear Reactions* (Abstract); *Om Atomkernereaktioner*, Overs. Dan. Vidensk. Selsk. Virks. Juni 1937 – Maj 1938, p. 32   **9**, 287–289

    Translation   **9**, 289

(109) *Magister Fritz Kalckar*, Politiken, 7 January 1938 **12**, 385–386

Translation **12**, 387–388

(110) *Nuclear Photo-effects*, Nature **141** (1938) 326–327 **9**, 297–299

(111) *Quantum of Action and Atomic Nucleus*; **9**, 301–317
Wirkungsquantum und Atomkern, Ann. d. Phys.
32 (1938) 5–19

Translation **9**, 318–329

(112) *Biology and Atomic Physics*, "Celebrazione del sec- **10**, 49–62
ondo centenario della nascita di Luigi Galvani",
Bologna – 18–21 ottobre 1937-XV: I. Rendiconto
generale, Tipografia Luigi Parma 1938, pp. 68–78

(113) *Resonance in Nuclear Photo-Effects*, Nature **141** **9**, 331–332
(1938) 1096–1097

(114) *Analysis and Synthesis in Science*, International En- **10**, 63–64
cyclopedia of Unified Science **1** (1938) 28

(115) *Science and its International Significance*, Danish **11**, 527–531
Foreign Office Journal, No. 208 (May 1938) 61–63

(116) *Symposium on Nuclear Physics, Introduction*, Brit. **9**, 333–335
Ass. Adv. Sci., Report of the Annual Meeting, 1938
(108th Year), Cambridge, August 17–24, London
1938, p. 381 (Abstract)

Nature (Suppl.) **142** (1938) 520–521 (Report) **9**, 336–337

(117) *Matter, Structure of*, Encyclopædia Britannica Book **12**, 49–52
of the Year 1938, pp. 403–404

(118) *The Causality Problem in Atomic Physics*, "New **7**, 299–322
Theories in Physics", Conference organized in collaboration with the International Union of Physics
and the Polish Intellectual Co-operation Committee,
Warsaw, May 30th – June 3rd 1938, International
Institute of Intellectual Co-operation, Paris 1939,
pp. 11–30

(119) [Two speeches for Rutherford, 1932], A.S. Eve,   **12**, 267–269
"Rutherford: Being the Life and Letters of the Rt
Hon. Lord Rutherford, O.M.", Cambridge University
Press, Cambridge 1939, pp. 361–363

(120) *Natural Philosophy and Human Cultures*,   **10**, 237–249
Congrès internal des sciences anthropologiques
et ethnologiques, compte rendu de la deuxième
session, Copenhagen 1938, Ejnar Munksgaard,
Copenhagen 1939, pp. 86–95

(121) *Reactions of Atomic Nuclei* (Abstract); *Om Atom-*   **9**, 339–340
kernernes Reaktioner, Overs. Dan. Vidensk. Selsk.
Virks. Juni 1938 – Maj 1939, p. 25

English version in Nature **143** (1939) 215   **9**, 340

(122) *Disintegration of Heavy Nuclei*, Nature **143** (1939)   **9**, 341–342
330

(123) *Resonance in Uranium and Thorium Disintegration*   **9**, 343–345
and the Phenomenon of Nuclear Fission, Phys. Rev.
55 (1939) 418–419

(124) *Mechanism of Nuclear Fission* (with J.A. Wheeler),   **9**, 359–361
Phys. Rev. **55** (1939) 1124

(125) *The Mechanism of Nuclear Fission* (with J.A.   **9**, 363–389
Wheeler), Phys. Rev. **56** (1939) 426–450

(126) *Nuclear Reactions in the Continuous Energy Region*   **9**, 391–393
(with R. Peierls and G. Placzek), Nature **144** (1939)
200–201

(127) *Matter, Structure of*, Encyclopædia Britannica Book   **12**, 53–54
of the Year 1939, pp. 409–410

(128) *The Fission of Protactinium* (with J.A. Wheeler),   **9**, 403–404
Phys. Rev. **56** (1939) 1065–1066

(129) *Writer and Scientist*; *Digter og Videnskabsmand*,   **12**, 461–463
"Festskrift til Niels Møller paa Firsaarsdagen 11.
December 1939", Munksgaard, Copenhagen 1939,
pp. 80–81
    Translation   **12**, 464–465

(130) *On the Fragments Ejected in the Disintegration of the*   **8**, 317–318
Uranium Nucleus; *Om de ved Urankernernes Søn-*
derdeling udslyngede Fragmenter, Overs. Dan. Vidensk. Selsk. Virks. Juni 1939 – Maj 1940, pp. 49–50
    Translation   **8**, 318

(131) *Meeting on 20 October 1939*; *Mødet den 20. Oktober*   **11**, 379–381
1939, Overs. Dan. Vidensk. Selsk. Virks. Juni 1939
– Maj 1940, pp. 25–26
    Translation   **11**, 382–383

(132) *The Theoretical Explanation of the Fission of Atomic*   **9**, 409–410
Nuclei (Abstract); *Den teoretiske Forklaring af Atom-*
kernernes Fission, Overs. Dan. Vidensk. Selsk. Virks.
Juni 1939 – Maj 1940, p. 28
    Translation   **9**, 410

(133) *Foreword*, C. Møller and E. Rasmussen, "The World"   **12**, 73–75
and the Atom," Allen & Unwin, London 1940, p. 9

(134) *Scattering and Stopping of Fission Fragments*, Phys.   **8**, 319–321
Rev. **58** (1940) 654–655

(135) *Successive Transformations in Nuclear Fission*, Phys.   **9**, 475–479
Rev. **58** (1940) 864–866

(136) *Velocity-Range Relation for Fission Fragments* (with   **8**, 323–326
J.K. Bøggild, K.J. Brostrøm and T. Lauritsen), Phys.
Rev. **58** (1940) 839–840

(137) *Meeting on 15 March [1940]*; *Mødet den 15. Marts*   **11**, 385–386
[1940], Overs. Dan. Vidensk. Selsk. Virks. Juni 1939
– Maj 1940, pp. 40–41
    Translation   **11**, 387

[19]

(138) *Meeting on 20 September 1940*; *Mødet den 20.*   **11**, 389–390
Septbr. 1940, Overs. Dan. Vidensk. Selsk. Virks. Juni
1940 – Maj 1941, pp. 25–26

    Translation   **11**, 391

(139) *Disintegration of Heavy Nuclei* (Abstract); *Tunge*   **9**, 481–482
Atomkerners Sønderdeling, Overs. Dan. Vidensk.
Selsk. Virks. Juni 1940 – Maj 1941, p. 38

    Translation   **9**, 482

(140) *Danish Culture. Some Introductory Reflections*;   **10**, 251–261
Dansk Kultur. Nogle indledende Betragtninger,
"Danmarks Kultur ved Aar 1940", Det Danske
Forlag, Copenhagen 1941–1943, Vol. 1, pp. 9–17

    Translation   **10**, 262–272

(141) *Recent Investigations of the Transmutations of Atomic*   **9**, 411–442
Nuclei; *Nyere Undersøgelser over Atomkernernes
Omdannelser*, Fys. Tidsskr. **39** (1941) 3–32

    Translation   **9**, 443–466

(142) *Kirstine Meyer, n. Bjerrum: 12 October 1861 – 28*   **12**, 425–429
September 1941; *Kirstine Meyer, f. Bjerrum: 12. Ok-
tober 1861 – 28. September 1941*, ibid., 113–115

    Translation   **12**, 430–431

(143) *Eighth Presentation of the H.C. Ørsted Medal*   **11**, 469–472
[K. Linderstrøm-Lang]; *Ottende Uddeling af H.C.
Ørsted Medaillen*, Fys. Tidsskr. **39** (1941) 175–177,
192–193

    Translation   **11**, 473–475

(144) *Velocity–Range Relation for Fission Fragments*, Phys.   **8**, 327–333
Rev. **59** (1941) 270–275

(145) *Professor Martin Knudsen*, Berlingske Aftenavis, 14   **12**, 289–291
February 1941

    Translation   **12**, 292–294

(146) *The University and Research*; *Universitetet og Forsk-*   **11**, 533–546
ningen, Politiken, 3 June 1941
    Translation   **11**, 547–552

(147) *Farewell to Sweden's Ambassador in Copenhagen*;   **12**, 481–482
Afsked med Sveriges Gesandt i København, Politiken,
15 November 1941
    Translation   **12**, 483

(148) *Mechanism of Deuteron-Induced Fission*, Phys. Rev.   **9**, 483–484
59 (1941) 1042

(149) *Analysis and Synthesis in Atomic Physics* (Abstract);   **7**, 323–324
Analyse og Syntese indenfor Atomfysikken, Overs.
Dan. Vidensk. Selsk. Virks. Juni 1941 – Maj 1942,
p. 30
    Translation   **7**, 324

(150) *Meeting on 30 January 1942*; *Mødet den 30. Januar*   **11**, 392–394
1942, Overs. Dan. Vidensk. Selsk. Virks. Juni 1941
– Maj 1942, pp. 32–34
    Translation   **11**, 395–396

(151) [Tribute to Bering], "Vitus Bering 1741–1941",   **12**, 237–243
H. Hagerup, Copenhagen 1942, pp. 49–53
    Translation   **12**, 244–246

(152) *Memorial Evening for Kirstine Meyer in the Society*   **11**, 493–496
for Dissemination of Natural Science; *Mindeaften for*
Kirstine Meyer i Selskabet for Naturlærens Udbre-
delse, Fys. Tidsskr. **40** (1942) 173–175
    Translation   **11**, 497–499

(153) *Foreword*; *Forord*, G. Gamow, "Mr. Tompkins   **12**, 77–80
i Drømmeland", Gyldendal, Copenhagen 1942,
pp. 7–8
    Translation   **12**, 81–82

[21]

(154) *Meeting on 13 November 1942 on the 200th Anniversary of the Establishment of the Academy*; *Mødet den 13. November 1942 paa 200-Aarsdagen for Selskabets Stiftelse*, Overs. Dan. Vidensk. Selsk. Virks. Juni 1942 – Maj 1943, pp. 26–28, 31–32, 36, 40–41, 44–48   **11**, 397–405

Translation   **11**, 407–414

(155) *Harald Høffding's 100th Birthday*; *Harald Høffdings 100-Aars Fødselsdag*, ibid., pp. 57–58   **10**, 323–324

Translation   **10**, 325

(156) *Science and Civilization*, The Times, 11 August 1945   **11**, 121–124

(157) *A Challenge to Civilization*, Science **102** (1945) 363–364   **11**, 125–129

(158) *Meeting on 19 October 1945*; *Mødet den 19. Oktober 1945*, Overs. Dan. Vidensk. Selsk. Virks. Juni 1945 – Maj 1946, pp. 29–31   **11**, 415–416

Translation   **11**, 417

(159) *Meeting on 19 October 1945*; *Mødet den 19. Oktober 1945*, Overs. Dan. Vidensk. Selsk. Virks. Juni 1945 – Maj 1946, pp. 31–32 [M. Knudsen's Retirement as Secretary of the Royal Danish Academy]   **12**, 295–296

Translation   **12**, 297

(160) *On the Transmutations of Atomic Nuclei* (Abstract); *Om Atomkernernes Omdannelser*, Overs. Dan. Vidensk. Selsk. Virks. Juni 1945 – Maj 1946, p. 31   **9**, 485–486

Translation   **9**, 486

(161) *On the Problem of Measurement in Atomic Physics*; *Om Maalingsproblemet i Atomfysikken*, "Festskrift til N.E. Nørlund i Anledning af hans 60 Aars Fødselsdag den 26. Oktober 1945 fra danske Matematikere, Astronomer og Geodæter, Anden Del", Ejnar Munksgaard, Copenhagen 1946, pp. 163–167   **11**, 655–661

Translation   **11**, 662–666

(162) *Humanity's Choice Between Catastrophe and Happier Circumstances*; *Menneskehedens Valg mellem Katastrofe og lykkeligere Kaar*, Politiken, 1 January 1946   **11**, 149–151

      Translation   **11**, 152–154

(163) *A Personality in Danish Physics*; *En Personlighed i dansk Fysik* [H.M. Hansen], Politiken, 7 September 1946   **12**, 325–327

      Translation   **12**, 328–330

(164) *Speech at the Memorial Ceremony for Ole Chievitz 31 December 1946*; *Tale ved Mindehøjtideligheden for Ole Chievitz 31. December 1946*, Ord och Bild **55** (1947) 49–53   **12**, 449–456

      Translation   **12**, 456–460

(165) *Meeting on 25 April 1947*; *Mødet den 25. April 1947*, Overs. Dan. Vidensk. Selsk. Virks. Juni 1946 – Maj 1947, pp. 53–54   **11**, 419–422

      Translation   **11**, 423–424

(166) *Newton's Principles and Modern Atomic Mechanics*, "The Royal Society Newton Tercentenary Celebrations, 15–19 July 1946", Cambridge University Press, Cambridge 1947, pp. 56–61   **12**, 219–225

(167) *Problems of Elementary-Particle Physics*, Report of an International Conference on Fundamental Particles and Low Temperatures held at the Cavendish Laboratory, Cambridge, on 22–27 July 1946, Volume 1, Fundamental Particles, The Physical Society, London 1947, pp. 1–4   **7**, 217–222

(168) *Atomic Physics and International Cooperation*, Address at Symposium of the National Academy of Sciences, Present Trends and International Implications of Science, Philadelphia, 21 October 1946, Proc. Am. Phil. Soc. **91** (1947), 137–138   **11**, 131–134

[23]

(169) *Meeting on 17 October 1947 in Commemoration of King Christian X; Mødet den 17. Oktober 1947 til Minde om Kong Christian X*, Overs. Dan. Vidensk. Selsk. Virks. Juni 1947 – Maj 1948, pp. 26–29   **11**, 425–429

  Translation   **11**, 430–432

(170) [Tribute to Rutherford], *Hommage à Lord Rutherford 7–8 Novembre 1947*, Pamphlet, World Federation of Scientific Workers 1948, pp. 15–16   **12**, 275–278

(171) *The Penetration of Atomic Particles through Matter*, Mat.–Fys. Medd. Dan. Vidensk. Selsk. **18**, No. 8 (1948)   **8**, 423–568

(172) *On the Notions of Causality and Complementarity*, Dialectica **2** (1948) 312–319   **7**, 325–337

(173) [Foreword], "Niels Bjerrum: Selected Papers, edited by friends and coworkers on the occasion of his 70th birthday on the 11th March 1949", Munksgaard, Copenhagen 1949, p. 3   **12**, 443–443

(174) *Greeting from Niels Bohr; Niels Bohr's Hilsen*, "Akademisk Boldklub [Academic Ball Club] 1939–1949", Copenhagen 1949, pp. 7–8   **11**, 675–678

  Translation   **11**, 679–680

(175) *Meeting on 11 March 1949; Mødet den 11. Marts 1949*, Overs. Dan. Vidensk. Selsk. Virks. Juni 1948 – Maj 1949, pp. 45–46   **11**, 433–434

  Translation   **11**, 435

(176) *Atoms and Human Knowledge; Atomerne og vor erkendelse*, Berlingske Tidende, 2 April 1949   **12**, 57–62

  Translation   **12**, 63–70

(177) *Niels Bohr on Jest and Earnestness in Science; Niels Bohr om spøg og alvor i videnskaben*, Politiken, 17 April 1949   **12**, 177–186

  Translation   **12**, 187–196

(178) *Professor Martin Knudsen Died Yesterday*; *Prof.* **12**, 299–300
Martin Knudsen død i gaar, Politiken, 28 May 1949
   Translation **12**, 301–302

(179) *Martin Knudsen 15.2.1871–27.5.1949*, Fys. Tidsskr. **12**, 303–307
47 (1949) 145–147
   Translation **12**, 308–311

(180) *He Stepped in Where Wrong had been Done: Obitu-* **12**, 433–434
ary by Professor Niels Bohr; *Han traadte hjælpende*
til hvor uret blev begaaet: Mindeord af professor
Niels Bohr [A. Friis], Politiken, 7 October 1949
   Translation **12**, 435

(181) *The Internationalist* [Albert Einstein], UNESCO **12**, 369–372
Courier **2** (No. 2, 1949), 1, 7

(182) *Discussion with Einstein on Epistemological Prob-* **7**, 339–381
lems in Atomic Physics, "Albert Einstein, Philoso-
pher–Scientist" (ed. P.A. Schilpp), The Library of
Living Philosophers, Vol. VII, Evanston, Illinois
1949, pp. 201–241

(183) [Obituary for M. Knudsen], Overs. Dan. Vidensk. **12**, 313–319
Selsk. Virks. Juni 1949 – Maj 1950, pp. 61–65
   Translation **12**, 320–324

(184) [*Discussion Remarks*], "Les particules élémentaires", **8**, 569–572
Rapports et discussions du huitème Conseil de
physique tenu à Bruxelles du 27 septembre au 2 oc-
tobre 1948, R. Stoops, Bruxelles 1950, pp. 107, 125–
127
Discussion of lectures by R. Serber ("Artificial
Mesons") and C.F. Powell ("Observations on the
Properties of Mesons of the Cosmic Radiation")

(185) *Some General Comments on the Present Situation in* **7**, 223–228
Atomic Physics, "Les particules élémentaires", *ibid.*,
pp. 376–380
Contribution to the general discussion

(186) *Open Letter to the United Nations, June 9th, 1950*,   **11**, 171–185
Schultz, Copenhagen 1950

(187) *Field and Charge Measurements in Quantum Electro-*   **7**, 211–216
dynamics (with L. Rosenfeld), Phys. Rev. **78** (1950)
794–798

(188) *Niels Bohr's deeply serious appeal*; *Niels Bohrs dybt*   **11**, 447–449
alvorlige appel, Politiken, 19 January 1951
Translation   **11**, 450–452

(189) *H.C. Ørsted*, Fys. Tidsskr. **49** (1951) 6–20   **10**, 341–356
Translation   **10**, 357–369

(190) *The Epistemological Problem of Natural Science*   **7**, 383–384
(Abstract); *Naturvidenskabens Erkendelsesproblem*,
Overs. Dan. Vidensk. Selsk. Virks. Juni 1950 – Maj
1951, p. 39
Translation   **7**, 384

(191) *Meeting on 2 February 1951*; *Mødet den 2. Februar*   **11**, 453–457
1951, ibid., 453–457
Translation   **11**, 458–461

(192) *Meeting on 19 October 1951*; *Mødet den 19. Oktober*   **11**, 437–438
1951, Overs. Dan. Vidensk. Selsk. Virks. Juni 1951
– Maj 1952, pp. 33–34

(193) *Meeting on 16 November 1951*; *Mødet den 16. No-*   **11**, 439–440
vember 1951, ibid., p. 39
Translation   **11**, 441

(194) *Statement by Professor Niels Bohr in the current af-*   **11**, 463–464
fairs programme on Danish National Radio, Mon-
day 4 February 1952; *Udtalelse af Professor Niels*
Bohr i radioens aktuelle kvarter, mandag den 4.
februar 1952, Videnskabsmanden: Meddelelser fra
Foreningen til Beskyttelse af Videnskabeligt Arbejde
6 (No. 1, 1952), p. 3
Translation   **11**, 465–467

(195) [Discussion remarks to papers by W. Kohn and B. Mottelson], "Report of the International Conference sponsored by the Council of Representatives of European States for Planning an International Laboratory and Organizing Other Forms of Co-Operation in Nuclear Research, Institute for Theoretical Physics, Copenhagen, June 3–17, 1952" (eds. O. Kofoed-Hansen, P. Kristensen, M. Scharff and A. Winther), pp. 16, 19

    Transcription   **9**, 527–529

(196) *Medical Research and Natural Philosophy*, Acta Medica Scandinavica (Suppl.) **142** (1952) 967–972   **10**, 65–72

(197) *On the Death of Hendrik Anthony Kramers*; *Ved Hendrik Anton Kramers død*, Politiken, 27 April 1952   **12**, 347–349

    Translation   **12**, 350–352

(198) *Hendrik Anthony Kramers †*, Ned. T. Natuurk. **18** (1952) 161–166   **12**, 353–360

(199) *Electron Capture by Swiftly Moving Ions of High Nuclear Charge* (Abstract); *Elektronindfangning af hurtigt bevægede Ioner med høj Kerneladning*, Overs. Dan. Vidensk. Selsk. Virks. Juni 1951 – Maj 1952, p. 49   **8**, 579–581

    Translation   **8**, 581

(200) [Discussion Contribution on Superconductivity], "Proceedings of the Lorentz Kamerlingh Onnes Memorial Conference, Leiden University 22–26 June 1953", Stichting Physica, Amsterdam 1953, pp. 761–762   **12**, 37–39

(201) *A Fruitful Lifework*; *Et frugtbart livsværk*, "Noter til en mand: Til Jens Rosenkjærs 70-aars dag" (eds. J. Bomholt and J. Jørgensen), Det Danske Forlag, Copenhagen 1953, p. 79   **12**, 473–474

    Translation   **12**, 475

(202) *Ninth Presentation of the H.C. Ørsted Medal*   **11**, 477–481
[A. Langseth]; *Niende Uddeling af H.C. Ørsted
Medaillen*, Fys. Tidsskr. **51** (1953) 65–67, 80
Translation   **11**, 482–485

(203) *[Preface]*; *[Forord]*, ...fra Thrige **6**, No. 1 (1953) 2–4   **11**, 555–558
Translation   **11**, 559–560

(204) *Speech Given at the 25th Anniversary Reunion of*   **10**, 223–232
the Student Graduation Class 21 September 1928;
Tale ved Studenterjubilæet, 1903–1928, Private print,
1953
Translation   **10**, 233–236

(205) *Address Broadcast on Danish National Radio, 16*   **11**, 681–684
October 1953; *Tale ved Statsradiofoniens udsendelse
den 16. oktober 1953*, "Det kongelige Teater,
Forestillingen lørdag den 17. oktober 1953", pp. 4–6
Translation   **11**, 685–688

(206) *Physical Science and the Study of Religions*, Studia   **10**, 275–280
Orientalia Ioanni Pedersen Septuagenario A.D. VII
id. Nov. Anno MCMLIII, Ejnar Munksgaard, Copenhagen 1953, pp. 385–390

(207) *The Rebuilding of Israel: A Remarkable Kind of Ad-*   **11**, 689–693
venture; *Israels Genopbygning: Et Æventyr af ejen-
dommelig Art*, Israel **7** (No. 2, 1954) 14–17
Translation   **11** 694–699

(208) *Electron Capture and Loss by Heavy Ions Penetrat-*   **8**, 593–625
ing through Matter (with J. Lindhard), Mat.–Fys.
Medd. Dan. Vidensk. Selsk. **28**, No. 7 (1954)

(209) *Address at the Opening Ceremony*, Acta Radiologica   **10**, 73–78
(Suppl.) **116** (1954) 15–18

(210) *Foreword*; *Forord*, "Johan Nicolai Madvig: Et min-   **12**, 247–248
deskrift", Royal Danish Academy of Sciences and
Letters and the Carlsberg Foundation, Copenhagen
1955, p. vii
    Translation   **12**, 249

(211) *Unity of Knowledge*, "Unity of Knowledge" (ed. L.   **10**, 79–98
Leary), Doubleday & Co., New York 1955, pp. 47–
62

(212) *Rydberg's discovery of the spectral laws*, Proceed-   **10**, 371–379
ings of the Rydberg Centennial Conference on
Atomic Spectroscopy, Lunds Universitets Årsskrift.
N.F. Avd. 2, Bd. 50, Nr 21 (1955) 15–21

(213) *Greater International Cooperation is Needed for*   **11**, 561–570
Peace and Survival, "Atomic Energy in Industry:
Minutes of 3rd Conference October 13–15, 1954",
National Industrial Conference Board, Inc., New
York 1955, pp. 18–26

(214) *Obituary*; *Mindeord* [Einstein], Børsen, 19 April   **12**, 373–374
1955
    Translation   **12**, 375

(215) *Albert Einstein 1879–1955*, Sci. Am. **192** (1955) 31   **12**, 377–378

(216) *The Physical Basis for Industrial Use of the Energy*   **11**, 571–584
of the Atomic Nucleus; *Det fysiske grundlag for in-*
dustriel udnyttelse af atomkerne-energien, Tidsskrift
for Industri (No. 7–8, 1955) 168–179
    Translation   **11**, 585–607

(217) *Physical Science and Man's Position*, Ingeniøren **64**   **10**, 99–106
(1955) 810–814

(218) *The Goal of the Fight: That We in Freedom May Look Forward to a Brighter Future*; *Kampens mål: At vi i frihed kan se hen til en lysere fremtid*, "Ti år efter", Kammeraternes Hjælpefond, Copenhagen 1955 ..... **11**, 701–702

Translation **11**, 703–704

(219) *My Neighbour*; *Min Genbo*, "Halfdan Hendriksen: En dansk Købmand og Politiker", Aschehoug, Copenhagen 1956, pp. 171–172 ..... **12**, 467–470

Translation **12**, 471–472

(220) *Atoms and Society*; *Atomerne og Samfundet*, "Den liberale venstrealmanak", ASAs Forlag, Copenhagen 1956, pp. 25–32 ..... **11**, 609–617

Translation **11**, 618–631

(221) *Mathematics and Natural Philosophy*, The Scientific Monthly **82** (1956) 85–88 ..... **11**, 667–672

(222) *A Shining Example for Us All*; *Et lysende forbillede for os alle* [H.M. Hansen], Politiken, 14 June 1956 ..... **12**, 331–332

Translation **12**, 333–334

(223) [Obituary for H.M. Hansen], Fys. Tidsskr. **54** (1956) 97 ..... **12**, 335–337

Translation **12**, 338

(224) *On Atoms and Human Knowledge*; *Atomerne og den menneskelige erkendelse*, Overs. Dan. Vidensk. Selsks. Virks. Juni 1955 – Maj 1956, pp. 112–124 ..... **7**, 395–410

English version in Dædalus **87** (1958) 164–175 ..... **7**, 411–423

(225) *Open Letter to the Secretary General of the United Nations, November 9th, 1956*, Private print, Copenhagen 1956 ..... **11**, 191–192

(226) *Autobiography of the Honorary Doctor*; *Selvbiografi af æresdoktoren*, Acta Jutlandica **28** (1956) 135–138 ..... **12**, 207–212

Translation **12**, 213–217

(227) *Greeting to the Exhibition from Professor Niels Bohr*; **11**, 633–635
Hilsen til udstillingen fra professor Niels Bohr, Elektroteknikeren **53** (1957) 363
    Translation **11**, 636

(228) [For the tenth anniversary of the journal "Nucleon- **12**, 83–84
ics"], Nucleonics **15** (September 1957) 89

(229) *The Presentation of the first Atoms for Peace Award* **11**, 637–644
to Niels Henrik David Bohr, October 24, 1957, Na-
tional Academy of Sciences, Washington, D.C. 1957,
pp. 1, 5, 18–22

(230) *Obituary*; *Mindeord*, "Bogen om Peter Freuchen" **12**, 477–478
(eds. P. Freuchen, I. Freuchen and H. Larsen), Frem-
ad, Copenhagen 1958, p. 180
    Translation **12**, 479

(231) *His Memory a Source of Courage and Strength*; *Hans* **12**, 445–446
minde en kilde til mod og styrke [N. Bjerrum], Poli-
tiken, 1 October 1958
    Translation **12**, 447

(232) *Quantum Physics and Philosophy – Causality and* **7**, 385–394
Complementarity, "Philosophy in the Mid-Century,
A Survey" (ed. R. Klibansky), La nuova Italia ed-
itrice, Firenze 1958, pp. 308–314

(233) *Professor Niels Bohr on Risø*; *Professor Niels Bohr* **11**, 645–647
om Risø, Elektroteknikeren **54** (1958) 238–239
    Translation **11**, 648–652

(234) *Preface* and *Introduction*, "Atomic Physics and Hu- **10**, 107–112
man Knowledge", John Wiley & Sons, New York
1958, pp. v–vi, 1–2

(235) *Physical Science and the Problem of Life*, "Atomic **10**, 113–123
Physics and Human Knowledge", John Wiley &
Sons, New York 1958, pp. 94–101

(236) [Foreword], J. Lehmann, "Da Nærumgaard blev   **12**, 393–394
børnehjem i 1908", Det Berlingske Bogtrykkeri,
Copenhagen 1958, p. i
Translation   **12**, 395–395

(237) [Greeting on the National Day of the Republic of Yu-   **12**, 227–228
goslavia], Slovenski Poročevalec, Ljubljana, 29 No-
vember 1958
Manuscript   **12**, 229–230

(238) *R.J. Bošković*, "Actes du symposium international   **12**, 231–234
R.J. Bošković, 1958", Académie Serbe des sci-
ences, Académie Yougoslave des sciences et des arts,
Académie Slovène des sciences et des arts, Belgrade,
Zagreb, Ljubljana 1959, pp. 27–28

(239) *Tenth and Eleventh Presentation of the H.C. Ørsted*   **11**, 487–489
Medal [J.A. Christiansen and P. Bergsøe]; *Tiende*
og ellevte Uddeling af H.C. Ørsted Medaillen, Fys.
Tidsskr. **57** (1959) 145, 158
Translation   **11**, 490–491

(240) [Tribute to Tesla], "Centenary of the Birth of Nikola   **12**, 235–236
Tesla 1856–1956", Nikola Tesla Museum, Belgrade
1959, pp. 46–47

(241) *Foreword*; *Forord*, "Hanna Adler og hendes Skole",   **12**, 397–402
Gad, Copenhagen 1959, pp. 7–10
Translation   **12**, 403–405

(242) *Foreword*; *Forord*, J.R. Oppenheimer, "Naturviden-   **12**, 85–86
skab og Livsforståelse", Gyldendal, Copenhagen
1960, pp. i–ii
Translation   **12**, 87

(243) *Foreword*, "Theoretical Physics in the Twentieth Cen-   **12**, 361–366
tury: A Memorial Volume to Wolfgang Pauli" (eds.
M. Fierz and V. Weisskopf), Interscience, New York
1960, pp. 1–4

(244) *Quantum Physics and Biology*, Symposia of the So-   **10**, 125–131
ciety for Experimental Biology, Number XIV: "Mod-
els and Analogues in Biology", Cambridge 1960,
pp. 1–5

(245) *Ebbe Kjeld Rasmussen: 12 April 1901 – 9 October*   **12**, 379–382
1959; *Ebbe Kjeld Rasmussen: 12. april 1901 – 9.*
oktober 1959, Fys. Tidsskr. **58** (1960) 1–2
     Translation   **12**, 383–384

(246) *The Connection Between the Sciences*, Journal Mon-   **10**, 145–153
dial de Pharmacie **3** (1960) 262–267

(247) *Physical Models and Living Organisms*, "Light and   **10**, 133–137
Life" (eds. W.D. McElroy and B. Glass), The Johns
Hopkins Press, Baltimore 1961, pp. 1–3

(248) *The Unity of Human Knowledge*, Revue de la Fonda-   **10**, 155–160
tion Européenne de la Culture, July 1961, pp. 63–66

(249) *Atomic Science and the Crisis of Humanity*; *Atomvi-*   **10**, 281–288
denskab og menneskehedens krise, Politiken, 20 April
1961
     Translation   **10**, 289–293

(250) *The Rutherford Memorial Lecture 1958: Reminis-*   **10**, 381–420
cences of the Founder of Nuclear Science and of
Some Developments Based on his Work, Proc. Phys.
Soc. **78** (1961) 1083–1115

(251) *Preface to the 1961 reissue*, "Atomic Theory and the   **12**, 89–90
Description of Nature", Cambridge University Press,
Cambridge 1961, p. vi

(252) *Meeting on 14 October 1960*; *Mødet den 14. oktober*   **11**, 443–445
1960, Overs. Dan. Vidensk. Selsk. Virks. Juni 1960
– Maj 1961, pp. 39–41

(253) *The Solvay Meetings and the Development of Quan-*   **10**, 429–454
tum Physics, "La théorie quantique des champs",
Douzième Conseil de physique tenu à l'Université
Libre de Bruxelles du 9 au 14 octobre 1961, Interscience Publishers, New York 1962, pp. 13–36

(254) *Foreword by the Danish editorial committee*;   **12**, 91–94
Den danske redaktionskomités forord, I.B. Cohen,
"Fysikkens Gennembrud", Gyldendals Kvantebøger,
Copenhagen 1962, pp. 5–6

     Translation   **12**, 95–96

(255) *The General Significance of the Discovery of the*   **12**, 279–282
Atomic Nucleus, "Rutherford at Manchester" (ed. J.B.
Birks), Heywood, London 1962, pp. 43–44

(256) *Address at the Second International Germanist*   **10**, 139–143
Congress [Copenhagen, August 22, 1960],
"Spätzeiten und Spätzeitlichkeit", Francke Verlag,
Bern 1962, pp. 9–11

(257) [Tribute to Russell], "Into the 10th Decade: Tribute   **12**, 389–390
to Bertrand Russell", Malvern Press, London [1962]

(258) *Light and Life Revisited*, ICSU Review **5** (1963)   **10**, 161–169
194–199

(259) *The Genesis of Quantum Mechanics*, "Essays 1958–   **10**, 421–428
1962 on Atomic Physics and Human Knowledge",
Interscience Publishers, New York 1963, pp. 74–78

(260) *Recollections of Professor Takamine*, "Toshio Taka-   **12**, 367–368
mine and Spectroscopy", Research Institute for Ap-
plied Optics, Tokyo 1964, pp. 384–386

(261) [Tribute to Weisgal], "Meyer Weisgal at Seventy"   **12**, 485–488
(ed. E. Victor), Weidenfeld and Nicolson, London
1966, pp. 173–174

CUMULATIVE INDEX FOR THE COLLECTED WORKS

INDEX (VOLUMES 1–12)

The following cumulative index has been created by merging the individual indexes from all twelve volumes. The result has then been edited as necessary. This has not been a trivial process, and some comments are in order.

The publication of the Niels Bohr Collected Works has taken thirty-four years to complete, and numerous people have been involved at different times. Although it was clear from the beginning that each volume should include a subject index, it was never foreseen that these indexes would ever be coordinated into one integrated index. As a result, each editor constructed his index according to his own preferences and idiosyncrasies. Subjects were emphasized differently, and even the index term used to denote a particular topic could vary widely.

In order both to keep the cumulation manageable and to ensure that the intentions of the each editor were not garbled or lost, we have as a general rule kept the original terms, introducing cross-references when necessary. The only exception is for names of persons, where the original editors' different spelling and varying use of initials have been standardized to ensure that there is only one entry for each person.

Over the years there have also been some changes of convention. Originally, "Bohr, Niels" was included as an index term. This soon proved impractical and these references are not included in the cumulative index. However, when the introduction to the index in only one or a few of the volumes identifies subjects that are not indexed because they appear throughout the volume(s), these terms are included in the cumulative index with the volume number followed by "—".

In the volumes first published (1–6 and 8–9) an italicized page number indicated a biographical note, whereas f and ff after a page number indicated that the subject continued to be discussed on one or more of the pages immediately following. Although these conventions were not upheld in the last published volumes (7 and 10–12), they have been included in the cumulative index as a help to the reader.

In the early volumes, the name of a person in a figure caption was only indexed if the person was also mentioned in the running text. This limitation does not apply for the indexes to the last published volumes, in which a p after a page number means that the subject stems from a figure caption, whereas an n after a page number means that it is mentioned only in a footnote (and not in the running text) on that page.

As in the individual volumes, subjects in the front material of these volumes (*i.e.*, pages with Roman numerals as page numbers) have been referred to in the cumulative index only when they are also mentioned elsewhere in the volume in question. Again following the indexes of the individual volumes, most institutions mentioned in the running text are indexed, whereas places are indexed only selectively.

Notwithstanding its imperfections, the editor is convinced that the cumulative index will be of invaluable help to the reader of this new complete edition of the Niels Bohr Collected Works.

Aardenne, G. van   **Vol. 3:** 609, 620
Aarhus, University of   **Vol. 11:** 513*n*
     Vol. 12: 4*n*, 14, 103, 210, 212, 215,
     217, 518
Aaserud, F.   **Vol. 10:** 6*n*, 20*n*, 21
Abelson, P.   **Vol. 9:** 365, 566, 659*f*
Abildgaard, P.C.   **Vol. 11:** 402, 412
     Vol. 12: 184, 193
Abraham, M.   **Vol. 2:** 128, 232
Absalon   **Vol. 10:** 255, 264
absorption   **Vol. 3:** 254–258, 368
     Vol. 7: 109, 157, 345, 348, 369
   coefficient   **Vol. 2:** 29–30, 238, 597
     Vol. 8: 40, 61–62, 64–65
   experiments on in Manchester   **Vol. 2:** 4,
     11–14, 624
   infrared   **Vol. 2:** 139, 152, 170, 230*ff*, 420,
     445, 446–448, 457, 565, 567, 631
   lecture on in Copenhagen   **Vol. 2:** 6–7,
     43–44, 624
   of α-rays   **Vol. 1:** xxiv, 555, 559, 573
   of charged particles (*see also* stopping of
     charged particles)   **Vol. 2:** 1–99, 103,
     331, 493, 495, 499–500, 515, 517, 551,
     552, 577*ff*, 624, 628, 629, 630
     Vol. 6: 76, 82, 131, 156, 375
   of γ-rays   **Vol. 2:** 374, 597
   of radiation (*see also* coupling between
     emission and absorption)   **Vol. 2:** 116,
     139, 152, 170, 175–179, 201, 222, 224,
     225, 230–232, 289, 338–339, 446–448,
     450, 452, 457–459, 584, 597
     Vol. 5: 8*f*, 11*f*, 21, 24, 32, 39, 44*ff*,
     65, 75, 85, 102*ff*, 112*ff*, 121*f*, 166, 169,
     179, 195, 275, 299, 303, 309*ff*, 317,
     320*ff*, 336, 347, 387, 398*ff*, 417, 420,
     492, 496, 506
     Vol. 6: 13, 40, 140, 382, 395*f*, 417, 460
     Vol. 8: 30, 31, 34, 216, 395, 399, 492,
     498*f*, 546, 675, 685, 694, 699*f*
   selective   **Vol. 12:** 20, 42
   spectra   **Vol. 4:** 446, 475
Academic Ball Club (Copenhagen)
     Vol. 11: 372, 676*p*, 677, 679
     Vol. 12: 478, 479
Académie des inscriptions et belles-lettres
     (Paris)   **Vol. 12:** 318, 323
Académie des sciences (Paris)   **Vol. 12:** 496
Academy of Sciences of the U.S.S.R.
     Vol. 11: 8, 10, 203, 204, 213, 217,
     223–225, 230, 298, 303, 736

Institute for Chemical Physics
     Vol. 11: 230*n*
   Institute for Physical Problems (Moscow)
     Vol. 11: 10, 223, 225, 229, 230,
     298, 333
   Institute for Physiology (Leningrad)
     Vol. 11: 198, 201
   P.N. Lebedev Institute of Physics
     Vol. 11: 230*n*
   Physico–Technical Institute (Leningrad)
     Vol. 11: 197, 201
   State Optical Institute (Leningrad)
     Vol. 11: 198, 201
Academy of Surgery (Copenhagen)
     Vol. 12: 126, 451, 456
Academy of Technical Sciences (Denmark)
     Vol. 11: 360, 363, 573, 585, 586,
     647, 650
     Vol. 12: 183, 184, 192, 193
accelerator (*see also* cyclotron; high-voltage
     generator/laboratory/plant)
     Vol. 10: 408, 409
     Vol. 11: 568
     Vol. 12: 84
   electromagnetic   **Vol. 11:** 531
Acheson–Lilienthal Report   **Vol. 11:** 66–68,
     157, 307–309, 325, 329
Acheson, D.   **Vol. 11:** 66, 68, 325, 327,
     329, 330
Achilles paradox   **Vol. 10:** 157
     Vol. 11: 669
acoustics   **Vol. 12:** 47, 327, 329
action
   at a distance   **Vol. 6:** 102
   classical/macroscopic   **Vol. 6:** 205, 214,
     233, 250, 261, 282, 381
   minimum quantity of   **Vol. 7:** 267, 317
   principle   **Vol. 6:** 204, 210
   principle of least   **Vol. 5:** 244, 247
action-angle variables (*see also* multiply
     periodic systems)   **Vol. 5:** 45,
     158–165, 277
     Vol. 6: 47, 49, 165*f*, 178,
     463–464, 465
activation energy for fission (*see also* critical
     deformation energy)   **Vol. 9:** 70, 433,
     459, 666
actors/spectators   **Vol. 6:** xxii, 196, 235, 253
     Vol. 10: xxxviii, xl, 60, 71, 97, 160,
     193, 229, 234, 351, 364, 478, 564, 571

[39]

Vol. 11: 134, 565
Vol. 12: 86, 87, 157, 170, 186, 196, 201, 205, 224, 463, 464
Adam, N.K. **Vol. 1:** 11
Adams, E.P. **Vol. 1:** 117, 118, 119
Vol. 2: 32, 33, 75
Vol. 8: 64*f*, 86, 102, 145
adaptation **Vol. 10:** 8, 10, 13, 14, 544, 548, 549, 553, 554, 566
Addendum (to Roosevelt) **Vol. 11:** 43–45, 47, 48, 69, 71, 113, 118, 119, 159, 178, 278, 279, 285, 320
addendum to Prize Essay **Vol. 1:** 4, 11
  translation of **Vol. 1:** 67–78
adiabatic
  hypothesis **Vol. 4:** 51
  invariance **Vol. 3:** 7, 11, 75, 87, 369, 405, 467, 469*ff*, 508, 538
    Vol. 10: 395, 439
  invariants **Vol. 1:** xxxii–xxxiii
  limit (*see also* cut-off distance/parameter)
    Vol. 8: 9, 15, 64, 109, 113, 132, 170, 174, 178, 255, 365, 367, 493*ff*, 502, 506–509, 517, 608, 635*f*
  principle/process/transformation
    Vol. 2: 341, 436–440, 465*ff*, 604, 631
    Vol. 5: 16, 182, 188, 197, 202, 224, 255, 262, 277, 278–279, 383, 385, 446, 450–451, 454, 455*f*, 504
    Vol. 6: 127, 154, 345, 389, 423
    Vol. 8: 9, 229, 236, 257, 309, 320, 329, 376, 378, 381*f*, 394, 398, 490*f*, 499, 508, 565, 572, 805
Adigard des Gautries, J. **Vol. 10:** 301*n*
Adler, D.B. **Vol. 12:** 120, 394, 395, 399, 403
Adler, Ellen, *see* Bohr, Ellen
Adler, H. **Vol. 1:** 526, 527, 536, 537, 572, 573, 584, 585
    Vol. 6: 24, 424, 449, 451
    Vol. 10: 36*p*
    Vol. 12: 120, 122, 123, 125, 396*p*, 399, 400, 401*p*, 402–405, 412, 420, 512, 513, 533
Adler, J. **Vol. 12:** 120, 394, 395, 399, 403
Adventures of a Danish Student **Vol. 10:** 159, 200, 232, 236, 256, 266, 480, 481, 483
Advisory Committee on Atomic Energy (U.K.)
    Vol. 11: 50
Aebi, M. **Vol. 7:** 328
Aeschylus **Vol. 10:** 189
aesthetics **Vol. 10:** 141, 345, 358

Ageno, M. **Vol. 9:** 74, 478
Aggerbeck, E.H. **Vol. 12:** 242, 246
agriculture **Vol. 12:** 84
Aharonov, Y. **Vol. 7:** 281*n*
air
  absorption in **Vol. 2:** 29–30, 36, 75–76, 78, 89
  ionization in **Vol. 2:** 87–89
Akademie der Wissenschaften (Göttingen)
    Vol. 11: 347, 440, 441
Alberto, A. **Vol. 11:** 563
alchemy **Vol. 11:** 97, 105, 139, 143, 531, 568, 577, 593, 646, 649
alchemy, modern **Vol. 10:** 86, 398, 449
Alder, K. **Vol. 10:** 556
Alexander, S. **Vol. 10:** 397
Alexandrine, Queen **Vol. 11:** 673*p*
Alfvén, H. **Vol. 8:** 765
Alikhanov, A.I. **Vol. 11:** 230
alkali metals **Vol. 4:** 62*ff*, 223*ff*
  arc spectra of **Vol. 4:** 383*ff*, 583*ff*, 636*ff*
  complex structure of **Vol. 4:** 596, 645
  spectra of (*see also* lithium, spectrum of)
    Vol. 2: 366, 370, 385, 387, 399, 405, 407
alkali spectra **Vol. 5:** 43*f*, 56, 122*f*, 147, 152, 168, 171, 251*f*, 272, 289, 343*f*, 398–399, 401, 407, 423, 424*ff*, 447, 451, 496, 501, 506
    Vol. 6: 385, 429*f*
alkaline earths
  arc spectra of **Vol. 4:** 600, 648
  spark spectra of **Vol. 4:** 588, 639
  spectra of **Vol. 5:** 251*f*
    Vol. 6: 385*f*, 395
Allen, H.S. **Vol. 2:** 265, 328, 377, 378, 379
    Vol. 6: 51
Allison, S.K. **Vol. 5:** 324, 496
alloys **Vol. 1:** 148, 343, 395
    Vol. 12: 24, 28
  conductivities of **Vol. 1:** 343
α-decay/particle/rays **Vol. 6:** 230–231, 246, 376, 378, 405*ff*
    Vol. 8: 210, 563
    Vol. 10: 383, 385, 387, 391, 397, 398, 403, 404, 408–412, 414, 443, 447, 449–451
α-disintegrations **Vol. 7:** 180, 190
α-particle, nature of **Vol. 2:** 189, 290
α-particles, stopping of, *see* stopping of charged particles

α-ray
  absorption of, *see* absorption of charged
    particles
  emission **Vol. 9:** 29*f*, 87, 95*ff*, 113, 131,
    146, 152*f*, 155*f*, 186, 189, 201, 204,
    210, 215*f*, 218, 220, 230, 248,
    249–252, 255*ff*, 263*f*, 277, 281, 308*f*,
    322, 337, 344, 361, 367, 374, 376, 415,
    426, 444, 453, 658*f*, 665
  ionization by **Vol. 2:** 476, 485
  origin of **Vol. 2:** 119–120, 212–214, 236,
    418, 423–425, 440–441
  scattering, *see* scattering of α-rays;
    scattering of charged particles
α-rays/particles **Vol. 11:** 574–578, 587–590,
    592–596
  Vol. 12: 42, 43, 60, 61, 66, 67, 145,
    256, 257, 260, 261, 265, 266, 268
aluminium, absorption in **Vol. 2:** 4, 11–14,
    30, 35–38, 68–70, 75, 76, 79–81, 83
Alvarez, L.W. **Vol. 9:** 365
Amaldi, E. **Vol. 8:** 790
  Vol. 9: 156, 478
Ambrosen, J.P. **Vol. 12:** 17*p*
America, lectures in **Vol. 3:** 45
  Science Club lecture, notes for
    Vol. 3: 45, 46
  Silliman lectures, notes for
    Vol. 3: 581–601
American Academy of Arts and Sciences
    Vol. 7: 287, 523
American Philosophical Society
    Vol. 11: 52, 133
American Physical Society **Vol. 11:** 291
  Chicago Meeting, 1923 **Vol. 5:** 6, 404, 487
  Cincinnati Meeting, 1923 **Vol. 5:** 4–5, 382
  Washington Meeting, 1924 **Vol. 5:** 496
  New York meeting, February 1939
    Vol. 9: 61, 64, 553
  Princeton meeting, June 1939 **Vol. 9:** 383
  Washington meeting, April 1939
    Vol. 9: 68–69, 359–361, 386, 663
American–Scandinavian Foundation
    Vol. 11: 531, 733
Ames, J.S. **Vol. 2:** 275
amplification **Vol. 10:** xl, 17–19, 33, 60, 61,
    68, 70, 86, 89, 91, 103, 104, 120, 122,
    128, 180, 207, 404, 519, 522–528,
    545, 549

amplitude function, *see* wave function
Ampère, A.M. **Vol. 5:** 129, 136
  Vol. 10: 209, 349, 362, 363
  Vol. 11: 530
analogy (*see also* correspondence with
  classical physics) **Vol. 2:** 113, 114,
    128, 172–175, 176–178, 243, 295, 325,
    326, 341, 344, 345, 353, 355, 358–360,
    363–364, 367, 368, 374, 382, 394,
    398–399, 404, 444–445, 446, 449, 453,
    526, 584
  Vol. 5: 36*ff*, 44*f*, 51, 53, 63*ff*, 81, 85,
    105*f*, 109*f*, 113, 116, 145–153, 183,
    198, 251*f*, 256, 263, 276, 279*f*, 288,
    311, 347, 358*f*, 367, 369, 394*ff*, 434,
    436, 448, 452, 506
  Vol. 6: xxiv, xxvi, 35, 49, 59, 61, 63,
    93, 95, 123, 141, 153, 155, 157, 158,
    167, 310, 316, 320, 397, 406, 434,
    466, 470
  Vol. 8: 9, 17, 27, 33, 36, 53, 121, 124,
    147, 223, 256, 290, 355, 371, 394, 396,
    399*f*, 451, 457, 476, 483, 492, 498*f*,
    504, 510, 515, 520, 531, 542*f*, 546,
    552*f*, 560*f*, 636, 672, 683, 689
  between material particles and radiation
    Vol. 6: 28, 70–71, 92, 116, 149, 264,
    287, 394, 423
  between optical spectra and X-ray spectra
    Vol. 12: 45
  between quantum theory and
    biology/psychology **Vol. 6:** 189, 196,
    204, 206, 212, 216*f*, 272, 298, 352
  between quantum theory and classical theory
    Vol. 12: 44
  between quantum theory and
    thermodynamics **Vol. 6:** 328*f*, 400,
    453, 469, 472*f*
  Hamilton's, between mechanics and optics
    Vol. 6: 111, 150, 204, 210, 264, 287
    Vol. 12: 47
analysis
  in atomic physics **Vol. 7:** 259, 323, 324,
    342, 379
  of experience **Vol. 7:** 263, 358, 376,
    522, 523
analysis/synthesis **Vol. 10:** xlv, 41, 44, 58,
    64, 83, 104, 109, 129, 143, 152, 159,
    175, 177, 181, 278, 316, 321, 446,
    584, 591

Vol. 11: 134, 658, 660, 663, 665, 672
Vol. 12: 103, 220, 224, 234, 363
anatomy **Vol. 10:** 54, 67, 91, 104, 117, 126, 130, 135, 150, 175, 179
ancient Greece, *see* Antiquity
Andersen, A. **Vol. 11:** 364, 708, 709
Andersen, H.C. **Vol. 10:** 256, 258, 265, 269
Vol. 11: 311, 313
Vol. 12: 152, 164, 181, 189
Andersen, H.H. **Vol. 8:** vi, 101
Andersen, H.N. **Vol. 10:** 250p
Andersen, M. **Vol. 11:** 358p
Vol. 12: 417, 418, 423, 424
Anderson, A. **Vol. 11:** 26, 305, 319, 322, 326, 329
Anderson, C.D. **Vol. 7:** 4
Vol. 8: 235, 810
Vol. 9: 126f, 166, 174, 216, 219
Vol. 10: 404, 450, 452
Vol. 11: 569
Anderson, H.L. **Vol. 9:** 61, 365, 379f, 384, 386
Anderson, J. **Vol. 11:** 15–17, 20–28, 30, 32–43, 45–47, 49, 50, 54, 55p, 58, 66, 67, 69, 71, 76, 78, 86, 89, 98, 99, 227, 231, 239, 242, 246, 250–256, 258, 261, 262, 267–269, 271–274, 276–278, 280, 283, 285–287, 290, 299, 304, 305, 319, 325–327, 329, 330
Andrade, E.N. da C. **Vol. 2:** 307, 308, 420, 501, 594, 597, 627
Vol. 4: 663
Vol. 8: 667, 672, 676
Vol. 10: 393
Vol. 12: 10, 111, 280p, 494
Anfinsen, C. **Vol. 10:** 168
angle variables **Vol. 3:** 95, 124
angular momentum (*see also* spin)
Vol. 3: 8, 13
Vol. 5: 52, 57, 146, 151, 272, 288f, 304, 339, 384, 386, 390, 393, 395, 412, 414, 423f, 426–432, 445, 450, 459f, 464–482, 499–500
Vol. 6: 52, 145, 158, 307, 310, 348f, 392, 394, 404
Vol. 8: 219, 294, 297, 635, 702f, 804
Vol. 12: 48
conservation of **Vol. 3:** 43, 100
for systems possessing axis of symmetry
Vol. 3: 101

including orbital moment **Vol. 9:** 11, 13, 30, 38, 42, 79, 80, 94, 106, 113, 135f, 138, 197f, 202, 236–237, 239, 255, 261, 317, 329, 337, 376ff, 380f, 392, 491, 493ff, 498, 500f, 511ff, 515ff, 524, 578f, 615, 634, 637, 668
quantization of
Vol. 2: 106–107, 108, 115, 120, 126, 129ff, 175, 180–185, 189ff, 216ff, 227, 233, 255, 258, 270, 313ff, 326–327, 336, 354, 382, 394, 396f, 425, 441–442, 447, 464ff, 475, 477, 504, 545f, 586, 588
Ann Arbor, *see* Michigan, University of
annihilation (*see also* creation/annihilation; pair creation) **Vol. 9:** 126, 415, 444
anomalous scattering, *see* scattering, anomalous
Anthropological and Ethnological Sciences, International Congress of, 1938
Vol. 10: 238p, 241, 249
Vol. 11: 11, 53, 338
anthropology (*see also* ethnography/ethnology)
Vol. 10: v, xxiii, xlix, 98, 102, 112, 219, 240, 245, 247, 253, 262
antiparticle **Vol. 10:** 404, 450, 454
antiproton **Vol. 9:** 12
Antiquity **Vol. 7:** 398, 412
Vol. 10: 68, 86, 102, 116, 126, 130, 135, 157, 174, 194, 207, 208, 297, 343, 357, 404
Vol. 11: 138, 141, 565, 611, 619, 646, 648, 669, 670, 694
Vol. 12: 12
antisymmetry **Vol. 7:** 68, 129
Aoki, H. **Vol. 9:** 239
Appel, J. **Vol. 3:** 660, 661
Vol. 10: 267n
Appleton, E.V. **Vol. 11:** 227, 730
a priori probability **Vol. 3:** 12, 75, 91, 92, 93, 142, 308, 313, 475
Arbejderbladet (Danish newspaper)
Vol. 11: 9
arc spectra **Vol. 2:** 387–388, 509, 559, 562
Vol. 4: 205, 278ff, 445, 474
archaeology **Vol. 10:** 253, 262, 300
Archimedes **Vol. 5:** 129, 136, 273
Vol. 10: 53, 141, 157, 208
Vol. 11: 669

argon **Vol. 8:** 29*f*, 236, 315, 320, 325, 331*ff*,
    478, 558, 575*ff*, 606*f*, 615*ff*, 621*f*,
    689, 692*f*, 720, 740, 743*ff*, 784
  Vol. 12: 45
Argonne School of Nuclear Science and
    Engineering **Vol. 11:** 630*n*
Aristotelian dynamics **Vol. 11:** 669
Aristotle **Vol. 10:** 3, 53, 55, 91, 104, 117,
    165, 174
arithmetic **Vol. 10:** 42, 511–513
Arley, N. **Vol. 11:** 358*p*
armaments race **Vol. 11:** 27, 32, 42, 234
arms control, *see* international
    control/regulation
Arndt, F. **Vol. 5:** *402*
Arneson, R.G. **Vol. 11:** 76, 321, 326, 330
Arnold, W. **Vol. 10:** 134
Arnold, W.A. **Vol. 9:** 559
Arnot, F.L. **Vol. 8:** 457, 566
Aron, R. **Vol. 7:** 411
Arrhenius, S.A. **Vol. 2:** 124
  Vol. 3: 583
  Vol. 4: 429, 467
  Vol. 6: 374
  Vol. 8: 727
  Vol. 12: 125, 438, 441
Arrøe, O.H. **Vol. 8:** 560, 566
art **Vol. 10:** 95, 96, 160, 254, 258–261, 263,
    269, 270, 272, 300, 567, 573
  Vol. 12: 159, 172, 359, 371, 413, 416,
    417, 421, 423
  and science **Vol. 12:** 127, 416, 423,
    462, 464
  of writing **Vol. 12:** 462, 464, 515
artificial transmutation, *see* transmutation,
    artificial
Arup, E. **Vol. 11:** 406*p*
Asmussen, D. **Vol. 10:** 312*p*
aspiration **Vol. 10:** 105, 280
Association of Engineers (Denmark)
  Vol. 11: 55, 294, 295, 324, 329, 583,
    606*n*, 731, 736
  Vol. 12: 534
Association of Physical Oceanography
  Vol. 12: 300, 302
Association of the Hard of Hearing (Denmark)
  Vol. 12: 327, 330, 332, 333
assumptions, Bohr's principal (*see also* angular
    momentum, quantization of; frequency
    condition, Bohr's; stationary state)

  Vol. 2: 104*ff*, 127, 137, 147, 167, 174*f*,
    182, 184–185, 189*ff*, 216*ff*, 232–233,
    255, 258, 293–294, 304, 308, 313–315,
    338, 341–342, 351, 354–355, 364,
    367–368, 392–398, 433*ff*, 460–461,
    464*ff*, 475, 529–530, 554*f*, 572, 579,
    582*f*, 604
Asther, N. **Vol. 12:** 129*p*
Aston's rule **Vol. 6:** 404
Aston, F.W. **Vol. 2:** 423
  Vol. 3: 38, 454
  Vol. 4: 431, 468, 658
  Vol. 6: 145, 378
  Vol. 8: 750
  Vol. 9: 94, 100, 154, 165, 174, 201,
    285, 306, 320
  Vol. 10: 399, 409
  Vol. 11: 566
  Vol. 12: 42, 342*p*
Astronomical Society (Denmark)
  Vol. 11: 488, 490
astronomy **Vol. 7:** 399, 413
  Vol. 10: 20, 30, 157, 241
  Vol. 11: 369, 539, 545, 548, 551, 657,
    662, 669
  Vol. 12: 233, 378
astrophysics **Vol. 10:** 399
  Vol. 12: 357
asymmetrical fission, *see* fission,
    symmetrical/asymmetrical
asymmetry, *see* symmetry
asymptotic agreement with classical theory, *see*
    correspondence with classical physics
Atkinson, R. **Vol. 10:** 409
Atom, Encyclopædia Britannica article
  Vol. 4: 657–663
atomic
  age **Vol. 11:** 359, 361, 569, 581, 601,
    646, 650
  Vol. 12: 487
  bomb **Vol. 11:** v, 3, 5, 12–19, 23, 24, 27,
    29, 32, 34, 37, 39, 42–44, 49–51, 53,
    55–57, 59, 60, 65, 68, 70, 74, 77, 80,
    83, 87, 88, 90, 92, 94, 97, 99, 100, 103,
    106, 107, 109, 110, 113–115, 138, 142,
    161, 164–166, 176–182, 189, 228, 229,
    233, 234, 242, 259, 261, 270, 289, 294,
    295, 299, 324, 328, 360, 361, 377, 580,
    581, 600, 601*n*, 614, 615, 617, 626,
    627, 630, 730

bomb, as challenge to civilization
    Vol. 11: 3, 51, 52, 54, 63, 70, 75, 79,
    113, 118, 123, 124, 127, 140, 143, 156,
    160, 168, 175, 182, 188, 192, 294, 295,
    362, 368, 611, 619, 643, 646, 650
bomb, implications of   **Vol. 11:** 5, 15, 16,
    18, 20, 22, 25, 35–37, 42, 49, 52, 77,
    87–89, 92, 98, 100, 105, 113, 156, 232,
    234, 235, 241, 242, 268, 269, 271, 276,
    282, 291, 296, 303, 321, 324, 328
bomb project (American–British)
    Vol. 11: 3, 15, 16, 18–28, 33, 36–38,
    40–42, 46, 47, 60, 63, 69, 86–109, 113,
    119, 139, 142, 143, 159, 166, 175–178,
    227n, 231–237, 240–242, 250, 256,
    257, 259, 261–263, 265, 266, 268, 273,
    278, 280–284, 291, 320, 361, 702, 703
    Vol. 12: 211, 216
bomb project (German)   **Vol. 11:** 16, 23,
    30, 32
bomb project (Soviet)   **Vol. 11:** 23, 24, 332
constitution/structure (*see also* atomic
    model)   **Vol. 5:** 17f, 34, 57, 63, 73,
    90ff, 103, 107, 124, 128, 130, 133, 136,
    138, 141, 145, 147f, 152, 156, 191,
    205, 221, 230, 238f, 241–248, 261,
    267, 272, 273–280, 300, 317, 329f,
    332, 335, 351, 371, 373, 406f, 422,
    431f, 446f, 451f, 462f, 493, 498, 512
    Vol. 6: 56, 75, 80, 82, 88, 94, 123f,
    136, 152, 155, 158, 204, 211, 224ff,
    237, 239, 242f, 262, 263, 266, 274,
    284, 285, 290, 299, 317, 359, 371,
    373ff, 381, 383, 385f, 388, 392, 395f,
    399f, 407, 457f
    Vol. 7: 12, 27, 43, 45, 60–64, 76, 79,
    84, 86, 92, 102, 120, 123–126, 134,
    136, 140, 141, 145, 152, 165, 171, 176,
    178–180, 183, 187–190, 197, 198, 215,
    220, 225, 249, 304, 305, 308, 343, 344,
    348, 350, 361, 362, 365, 469, 510,
    511, 520
    Vol. 8: 3ff, 11, 17f, 20, 22, 24, 53, 71,
    75ff, 84f, 90ff, 99f, 105, 109, 113,
    118, 171, 174, 179, 184ff, 189ff, 197,
    199, 207, 223, 289, 427, 465, 497, 603,
    657, 668, 683, 686, 694, 727, 729,
    766f, 769ff, 779ff, 784
    Vol. 10: 68, 69, 86, 102, 111, 130, 136,
    148, 151, 165, 166, 177, 208, 211, 214,
    332, 337–339, 373, 375, 377, 379, 384,

386, 387, 390, 393, 394, 396, 397,
    403–405, 408, 424, 427, 431, 434–437,
    443, 446, 453, 509–513
    Vol. 11: 52, 90, 101, 123, 133, 138,
    142, 175, 372, 511, 514, 520, 522,
    529–531, 539, 548, 566, 567, 611–614,
    619, 620, 622, 626, 641, 659, 664
    Vol. 12: 4, 10, 11, 20, 42–47, 51, 60,
    65, 66, 84, 138–141, 145, 149, 150,
    199, 203, 209, 213, 221, 252, 255–257,
    260, 261, 265, 281, 288, 326, 329, 340,
    341, 344–346, 356, 368
core   **Vol. 3:** 509, 539
    Vol. 5: 52, 395, 411ff, 423f, 427f
energy, *see* nuclear energy
energy, as power source   **Vol. 11:** 13, 18, 50,
    57, 66, 68, 75, 78, 87, 92, 103, 106, 109,
    113, 114, 117, 140, 143, 150, 152, 157,
    163, 166, 175–177, 179–181, 183, 233,
    234, 243, 244, 246, 294, 295, 308, 309,
    311–313, 321, 324, 328, 333, 359–361,
    363–367, 369, 370, 562, 565, 570,
    573, 579, 580, 584–586, 597, 599, 607,
    611, 616–619, 627, 629–631, 641, 646,
    647, 650, 707, 708, 716–718, 720, 731
    Vol. 12: 84, 211, 216
energy, in physics   **Vol. 11:** 29, 50, 56, 60,
    63, 123, 229, 333, 568, 575, 579–583,
    590, 598, 600–605, 614, 624, 635, 636,
    646–649, 652
    Vol. 12: 43, 61, 68, 199, 203
model/picture (*see also* atomic
    constitution/structure)   **Vol. 5:** 36f,
    57, 64, 90ff, 130f, 138f, 146, 151,
    191f, 205, 224, 239, 251f, 257, 264,
    304, 504
    Vol. 6: 52, 226, 239, 359, 376, 377,
    379, 381, 383ff, 392
    Vol. 7: 33, 34, 59, 60, 63, 114,
    123–126, 160, 197, 203, 209, 220, 262,
    303, 308, 347, 350, 358, 375, 469, 470
    Vol. 10: 333, 384–390, 392, 395, 399,
    401, 407, 434, 436–438, 496n, 510, 511
    Vol. 11: 356, 510p, 511, 512, 520,
    521, 567
    Vol. 12: 10, 44, 356, 357
momentum   **Vol. 11:** 660, 665
number (*see also* periodic system of
    elements)   **Vol. 2:** 120, 189, 198ff,
    289–290, 314, 393, 419, 440, 530, 589
    Vol. 3: 204, 364

Vol. 4: 86, 195, 263, 346

Vol. 5: 180*f*, 196*f*, 276, 423, 427

Vol. 6: 224*f*, 228, 237*f*, 243, 379, 385, 387*ff*

Vol. 8: 7, 122, 220, 255, 263, 265, 270, 274, 279, 283, 326, 360*f*, 462, 473, 532*f*, 546, 558, 591, 599, 605, 607, 610, 615, 617, 623, 632, 653, 707, 725, 766*f*, 784

Vol. 9: 67, 135, 155, 193, 197, 231, 236, 274, 280, 306*f*, 311, 317, 320*f*, 324, 329, 344*f*, 350*ff*, 357, 365, 367*f*, 371*f*, 380, 415, 417, 428, 431, 440, 445*f*, 455, 457, 465, 478, 553, 593*f*

Vol. 10: 376, 384–387, 393, 394

Vol. 11: 613, 614, 621, 624

Vol. 12: 43–46, 61, 67

phenomena **Vol. 7:** 36, 219, 297, 305, 307, 308, 318, 342–344, 346–348, 358, 362, 364, 368, 373, 375

Vol. 11: 134, 567, 661, 665

Vol. 12: 47, 51, 156, 168, 185, 194, 222, 378

physics (*see also* nuclear physics)

Vol. 7: 216, 224, 279, 335, 341, 348, 351, 364, 376, 378–380, 398, 412, 504, 508

Vol. 11: 13, 36, 52, 62, 81, 90, 119, 133, 134, 137, 141, 184, 217, 223, 225, 311, 313, 356, 361, 370, 511, 520, 529, 530, 556, 559, 569, 574, 581, 587, 588, 601, 613, 622, 647, 650, 651, 660, 665

Vol. 12: 12, 13, 61, 62, 68, 70, 75, 86, 87, 105, 116, 157, 159, 160, 169, 172, 173, 184–186, 193–196, 199, 201, 203, 205, 210, 211, 214, 216, 217, 223, 224, 228, 229, 348–351, 355, 356, 359, 382, 384, 534

power stations **Vol. 11:** 234, 364, 573, 585, 616, 617, 627–630

properties **Vol. 12:** 61, 62, 68, 70, 79, 81, 156, 168, 209, 214, 222

radius, *see* dimensions, atomic

spectra, *see* spectra, atomic

stability (*see also* stability of atoms/stationary states) **Vol. 7:** 165*n*, 343, 415

Vol. 11: 512, 521, 567, 611–613, 619, 621, 622, 649

Vol. 12: 43, 59, 61, 63, 67, 68, 222, 257, 261, 282, 364

states **Vol. 11:** 512, 521, 567

state/transition **Vol. 7:** 104, 105, 153, 154, 174, 176, 177, 180, 185, 187, 190, 206, 305, 307, 332, 343–346, 348, 402

statistics **Vol. 10:** 519, 522

theory **Vol. 11:** 123, 203, 204, 355, 371, 529, 658, 663

Vol. 12: 10, 12, 43, 47, 51, 54, 60, 64, 84, 108, 114, 115, 118, 125, 156–158, 168, 170, 171, 224, 344, 345, 356, 438, 439, 441, 508–510

theory, fundamental assumptions of the **Vol. 3:** 569*ff*

volume **Vol. 2:** 105, 109, 137, 148, 208–209, 236, 261, 420, 530, 579

Vol. 8: 79, 95

weight (*see also* mass number)

Vol. 6: 375, 377*f*, 387, 404

Vol. 8: 10, 45*f*, 52*f*, 60, 66*ff*, 77, 79, 81*ff*, 87*ff*, 93, 95, 97*ff*, 103*f*, 133, 147, 150, 153, 163, 172, 176, 179, 669*ff*, 725, 752

Vol. 9: 15, 94, 155, 158, 164*ff*, 168, 170, 172*ff*, 177*f*, 189, 215, 218, 262, 365, 367*f*, 376, 379, 385, 417, 446, 470, 473, 593*f*, 652

Vol. 11: 575, 576, 590, 591, 613, 614, 616, 621–625, 627

Vol. 12: 42, 60, 64, 65, 252

atomic constitution, ideas of **Vol. 3:** 583, 584

quantum theory of **Vol. 3:** 587*ff*

Atomic Energy Authority (U.K.)

Vol. 11: 324, 328, 616, 629

Atomic Energy Commission (Denmark)

Vol. 11: viii, 363–366, 369, 573, 584, 586, 607, 616, 617, 629, 630, 646–648, 651, 652, 653*p*

Vol. 12: 210, 212, 215, 217, 382, 384

Atomic Energy Commission (U.S.)

Vol. 11: 325*n*, 361, 363, 562, 616, 629, 710*n*, 711*n*, 713*n*

Atomic Energy Committee (Denmark)

Vol. 11: 360, 586*n*

Atomic Energy Conference, Third Annual, *see* Peaceful Uses of Atomic Energy, New York Conference on

Atomic Energy Research Establishment (U.K.)

Vol. 11: 360, 616, 629

Vol. 12: 84

atomic model
  Bohr's **Vol. 8:** 4, 9, 18–19, 21, 24, 133,
    670, 720, 779*ff*
  Langmuir's **Vol. 8:** 780, 782
  Lewis' **Vol. 8:** 780, 782
  Nagaoka's **Vol. 2:** 127, 549
  Rutherford's (*see also* Rutherford scattering)
    Vol. 1: xxii, 109
    Vol. 2: 5*f*, 18, 32*ff*, 38–39, 46, 60, 63,
      68, 96*ff*, 133–134, 136*ff*, 146*ff*, 161*ff*,
      188*ff*, 215*ff*, 232, 280, 289–292, 306,
      308, 312*ff*, 352, 377, 393*ff*, 418*ff*, 440,
      447, 473, 528, 529, 579, 624
    Vol. 8: 4*ff*, 50, 64*ff*, 70*f*, 77, 87, 89,
      92–93, 102*ff*, 109*f*, 113, 130, 133, 138,
      145, 163, 170*ff*, 183, 186, 188, 191,
      197, 199, 490, 668, 720, 818
  Thomson's (*see also* Thomson's theory)
    Vol. 1: xxiii
    Vol. 2: 125, 127, 136, 146, 162,
      267–268, 529, 588*f*
    Vol. 8: 76–77, 92, 169, 172, 177, 186,
      191, 780, 782
atomic models, recent **Vol. 4:** 276, 277
atomic physics, program of **Vol. 3:** 584*ff*
atomic problems in general **Vol. 3:** 36, 37,
    397–414
atomic structure
  and physical and chemical properties
    Vol. 4: 181–340
  application of the quantum theory to
    Vol. 3: 455–499, 501–558
  fundamental postulates **Vol. 3:** 560–574
  theoretical conceptions of
    Vol. 4: 486*ff*, 520*ff*
  theory of series spectra **Vol. 3:** 502–531,
    532–558
atomism **Vol. 7:** 399, 413
    Vol. 10: 53, 55, 91, 104, 165
    Vol. 12: 42, 221, 222
atoms
  adiabatic invariants **Vol. 1:** xxxii–xxxiii
  building up of **Vol. 3:** 600*ff*
  constitution/structure of (*see also the
    element in question*) **Vol. 1:** xxi–xxiv,
    xxxiii–xxxiv, 109, 555, 559
    Vol. 4: 7, 44*ff*, 100*ff*
  correspondence principle
    Vol. 1: xxviii–xxix, xxxiv–xxxvi
  effect of external fields on spectra
    Vol. 3: 440*ff*

  formation by successive capture of electrons
    Vol. 4: 76, 100*ff*, 204*ff*, 217*ff*, 246,
      277*ff*, 445, 474, 578, 632
  multiply periodic systems
    Vol. 1: xxxii–xxxiv
  on the knowledge of **Vol. 4:** 7, 44–46
  our present knowledge of **Vol. 4:** 14,
    83–89
  periodic system of elements
    Vol. 1: xxii, xxxiv–xxxv
  quantum mechanics **Vol. 1:** xxxvii–xxxviii
  quantum postulates **Vol. 1:** xxv–xxx
  stability of **Vol. 4:** 438*ff*, 468*ff*
  with several electrons **Vol. 3:** 377,
    509*ff*, 538*ff*
    Vol. 4: 372*ff*
Atoms for Peace
  Award **Vol. 10:** 157, 192
    Vol. 11: 367, 368*p*, 638, 641, 642, 716,
      720, 722, 735
    Vol. 12: 228*p*, 229*p*
  Programme **Vol. 11:** 83, 360, 366, 619*n*
attenuation of radiation, *see* absorption of
    radiation
Attlee, C. **Vol. 11:** 50, 325, 329
Aubel, E. van **Vol. 3:** 361
Auger effect **Vol. 8:** 572
Auger, P. **Vol. 9:** 263
    Vol. 11: 298
    Vol. 12: 118, 494–496
Aurén, T.E. **Vol. 2:** 620
    Vol. 3: 653, 654
Austauscheffekt, *see* exchange effect
autocatalysis **Vol. 11:** 270
automata/machines **Vol. 10:** 12, 117, 130,
    135, 136, 151, 167, 179, 198
average energy/velocity loss, *see* mean
    energy/velocity loss
Avery, O.T. **Vol. 10:** 24
Axel, Prince of Denmark **Vol. 11:** 289
axial symmetry, of field **Vol. 3:** 120
  systems possessing **Vol. 3:** 101
Axis Powers **Vol. 11:** 18*n*, 28, 92

Bachelard, G. **Vol. 7:** 328
Back, E.E.A. **Vol. 1:** xxxi
    Vol. 2: 362, 365, 366
    Vol. 3: 164, 177, 184, 280, 281, 445,
      526, 554
    Vol. 4: 216, 379, 573, 629

Vol. 10: 338
Vol. 12: 340, 341
bacteria/bacteriology **Vol. 10:** 24, 55, 554
bacteriological science **Vol. 11:** 166
bacteriophage **Vol. 10:** 23, 24, 552, 555
Baedeker, K. **Vol. 1:** 231, 242, 244, 344,
    353, 355, 398
Baeyer, O. von **Vol. 2:** 37, 79
    Vol. 8: 69, 76, 91, 149
Bagge, E. **Vol. 8:** 523, 566
Baggesen, J. **Vol. 10:** 256, 266
Bahr, E. von **Vol. 2:** 339, 446, 457, 565, 568
    Vol. 3: 10, 190, 192
    Vol. 4: 437, 470
Bainbridge, K.T. **Vol. 9:** 201
Baker, J. (Aage Bohr's codename)
    Vol. 11: 264, 307, 310
Baker, N. (Niels Bohr's codename)
    Vol. 11: 16, 38, 39, 262, 263, 307, 310
Bakerian Lecture (by A. Fowler)
    Vol. 2: 326, 504*ff*
Bakerian Lecture (E. Rutherford) **Vol. 10:** 408
Ballinger, M. **Vol. 11:** 563
Balmer formula/series/spectrum
    Vol. 2: 110*ff*, 121, 132, 167*ff*, 177, 199,
    202–203, 233, 270, 280, 283, 294–295,
    296, 327–329, 333, 351, 354, 359, 364,
    368, 374, 377–380, 385–388, 395,
    399*f*, 468, 504*ff*, 546, 588, 626,
    629, 630
    Vol. 3: —
    Vol. 5: 157, 161*f*, 167, 170, 225, 239,
    274, 276*f*, 280, 338, 406, 445, 447,
    450*f*, 472, 474, 498
    Vol. 6: 383*ff*, 388
    Vol. 10: 332, 374, 378
    Vol. 12: 42, 44, 45
Balmer, J.J. **Vol. 2:** 167, 233, 270, 280, 283,
    351, 368, 630
    Vol. 3: 71, 207, 246, 251, 373, 411,
    423, 582
    Vol. 4: 434, 469, 659
    Vol. 5: 274
    Vol. 6: 228, 242, 382
    Vol. 10: 58, 374, 388
    Vol. 12: 42
Balmer–Rydberg constant, *see* Rydberg
    constant
Bancroft, F.W. **Vol. 1:** 520, 521
band spectra **Vol. 5:** 187, 202, 349,
    384, 386*ff*

Vol. 6: 395–398, 404, 439*f*
Vol. 9: 80, 231, 299, 316, 328, 611
Vol. 12: 48
Bang-Jensen, P. **Vol. 11:** 71, 326, 327, 330
Bang, F. **Vol. 10:** 224*p*
Bang, J. **Vol. 12:** 16
Banks, J. **Vol. 11:** 445
Barcroft, J. **Vol. 11:** 445
Bardeen, J. **Vol. 12:** 9
barium **Vol. 11:** 579, 580, 598, 599
Barker, E.F. **Vol. 5:** 29, 384, 386, 390, 392
Barkla radiation **Vol. 10:** 386, 387, 393, 394
Barkla, C.G. **Vol. 2:** 189, 413, 580
    Vol. 3: 361, 583, 584, 591
    Vol. 4: 268, 461, 481
    Vol. 6: 386
    Vol. 8: 169, 173, 177
    Vol. 10: 386
Barnes, A.H. **Vol. 8:** 214, 698
Barnes, E.L. **Vol. 1:** 106, 107, 546, 547,
    548, 549
Barnes, G. **Vol. 11:** 255, 257, 268, 271, 274
Barnett, J.S. **Vol. 10:** 339
Barres, J. **Vol. 2:** 618
barrier penetration (*see also* Gamow, G.)
    Vol. 6: 230–231, 246, 405
    Vol. 8: 344–345, 349–350, 371, 456,
    652–653
    Vol. 9: 110, 155–156, 165, 174, 185*ff*,
    203, 210, 249–257, 264, 281, 308, 310,
    314*f*, 322, 324, 326*ff*, 374*f*, 426, 430,
    453, 456, 551, 586*f*, 658, 666
Barschall, H.H. **Vol. 9:** 382, 385, 397, 404
Bartlett, A.A. **Vol. 5:** 3
Baruch, B. **Vol. 11:** 67, 68, 156, 158, 325, 329
Barzin, M. **Vol. 7:** 328
Bateman, H. **Vol. 2:** 64
    Vol. 5: 339
    Vol. 8: 134
Bateson, W. **Vol. 11:** 445
Battaglia, F. **Vol. 10:** 82
Bauer, E. **Vol. 7:** 32*p*, 301
Bayard, P. **Vol. 9:** 638, 640
Bayer, R. **Vol. 7:** 328
Bayliss, W.M. **Vol. 11:** 445
Baym, G. **Vol. 12:** vii, 4*n*, 25
Beament, J.W.L. **Vol. 10:** 126
Beatty, R.T. **Vol. 2:** 236
Beck, G. **Vol. 8:** 329, 405, 556, 566, 709*f*
    Vol. 9: 9, 10, 100, 137–138, 140, 202
    Vol. 12: 17*p*

[47]

Becker, R.　**Vol. 3:** 182, 183
　Vol. 5: 46
Beckmann Hansen, P.　**Vol. 11:** 358*p*
Becquerel, A.H.　**Vol. 2:** 119
　Vol. 10: 102, 385
　Vol. 11: 565, 646, 649
　Vol. 12: 277
Beethoven, L. van　**Vol. 11:** 376
　Vol. 12: 359
behaviourism　**Vol. 10:** xli, 92, 105, 545, 546,
　549, 550
Belinfante, F.J.　**Vol. 12:** 38*p*
Bell, G.I.　**Vol. 8:** 592, 599, 610*ff*, 625
Bendt-Nielsen, B.　**Vol. 11:** 358*p*
Benedicks, C.　**Vol. 3:** 653, 654, 655
benevolence　**Vol. 10:** 97
Benzon, A. von　**Vol. 10:** 74
Berg, G.J. van den　**Vol. 12:** 38*p*
Bergengren, J.　**Vol. 4:** 499, 532
Bergmann series　**Vol. 3:** 253
Bergsøe, P.　**Vol. 9:** 626, 628, 633, 635
　Vol. 11: 354, 488, 490
　Vol. 12: 100, 127
Bergson, H.　**Vol. 10:** 4*n*
Bergwitz, K.　**Vol. 8:** 668
Bering Strait　**Vol. 12:** 106, 239, 243,
　244, 246
Bering-Liisberg, A.S.　**Vol. 12:** 148
Bering, V.　**Vol. 12:** 106, 107, 238–241,
　244–246, 527, 528
Berkeley (*see also* California, University of)
　Vol. 8: 739
　Vol. 9: 39, 350, 353
　Vol. 12: 84, 386, 388
Berkeley, G.　**Vol. 10:** xlviii
Berle, A.　**Vol. 11:** 328, 331
Berlème, A.　**Vol. 2:** 618
　Vol. 3: 24, 286, 295
　Vol. 10: 224*p*
　Vol. 11: 512, 521
　Vol. 12: 199*n*
Berlin　**Vol. 5:** 24, 27, 48, 76, 228*f*, 235, 301,
　303, 308, 352, 365*ff*, 377*f*, 459, 461,
　463, 478, 481, 508
　Vol. 6: xii, 24, 46, 48, 317, 421, 456,
　463, 465
　Vol. 8: 221, 660*f*, 681, 688*f*, 702*ff*,
　708, 712*f*, 791
　Academy　**Vol. 6:** 21*f*, 317, 419*f*, 480
　Lecture, Bohr's, 1930　**Vol. 6:** 317, 480

Berlin, lecture in, 1920　**Vol. 3:** 21*ff*
　translation of　**Vol. 3:** 241–282
Berliner, A.　**Vol. 2:** 618
　Vol. 5: 455*f*
　Vol. 10: 524–526
Berlingske Aftenavis (Danish newspaper)
　Vol. 11: 9, 61, 206–208, 211
Berlingske Tidende, *see* Berlingske Aftenavis
Bernays, P.　**Vol. 7:** 328
Bernhard, Prince of the Netherlands
　Vol. 10: 159*p*
Bernoulli, A.L.　**Vol. 2:** 236
beryllium　**Vol. 11:** 577, 578, 593, 596,
　614, 624
beryllium atom, structure of　**Vol. 2:** 204–205
Berzelius, J.J.　**Vol. 4:** 429, 467
　Vol. 10: 366*p*
　Vol. 11: 354
Besso, M.　**Vol. 6:** 8
β-decay/particles/radiation/rays/spectrum
　Vol. 5: 94, 95, 216
　Vol. 6: 25, 309, 376, 378, 404, 406*f*,
　425, 442*f*, 445, 446–447
　Vol. 8: 17, 211, 669, 712*f*, 734, 736,
　754, 764, 793
　Vol. 10: 385, 387, 393, 411, 412,
　450, 451
　Vol. 11: 575, 578–580, 589, 590, 595,
　597, 599, 600
　Vol. 12: 12, 42, 43, 54, 60, 66, 256,
　260, 286, 287, 365
β-particles
　absorption of, *see* absorption of charged
　　particles
　secondary　**Vol. 5:** 5, 75*ff*, 184, 199, 336
　stopping of, *see* stopping of charged particles
β-ray
　emission　**Vol. 9:** 5–11, 13*f*, 55, 57, 68,
　　85–89, 96–97, 103, 113, 117, 125,
　　131*f*, 136, 137–140, 152*f*, 166, 174*f*,
　　215*f*, 218*ff*, 274*f*, 280, 295*f*, 306–307,
　　309, 321, 323, 344*f*, 350, 353,
　　355–357, 365*ff*, 378, 384*ff*, 415, 417,
　　420, 431, 435, 444*ff*, 448, 457, 461,
　　547*f*, 555, 557, 567, 605–606, 676, 679
　spectrum (*see also* probability distribution of
　　energy loss)　**Vol. 2:** 78*ff*
β-rays　**Vol. 7:** 181, 182, 191, 479
　origin of　**Vol. 2:** 119–120, 179, 212–214,
　　236–238, 418, 423–425, 428, 440–441,
　　581, 584, 586, 588

Bethe's stopping formula, *see* stopping formula, Bethe's
Bethe, H.A.   **Vol. 1:** xxiv
  Vol. 2: 10
  Vol. 7: 48
  Vol. 8: 36, 208, 214*f*, 217*f*, 220, 225, 239, 256*f*, 269, 278, 304, 330, 407, 428, 502, 519, 521, 523*f*, 549*f*, 566*f*, 635, 638, 700*f*, 811
  Vol. 9: 16*f*, 19, 24, 29*f*, 31, 32, 38*f*, 146, 181, 194, 209, 227, 248, 250, 259*f*, 264, 312, 325, 370, 377*f*, 385, 387*f*, 484, 491*f*, 507, 534, 539–540, 609, 613, 642*f*
  Vol. 10: 387, 417, 455*p*
  Vol. 11: 568
Bevan, P.V.   **Vol. 2:** 337, 448, 449, 564, 566, 604
  Vol. 5: 44
  Vol. 10: 401
Bialobrzeski, C.   **Vol. 7:** 260, 301
Bianco Luno   **Vol. 12:** 107*n*
bibliography to "On the Constitution..."
  Vol. 2: 117, 235–239, 626
Bieberbach, L.   **Vol. 10:** 471, 472
Biggs, H.T.   **Vol. 2:** 597, 616, 619, 620
Biilmann, E.   **Vol. 3:** 286, 295
  Vol. 11: 406*p*
Bijl, A.   **Vol. 8:** 728*ff*, 732
Bijl, H.J. van der   **Vol. 3:** 239
Bille, C.S.A.   **Vol. 12:** 152, 164
Billgren, P.   **Vol. 12:** 400, 404
billiard-ball model of nucleus   **Vol. 9:** 158, 167–168, 176, 207–208
binding
  energy (*see also* ionization potential)
    Vol. 8: 512, 603*f*, 610, 622*f*, 792
    Vol. 9: 12, 15, 22, 29, 34, 72, 76*f*, 93, 112–113, 145, 153, 169, 177, 188*f*, 193, 208, 210, 234, 258, 277, 285, 311, 313, 324, 326, 342, 345, 357, 361, 367, 369*f*, 375*f*, 381, 383, 385*ff*, 404, 407*f*, 419*ff*, 423*f*, 426, 428*ff*, 432, 434, 438, 440, 448*f*, 451*ff*, 455*f*, 458, 460, 463, 465, 470, 478, 484, 490, 507, 523–524, 560, 575*f*, 618, 646*f*
  force   **Vol. 8:** 6*ff*, 15, 51*ff*, 78*ff*, 93*ff*, 109, 113, 117, 131*f*, 136, 142, 154, 156, 170*f*, 174*f*, 178*f*, 253, 264, 272, 281, 371, 378, 411*f*, 438, 488*ff*, 493, 499, 504*f*, 511, 518, 524, 532, 611, 635

binding of electrons by atoms
  first two electrons   **Vol. 4:** 287*ff*
  third electron   **Vol. 4:** 140*ff*, 291*ff*
  fourth to sixth electrons
    Vol. 4: 146*ff*, 293*ff*
  seventh to tenth electrons   **Vol. 4:** 296, 297
  eleventh electron   **Vol. 4:** 153*ff*, 297*ff*
  twelfth to eighteenth electrons
    Vol. 4: 159*ff*, 301*ff*
  nineteenth electron   **Vol. 4:** 302, 399, 458, 480
biochemical science   **Vol. 11:** 166
biochemistry   **Vol. 7:** 408, 421
  Vol. 10: 23, 150, 152, 166, 168, 169, 180, 214, 466–469, 471, 472, 484, 497, 500, 558, 560
biology   **Vol. 6:** xxi, 196, 206, 216, 234–235, 251–253, 272–273, 274–275, 298–299, 300–301, 352
  Vol. 7: 336, 376, 408, 420, 421, 450, 453
  Vol. 8: 246, 655*f*, 714, 716, 731*ff*, 735
  Vol. 9: 167, 175, 216, 220, 599*ff*, 641, 643
  Vol. 10: —
  Vol. 11: 133, 134, 337, 516, 524, 569
  Vol. 12: 105, 157, 161, 170, 174, 201, 205, 210, 212, 215, 217, 502, 504
  radiation   **Vol. 12:** 502, 504
biophysics   **Vol. 7:** 408, 421
  Vol. 10: 21, 22, 166, 468, 469
  Vol. 11: 337
Biot, J.B.   **Vol. 10:** 349, 362
Birmingham   **Vol. 8:** 259
  Vol. 9: 49*f*, 612*f*, 615, 617, 619, 621, 670
Birmingham meeting, *see* British Association for the Advancement of Science
Birmingham, University of   **Vol. 11:** 87*n*
Birtwistle, S.   **Vol. 4:** 663
bismuth   **Vol. 11:** 575, 590
Bjerge, T.   **Vol. 8:** 235
  Vol. 9: 28, 263, 560, *629–630*, 631, 633, 635, 638, 640, 652
  Vol. 11: 358*p*, 518, 526
Bjerknes, C.A.   **Vol. 5:** *483*
Bjerknes, V.F.K.   **Vol. 1:** 422, 427
  Vol. 5: *483*, 485
  Vol. 8: 197, 199
Bjerrum, E.   **Vol. 12:** 497, 498
Bjerrum, J.P.   **Vol. 12:** 125

Bjerrum, N.   **Vol. 1:** 98, 101, 125
  Vol. 2: 224, 338, 339, 341, 437, 446,
  447, 452, 457, 565, 567, 568, 582, 631
  Vol. 3: 10, 189, 190, 191, 192, 233,
  329, 595, 664
  Vol. 4: 436, 453, 470, 478
  Vol. 5: 132, *139*, 305, 491*f*
  Vol. 6: 308, 396, 442, 444
  Vol. 7: 446
  Vol. 8: 704*f*
  Vol. 9: 576*f*, 627*f*
  Vol. 10: 274*p*, 386, 417
  Vol. 11: 406*p*, 525
  Vol. 12: 125, 126, 438, 438*p*, 439–441,
  444, 446, 447, 448*p*, 452, 453, 457,
  458, 496–498, 532
Bjørnson, B.   **Vol. 10:** 303
black-body
  energy fluctuations   **Vol. 7:** 73, 75, 98, 104,
    112, 120, 132, 134, 149–151, 154, 159,
    164, 486
  radiation (*see also* heat radiation; Planck's
    formula/law; thermal radiation)
    Vol. 2: 126, 274, 279–280, 286–289,
    292–293, 312, 351, 393, 434, 445, 450,
    474, 584, 589, 627
    Vol. 5: 45, 103*f*, 132, 139, 274, 275
    Vol. 6: 10, 203, 208, 227, 241, 382,
    398, 423
    Vol. 7: 34, 279, 306, 342, 344*p*, 347,
    400, 413, 414, 470
    Vol. 8: 27
    Vol. 9: 45, 108, 262, 293*f*, 298, 303,
    313, 315*f*, 318, 326*ff*, 335, 419, 423,
    426–428, 448, 451, 454, 501, 585, 587
Blackett, P.M.S.   **Vol. 7:** 32*p*, 468
  Vol. 8: 210, 223, 231, 634, 642,
  649–650, 786, 811
  Vol. 9: 134, 135*f*, 166, 174, 216, 219
  Vol. 10: 399, 404, 412, 450
  Vol. 11: 569
Blackman, M.   **Vol. 9:** 607*f*
Blanc, A.C.   **Vol. 7:** 328
Blaton, J.   **Vol. 8:** 435, 566
Blau, M.   **Vol. 9:** 425, 453, 586*f*
Blinkenberg, C.   **Vol. 11:** 406*p*
Bloch's stopping formula, *see* stopping
    formula, Bloch's
Bloch, C.E.   **Vol. 12:** 528
Bloch, F.   **Vol. 1:** xxiv
  Vol. 2: 10

  Vol. 6: 25, 425
  Vol. 7: 445, 465, 467
  Vol. 8: 216*ff*, 224, 238*f*, 253, 256, 269,
  278, 305, 330, 372, 407, 438, 504, 527,
  566, 636, 642, 650–656, 700*ff*, 704*f*,
  714, 716
  Vol. 9: 10, 14, 25*f*, 31, 32, 38, 44, 194,
  534, 540–544, 596*f*, 600*ff*, 606, 608*f*
  Vol. 10: 448, 534, 535
  Vol. 11: 698*n*
  Vol. 12: 5–8, 17*p*, 22–25, 27, 28, 30,
  33, 38*p*, 498, 500, 525
Bocciarelli, D.   **Vol. 9:** 478
Bôcher, M.   **Vol. 1:** 217, 331
Bock, J.C.   **Vol. 11:** 406*p*
Bodelsen, M.   **Vol. 12:** 402, 405
Boer, J. de   **Vol. 12:** 38*p*
Boerhaave, H.   **Vol. 12:** 431*n*
Bøggild, J.K.   **Vol. 1:** xxvii
  Vol. 2: 10
  Vol. 7: 396*p*
  Vol. 8: 233, 235, 237, 244, 246*f*, 255,
  262, 265, 318, 320, 323–326, 328, 331*f*,
  375, 405*f*, 467, 486, 554*f*, 560, 566,
  571, 576*f*, 585, 587, 731*f*, 734*f*, 749
  Vol. 9: 28, 413, 443, 621
  Vol. 11: 579, 598
Bøggild, M.   **Vol. 12:** 106, 241, 246
Bohm, D.   **Vol. 7:** 281*n*
Bohr–Bøggild–Brostrøm–Lauritsen paper
  Vol. 8: 237, 323–326, 331, 406
"Bohr Festspiele", Göttingen 1922
  Vol. 5: 23, 92, 407
Bohr–Kalckar papers   **Vol. 9:** 18*ff*, 25, 27–34,
  39, 49, 156, 211, 223–264, 299, 306,
  320, 369, 413, 443, 489, 505, 539,
  572*f*, 580*ff*, 595*f*, 600*f*, 610, 676*f*
Bohr–Kramers–Slater theory (*see also* Slater,
  J.C.)   **Vol. 6:** 37, 39, 60, 62, 65, 70,
  115, 139*f*, 148, 191, 417, 431
  Vol. 8: 31, 36, 206
  application/elaboration of   **Vol. 5:** 12,
    22–23, 31, 40–41, 46, 65–67, 110–118
  experimental test of, *see* Bothe–Geiger
    experiment; Compton–Simon
    experiment
  reception of   **Vol. 5:** 23–38, 74, 78, 84,
    86–88, 299, 308–309, 345*f*, 351, 362,
    380, 383, 385, 387*f*, 390, 392–393,
    402–403, 415–421, 440–441, 443,
    490–491

Bohr–Lindhard paper **Vol. 8:** 258–265, 581, 584*f*, 587, 592, 593–625, 823*f*

Bohr–Peierls–Placzek papers **Vol. 9:** 47, 48–52, 80, 332, 340, 377, 391–393, 487–519, 612–621, 659, 663, 678*f*

Bohr–Rosenfeld paper **Vol. 7:** 7*n*, 8, 11, 12, 18, 34*n*, 45, 189*n*, 197*n*, 230*n*, 235*n*, 324, 483*n*

Bohr–Wheeler papers **Vol. 9:** 67, 68–74, 81–83, 345, 359–389, 398, 399–404, 413, 443, 484, 551, 561*f*, 657–671, 679*f*

Bohr's stopping formula, *see* stopping formula, Bohr's

Bohr, Aage **Vol. 3:** 25
   Vol. 6: vii, 330, 414, 449, 469
   Vol. 7: 522, 523
   Vol. 8: vi, 212, 246*f*, 250*f*, 259, 262, 336, 358, 374, 392, 410, 420, 428, 435, 496, 503, 527, 566, 584, 646*f*, 649, 731, 733*ff*, 737*f*, 765, 822*ff*
   Vol. 9: vi, 81, 680
   Vol. 10: 25*n*, 164, 476, 487, 556, 558, 584, 585, 587–589, 591, 594
   Vol. 11: ii*p*, vii, 12*n*, 14*n*, 15, 16, 20, 22, 25, 27, 33, 35, 36, 42, 43, 57, 60, 78*n*, 79, 99, 238, 243, 264, 265, 287, 288, 290, 307, 310, 324, 326–330, 333, 340*n*, 375, 710, 711, 730–733
   Vol. 12: vii, 209, 211, 213, 216, 406*p*, 411, 412, 420, 421, 496–498, 515, 529–531, 533, 534

Bohr, Ann **Vol. 7:** 396*p*

Bohr, Christian (Niels Bohr's father)
   Vol. 1: xviii, xx, 4, 496, 497, 510, 511, 512, 513, 540, 541
   Vol. 6: xvii*ff*
   Vol. 10: ix, xxviii, 4, 5, 114*p*, 117, 298, 301, 312*p*, 313, 319, 543, 548
   Vol. 11: 6, 679*n*
   Vol. 12: 108, 120, 121, 123, 126, 135, 136, 138, 140, 144, 149, 150, 153, 157, 164–166, 170, 198, 202, 209, 212, 213, 217

Bohr, Christian (Niels Bohr's son)
   Vol. 6: 24, 424
   Vol. 8: 652, 704*f*
   Vol. 11: 218*n*
   Vol. 12: 122, 125, 127, 209, 213, 406*p*, 409, 410*p*, 411–418, 419*p*, 420–424, 497, 498, 515

Bohr, Ellen **Vol. 1:** xviii, 106, 496, 497, 498, 499, 500, 501, 502, 503, 504, 505, 508, 509, 522, 523, 532, 533, 534, 535, 538, 539, 540, 541, 542, 543, 544, 545, 558, 559, 570, 571
   Vol. 2: 520–521
   Vol. 5: 328, 503
   Vol. 6: xviii, 451
   Vol. 8: 74, 818
   Vol. 12: 120, 135, 136, 138, 140, 144, 149, 150, 209, 213, 412, 420, 524

Bohr, Erik **Vol. 8:** 336, 734*f*, 823
   Vol. 9: 58, 356, 560, 622*ff*, 627, 629*ff*, 660*f*, 679
   Vol. 11: 13*n*, 358*p*, 728
   Vol. 12: 209, 213, 406*p*, 411, 412, 420, 421, 496–498, 515, 526

Bohr, Ernest **Vol. 7:** 522
   Vol. 10: 584, 591, 592
   Vol. 12: vii, 105, 209, 213, 406*p*, 411, 412, 420, 421, 496–498, 515

Bohr, Hans **Vol. 6:** 24, 424
   Vol. 7: 396*p*, 458, 459
   Vol. 8: 822
   Vol. 9: 58, 60, 64, 550, 564, 579*ff*, 622, 624
   Vol. 10: 583
   Vol. 12: vii, 117*p*, 209, 213, 388*n*, 406*p*, 411, 412, 420, 421, 496–498, 515

Bohr, Harald (Niels Bohr's brother)
   Vol. 1: xviii, xx, 4, 10, 93, 94, 95, 96, 99, 102, 104, 107–109, 114, 495–586
   Vol. 2: 4, 5, 6, 8, 103, 104, 108, 126, 129, 520, 521, 537, 557, 558, 560, 561, 613
   Vol. 3: 5, 34, 627, 652
   Vol. 5: 219, 308, 328, 435, 437, 459, 461*ff*, 465*f*
   Vol. 6: xvii, xxiii, xxiv, 28, 42, 308, 435, 436
   Vol. 7: 31*n*
   Vol. 8: 4*f*, 7*ff*, 11, 14, 74, 224, 642, 687, 750, 818
   Vol. 10: xxix, xxx*p*, 36*p*, 114*p*, 221, 299*n*, 300, 460
   Vol. 11: 239, 339, 371, 372, 406*p*, 460*n*, 676*p*
   Vol. 12: 4, 128, 198, 202, 210, 215, 399, 403, 452, 457

Bohr, Harald (Niels Bohr's son)
   Vol. 12: 411, 412, 420, 421

Bohr, Jenny   **Vol. 1:** 496, 497, 558, 559
Bohr, Margrethe   **Vol. 1:** xxvi, 495, 536, 537
　　　Vol. 4: 683
　　　Vol. 5: 156, 230, 237, 240, 299*f*, 304,
　　　307*f*, 327*ff*, 342, 348, 355*f*, 380, 384,
　　　386*f*, 389, 408*ff*, 435, 437, 459, 461,
　　　488*f*, 494, 503, 512
　　　Vol. 6: xxii, xxv, 7, 12, 24, 58, 90, 422,
　　　424, 459, 461, 469, 478, 480
　　　Vol. 7: 261*p*, 287*p*, 396*p*, 476*p*
　　　Vol. 8: 5, 74, 116, 162, 268, 659, 662,
　　　665, 670, 676, 684, 692, 702*f*, 706,
　　　715, 717*f*, 727, 729, 735*ff*, 750, 755*f*,
　　　763, 818*ff*
　　　Vol. 9: 60, 68, 556, 564, 632, 634, 636,
　　　638*f*, 652, 667, 670
　　　Vol. 10: xxx*p*, xxxii, 288*p*, 299*n*, 301,
　　　302, 387, 396, 397, 411, 414, 464, 465,
　　　479, 481, 484, 487, 489, 494, 516, 517,
　　　543, 546, 547, 550, 551, 556, 558–561,
　　　567, 572, 582, 584, 586, 587, 589–593
　　　Vol. 11: 12, 28, 29, 49, 65*n*, 136*p*, 214,
　　　218, 220, 222, 228, 230, 238, 295, 297,
　　　299, 303, 308, 318, 319, 322, 326, 330,
　　　340, 361, 375, 376, 711, 712, 731, 736
　　　Vol. 12: 105, 107, 110, 117*p*, 122,
　　　129*p*, 138, 140, 144, 149, 150, 154*p*,
　　　167*p*, 209, 213, 368, 388*n*, 391*p*, 411,
　　　412, 414, 418, 420, 422, 424, 470, 472,
　　　496–498, 506, 507, 511–515, 524, 526,
　　　529, 531–533
Bohr, Marietta   **Vol. 11:** 327, 330
Bohr, Ulla   **Vol. 1:** 495
Bollnow, O.F.   **Vol. 5:** 303
Bologna meeting, 1937   **Vol. 9:** 44
Boltwood, B.B.   **Vol. 6:** 377
Boltzmann Festschrift   **Vol. 2:** 64
Boltzmann statistics   **Vol. 6:** 143
Boltzmann's *H*-theorem (*see also* entropy;
　　　irreversibility; second law of
　　　thermodynamics)
　　　Vol. 6: 320*ff*, 438, 440
Boltzmann, L.   **Vol. 1:** xvii, 135, 192, 201,
　　　242, 290, 321, 353, 394, 425, 430, 440,
　　　449, 451, 464
　　　Vol. 2: 64, 436, 454
　　　Vol. 3: 73, 75, 76, 208, 470, 473, 474
　　　Vol. 5: 273, 274
　　　Vol. 6: 320, 322*ff*, 327, 359, 400, 452
　　　Vol. 7: 342
　　　Vol. 10: 104, 432, 442

　　　Vol. 11: 670
　　　Vol. 12: 372
Bolyai, J.   **Vol. 12:** 372
Bonaparte, H.   **Vol. 7:** 282*n*
Bonaparte, N.   **Vol. 10:** 397
　　　Vol. 11: 26
bonds, homopolar/polar   **Vol. 12:** 51
Bonhoeffer, K.F.   **Vol. 6:** 397
Bonner, T.W.   **Vol. 9:** 478
Bonnesen, M.   **Vol. 11:** 9, 205*n*, 349
　　　Vol. 12: 186, 189*n*, 196
Bonnet, H.   **Vol. 7:** 302
"Book, The"   **Vol. 10:** v, vi, 477
Booth, E.T.   **Vol. 9:** 61, 365, 380, 385, 401,
　　　404, 441, 465, 469
Borch, M.   **Vol. 3:** 287, 296
　　　Vol. 11: 517, 525
Borelius, G.   **Vol. 1:** xxvii
　　　Vol. 12: 38*p*
Borelli, A.   **Vol. 10:** 54, 67
Born approximation   **Vol. 8:** 207*ff*, 214, 216,
　　　218, 230, 231, 270, 273, 279, 282, 288,
　　　289*f*, 293, 296*f*, 304*f*, 330, 343*f*,
　　　348, 363*f*, 372, 379*f*, 445*f*, 500*ff*,
　　　528, 536*ff*, 627, 629, 631, 635*ff*, 663,
　　　699*ff*, 793, 799, 811, 813
Born, H.   **Vol. 5:** 26, 299*ff*, 304, 308, 310,
　　　313, 349
Born, I.   **Vol. 5:** 26
Born, M.   **Vol. 1:** xxxvii, 566, 567
　　　Vol. 2: 127, 331
　　　Vol. 3: 22, 23, 42, 467, 650, 656,
　　　657, 664
　　　Vol. 4: 4, 7, 8, 24, 38, 39, 48, 57, 68,
　　　95, 105, 203, 276, 392, 593, 643,
　　　669–673, 693, 694, 700, 710
　　　Vol. 5: 23*f*, 26, 46, 53, 57, 72*f*, 76*f*,
　　　83*ff*, 156, 188*f*, 203, 220, 223, 225,
　　　238, 251*f*, 258*f*, 265*f*, 280, 294,
　　　299–313, 338, 345*f*, 348*f*, 352, 355*f*,
　　　362*ff*, 368*ff*, 396, 405, 440, 443,
　　　497*f*, 501
　　　Vol. 6: 3*f*, 8*f*, 12, 30, 36, 37*ff*, 52, 71,
　　　95*ff*, 103, 124*f*, 127, 137–139, 142,
　　　153, 164*f*, 170, 230, 245, 393, 415,
　　　417, 422*f*, 459
　　　Vol. 7: 4, 221, 262, 283, 308, 347
　　　Vol. 8: 30*ff*, 36, 207*f*, 343, 348, 637,
　　　642, 656–663, 678, 688, 801, 803
　　　Vol. 9: 21, 23, 28, 305, 319, 607*f*,
　　　617, 653

Vol. 10: xxiv, 212, 400, 401, 403, 426, 427, 441, 442, 454, 514, 515
Vol. 11: 298, 567
Vol. 12: 101*n*, 363, 530
Bose statistics **Vol. 7:** 220, 221, 227, 228
Bose–Einstein statistics **Vol. 6:** 50, 144, 397, 398, 404*f*, 466
Vol. 9: 94*f*, 117
Vol. 8: 211, 437, 442, 755
Vol. 10: 403, 444
Bose, E. **Vol. 1:** 158
Bose, S.N. **Vol. 6:** 144
Vol. 10: 442
Bošković Institute (Zagreb) **Vol. 12:** 104, 234, 516, 517, 532
Bošković, R.J. **Vol. 12:** 104, 105, 228, 229, 233, 234, 516, 517, 532
boson **Vol. 10:** 443, 452
botany **Vol. 10:** 148
Bothe–Geiger experiment **Vol. 5:** 22, 34, 36*f*, 68*f*, 75–80, 81*f*, 86*f*, 89*f*, 178, 190, 194, 204, 210*f*, 215, 221, 276, 303–304, 306, 328, 351, 352–354, 362*f*, 416, 420, 440, 443, 449, 453, 484*f*, 499
Vol. 6: 17, 102, 115, 139*f*, 148, 185
Vol. 8: 36, 206
Bothe, W. (*see also* Bothe–Geiger experiment)
Vol. 5: 5, 37, 68*f*, 75*ff*, 87, 89*f*, 95, 178, 190, 194, 204, 210*f*, 215, 221, 276, 303, 352, 440, 443, 499
Vol. 6: 139, 185
Vol. 7: 32*p*
Vol. 8: 255, 483*f*, 547, 566, 809
Vol. 9: 13, 43*f*, 50, 100, 134, 262, 298*f*, 314*f*, 327*f*, 336*f*, 389, 393, 542*f*, 602, 604
Vol. 10: 410
Vol. 11: 243, 244, 264, 265
Vol. 12: 6*n*
Bouligand, G. **Vol. 7:** 328
Bowen, I.S. **Vol. 10:** 392
Boyd Orr, Lord J. **Vol. 11:** 733
Braae, J. **Vol. 12:** 402, 404
Brackett, F. **Vol. 3:** 424, 583
Brackett series **Vol. 2:** 116, 169
Bradt, H.L. **Vol. 8:** 624*f*
Bragg's law **Vol. 5:** 121, 323
Vol. 8: 82, 98, 110, 113, 669*ff*, 752
Bragg's law of absorption **Vol. 2:** 105, 137, 148, 500

Bragg, W.H. **Vol. 2:** 28, 29, 336, 412, 428, 429, 578
Vol. 3: 362, 363, 395, 591
Vol. 4: 442, 473, 659
Vol. 6: 387
Vol. 8: 60*f*
Vol. 10: 387, 436
Vol. 11: 445
Vol. 12: 287
Bragg, W.L. **Vol. 2:** 428, 429
Vol. 3: 32, 361, 362, 591
Vol. 4: 442, 473, 659
Vol. 6: 36, 37, 100, 387, 415
Vol. 8: 223, 228, 642, 664, 698, 753, 794, 798, 804
Vol. 10: 330, 387, 435, 436, 441, 455*p*
Vol. 12: 43, 287
Brahe, T. **Vol. 5:** 49
Vol. 10: 257, 267, 354, 368
Vol. 11: 544, 545, 551
Vol. 12: 101*n*, 371, 529
brain **Vol. 10:** xxv, xliii, xliv, xlvii, 8, 19, 94, 105, 117, 169
branching of tracks **Vol. 8:** 155, 212, 225, 236*f*, 255, 312, 320*f*, 325*f*, 330, 371, 375*f*, 405*f*, 426, 464–469, 555, 678, 786, 793
branching rule **Vol. 5:** 289, 330
Brandes, G. **Vol. 10:** 257, 266, 303
Vol. 11: 311, 313
Vol. 12: 101*n*, 528
Brandt, F. **Vol. 10:** 36*p*, 298*n*, 301*n*, 325*n*
Brecht, B. **Vol. 12:** 14
Breit–Wigner formula **Vol. 9:** 29, 32, 48, 246*f*, 312, 325, 377, 379, 380, 388, 491, 507*f*, 510, 539, 613*ff*
Breit, G. **Vol. 3:** 483, 487
Vol. 4: 361
Vol. 5: 45, 46, 145, 148, 495
Vol. 6: 181*ff*
Vol. 8: 797
Vol. 9: 29, 31, 32, 47, 189, 209, 246, 312, 325, 377, 491, 507*f*, 510, 539
Bretscher, E. **Vol. 8:** 808
Briand, A. **Vol. 8:** 673
Brickwedde, F.G. **Vol. 9:** 56
Bridgman, P.W. **Vol. 7:** 411
Vol. 10: 440
Brillouin, L.N. **Vol. 3:** 361, 362, 363
Vol. 6: 36, 37, 318, 333, 336, 339, 347, 415

[53]

Vol. 7: 260, 301
Vol. 12: 17*p*
Brinkman, H.C. **Vol. 8:** 205*f*, 256, 530,
536*f*, 566, 789
Bristol Lecture, Bohr's, 1931 **Vol. 6:** 317,
319, 361–370, 481
Bristol, University of **Vol. 11:** 569
British Association for the Advancement of
Science **Vol. 2:** 124–125, 420, 532,
586, 587–588
Vol. 6: 7, 317, 355–356, 374
Vol. 8: 787
Vol. 9: 47, 333–337, 678
Liverpool Meeting, 1923 **Vol. 3:** 44, 45,
575–579
Vol. 5: 9, 487, 492
Southampton Meeting, 1925 **Vol. 5:** 339
Sydney Meeting, 1914 **Vol. 12:** 285, 288
British Broadcasting Corporation (BBC)
Vol. 10: 172, 587
Vol. 11: 376, 735
Britten, R. **Vol. 10:** 166
Broch, E.K. **Vol. 9:** 412
Brode, R.B. **Vol. 5:** 364*f*
Broek, A.J. van den **Vol. 2:** 118, 119, 120,
133, 189, 236, 290, 314, 352, 393, 419,
551, 552, 591
Vol. 3: 204, 242, 584
Vol. 4: 45, 195, 268, 422, 432, 468, 659
Vol. 8: 171, 175, 179
Vol. 10: 386, 417
Vol. 12: 43
Broer, L.J.F. **Vol. 12:** 38*p*
Broglie, L. de (*see also* de Broglie wave)
Vol. 1: xxxvii
Vol. 5: 81, 193, 206, 210*f*
Vol. 6: 31, 36, 37, 39, 53, 60, 62, 64,
70*f*, 77*f*, 84*f*, 92, 96, 102, 116*f*, 126,
137, 149, 151, 153, 184, 229, 245,
264*f*, 287*f*, 333–334, 359, 393*f*, 415,
417, 439, 441
Vol. 7: 32*p*, 260, 267, 300, 301, 309,
329, 347
Vol. 9: 134, 305, 319
Vol. 10: 212, 402, 426, 440, 442
Vol. 11: 567
Vol. 12: 11, 47, 358
Broglie, M. de **Vol. 2:** 597
Vol. 3: 32, 361, 362, 363, 368, 615
Vol. 4: 252, 319, 323
Vol. 10: 438

Bromberg, J. **Vol. 9:** vi, 21
Brøndal, V. **Vol. 10:** 299*n*, 300
Bronk, D.W. **Vol. 10:** 82
Vol. 11: 325, 329
Brønsted, J.N. **Vol. 2:** 614
Vol. 3: 292, 300
Vol. 8: 727, 729
Vol. 11: 406*p*
Vol. 12: 125, 438, 440
Brooke, R. **Vol. 10:** 169*n*
Brookhaven National Laboratory
Vol. 11: 569, 711*n*
Brorfelde astronomical observatory
Vol. 11: 460*n*
Brostrøm, K.J. **Vol. 2:** 10
Vol. 8: 234, 235, 237, 255, 318, 320,
323–326, 328, 331*f*, 375, 405*f*, 467,
486, 555, 566, 585, 587,
731*f*, 749, 795
Vol. 9: 28, 413, 443
Vol. 11: 358*p*
Brown **Vol. 4:** 37, 688
Brownian motion **Vol. 5:** 27, 273*f*
Vol. 12: 372
Brubaker, W. **Vol. 9:** 478
Brümmer **Vol. 1:** 65
Brunings, J.H.M. **Vol. 8:** 407, 540, 556, 566
Bruns, C. **Vol. 3:** 307
Brussels **Vol. 6:** 35, 37, 50, 99, 191, 310*ff*,
338, 415, 431, 467
Vol. 8: 220, 257, 569, 702*ff*, 712*f*
Brussels, University of **Vol. 12:** 534
Buch-Andersen, E. **Vol. 4:** 670
Vol. 8: 692
Vol. 12: 17*p*
Bucherer, A.H. **Vol. 2:** 199
Buchner, E.H. **Vol. 2:** 630
Buddha **Vol. 6:** xx
Vol. 10: 60, 502, 504
Vol. 12: 158, 171
building-up principle **Vol. 5:** 289, 332, 340,
424, 428
Bukharin, N. **Vol. 11:** 8–10, 214, 215,
217, 218
Bull, F. **Vol. 10:** 369*n*
Bulletin of the Atomic Scientists **Vol. 11:** 81,
328, 331
Bunsen, R.W. **Vol. 4:** 446, 475
Burbury, S.H. **Vol. 1:** 424, 425, 429, 430
Burger, H.C. **Vol. 5:** 251*f*, 279

[54]

Burgers, J.M.   **Vol. 3:** 7, 10, 83, 87, 88, 96, 124, 146, 159, 191, 192, 317, 333, 348, 355, 372, 409, 430, 463, 467, 470, 475, 639, 640
  Vol. 4: 357, 358, 359
Burrau, C.   **Vol. 1:** 540, 541
Bury, C.R.   **Vol. 4:** 244, 315, 316, 459, 463, 480, 482, 489, 522
  Vol. 6: 390
Bush, V.   **Vol. 8:** 788
  Vol. 11: 15, 38, 40–42, 45–48, 54, 66, 262, 267–269, 271, 275–278, 280, 285–287, 292, 300, 308, 325, 329
Byrnes, J.F.   **Vol. 11:** 68
Byskov, J.   **Vol. 10:** 250p, 271p

"C" (head of British intelligence) (*see also* Menzies, S.)
  Vol. 11: 25, 38, 263
Cabrera, B.   **Vol. 7:** 32p
  Vol. 10: 446
  Vol. 12: 376p
cadmium   **Vol. 11:** 578, 595
  spectrum of   **Vol. 2:** 409
calculation device   **Vol. 10:** 123, 130, 135, 167, 499
calculus   **Vol. 10:** 142, 157, 175
  Vol. 11: 669
Calhoun, R.L.   **Vol. 10:** 82
California Institute of Technology (Pasadena)
  Vol. 12: 386, 388
California, University of (*see also* Berkeley)
  Vol. 2: 333
  Vol. 11: 564
Calvin, J.   **Vol. 10:** 555
Cambridge   **Vol. 2:** 3
  Vol. 5: 7, 9, 54, 68, 225, 300, 315f, 338, 341f, 441, 487ff, 492, 502
  Vol. 6: xii, xv, 44f, 277f, 317, 357, 359f, 377, 438, 449f, 457f, 480
  Vol. 8: 4f, 28f, 203, 210, 223, 228, 647, 706, 709f, 715f, 754f, 785f, 806f
  Vol. 9: 6, 19, 21, 32, 47, 53, 63, 216, 220, 333, 550, 555, 567, 578ff, 610, 630, 651, 652
  Vol. 10: 383, 384, 386, 409, 410, 414, 426, 449, 450, 495
  Lectures, Bohr's, 1922   **Vol. 6:** 458

Lecture, Bohr's, 1925   **Vol. 5:** 82, 88f, 488f
  Vol. 6: 458
Lectures, Bohr's, 1930   **Vol. 6:** 317, 480
Lecture, Bohr's, 1947   **Vol. 6:** 449f
University of   **Vol. 11:** v, vii, ix, 10, 212, 218–220, 327, 330, 530
  Vol. 12: 108, 160, 173, 180, 188, 209, 213
Cambridge Philosophical Society lecture
  Vol. 1: 106ff
Cambridge University Press   **Vol. 12:** 13, 15, 90, 509, 510
Campbell, N.R.   **Vol. 4:** 10, 72, 73, 80
  Vol. 6: 4, 26, 28f, 31f, 69, 131, 156, 478
Campbell, R.   **Vol. 11:** 19, 20, 34, 38, 42, 43, 86, 88, 232, 233, 235, 250–256, 258, 261, 263, 270, 271, 273, 280, 283
canonical amplitudes/equations/ transformations/variables   **Vol. 7:** 64, 67, 126, 128, 292, 306, 307, 332, 347
canonical distribution/ensemble   **Vol. 6:** 325f, 467, 469f
capillarity (*see also* surface tension/energy)
  Vol. 9: 36, 70, 82, 259, 261, 373, 567, 669, 671
Captain's Room   **Vol. 12:** 153, 164
capture of electrons by α-particles (*see also* electron capture by α-particles)
  Vol. 6: 376, 458
capture of mesons   **Vol. 8:** 571–572, 765
capture/loss of electrons (*see also* electron capture)   **Vol. 8:** 25, 35f, 203–206, 214, 232, 236, 238, 241, 244ff, 256, 258–265, 320, 325, 329, 333, 358, 362–364, 375, 377ff, 381–382, 383, 385, 386–390, 405ff, 417, 426, 509, 528, 529–545, 548f, 556, 559, 573–581, 586, 588, 591f, 593–625, 637, 677f, 698, 750, 761, 787, 811, 823f
carbon   **Vol. 11:** 577, 581, 583, 593, 602, 605, 612, 614, 620, 624
  Vol. 12: 54
Careus, E.G.   **Vol. 10:** 566
Cario, G.   **Vol. 4:** 695, 697, 699, 702
  Vol. 8: 685, 688ff
Carlsberg Breweries   **Vol. 11:** 339, 341, 474n, 517n
  Vol. 12: 127, 128, 469–472, 533

Carlsberg Foundation **Vol. 3:** 24, 68, 184, 288, 296
  Vol. 5: xi, 134, 141
  Vol. 9: vii, 217, 221, 632f
  Vol. 10: 324, 325, 356, 369
  Vol. 11: ix, 7, 339, 343, 346, 351, 384p, 400, 401, 403, 410, 411, 413, 456, 457, 460, 461, 464, 466, 467, 471, 474, 512, 516, 517n, 521, 524, 647, 650
  Vol. 12: 107, 108, 127, 152, 162, 164, 175, 182, 190, 209–211, 213–216, 248, 249, 469, 471
Carlsberg Foundation fellowship **Vol. 1:** 103
Carlsberg Laboratory **Vol. 11:** 471, 474, 516, 524
  Vol. 12: 127
  Chemistry Division **Vol. 11:** 297n, 341, 353, 471, 474
  Vol. 12: 124
  Physiology Division **Vol. 11:** 474n
Carlsberg Mansion **Vol. 7:** 474, 477
  Vol. 10: 20, 21p, 36p, 302, 380p, 493, 582
  Vol. 11: 29, 64, 69, 79, 81, 82, 186p, 227, 326, 327, 329, 330, 340, 710, 713, 726, 731
  Vol. 12: 109, 110, 123, 127, 152, 153p, 154p, 163, 164n, 165p, 167p, 179p, 187p, 198, 202, 418, 424, 469, 471, 509, 512, 525
Carnap, R. **Vol. 10:** xli, 15
Carroll, L. **Vol. 12:** 14n, 79, 81, 518n
Casimir, H.B.G. **Vol. 6:** 304, 307, 332, 426f, 430, 479f
  Vol. 7: 520
  Vol. 8: 224, 709f, 754
  Vol. 9: 28, 597f, 676
  Vol. 12: 8, 38p, 358
Cassidy, D.C. **Vol. 5:** x, 94
Catalan, M.A. **Vol. 3:** 180
  Vol. 4: 317, 402, 606, 652
  Vol. 10: 438
Catalski, *see* Katzir-Katchalsky, A.
cathode rays **Vol. 2:** 18, 19, 28, 36–38, 88, 89, 190, 236, 278, 315–316, 333, 494, 495, 530, 546
  Vol. 6: 224, 237, 386
  Vol. 8: 10, 11, 20, 50f, 60f, 68–70, 75f, 83, 89, 91f, 98, 105, 158f, 722, 756
  Vol. 11: 658, 663
  Vol. 12: 42, 60, 66, 252

causality (*see also* conservation of energy and momentum; coupling between emission and absorption; determinism, probability) **Vol. 5:** 13, 22, 24, 26f, 42, 68, 93, 106ff, 114, 402, 490
  Vol. 7: 24, 260, 271, 272, 293, 299–322, 325–337, 342–344, 346, 352, 376, 384–394, 461, 462, 521, 522
  Vol. 9: 97, 114, 305, 320, 583, 585, 599, 601
  Vol. 10: xxvi, xxxv, xxxvi, xlv, 4, 7, 8, 17, 19, 20, 30, 33–35, 40, 41, 43–45, 47, 58, 59, 62, 69, 70, 85, 93, 97, 105, 127, 128, 130, 142, 148, 157, 158, 174–178, 180, 182, 195, 208, 209, 242, 244, 276, 277, 279, 284, 285, 290, 318, 322, 340, 405, 406, 453, 467–469, 496, 502, 504, 517–526, 539, 540, 542, 557, 573, 582–584, 587, 588
  Vol. 11: 134, 404, 414, 540, 541, 549, 661, 665, 670, 671
  Vol. 12: 79, 81, 156, 157, 168–170, 220–224, 233, 234
caustic surface **Vol. 1:** 17
Cavendish
  Dinner **Vol. 10:** 383, 384
  Laboratory **Vol. 1:** xxi, 525, 531
  Vol. 2: 3, 597
  Vol. 5: 335, 488
  Vol. 6: 317, 360, 405, 458, 480
  Vol. 8: 4, 204, 223, 706, 709
  Vol. 9: 95, 216, 220, 336, 630
  Vol. 10: 383, 384, 399, 400, 404, 408, 410–414
  Vol. 11: 87n, 212, 217, 218, 241n, 566, 568, 646, 649
  Vol. 12: 108–110, 138, 140, 144, 149, 150, 198, 202, 265, 266, 274
  professorship **Vol. 10:** 398, 399
cavity radiation, *see* thermal radiation
Cayley, A. **Vol. 5:** 259, 266, 280, 312
  Vol. 11: 445
Ceccato, S. **Vol. 7:** 328
cell **Vol. 10:** xlviii, 13, 24, 33, 70, 150, 166–169, 214, 485, 486, 544, 545, 548, 549, 551, 553, 554, 559, 560
central field **Vol. 3:** 122
  orbit **Vol. 3:** 515, 535
  perturbations **Vol. 3:** 266ff
central motion **Vol. 4:** 58ff, 108ff, 207, 269, 285, 359

central systems   **Vol. 4:** 93
Čerenkov, P.A.   **Vol. 8:** 497, 566
cerium   **Vol. 11:** 580, 599
CERN   **Vol. 10:** 546, 547, 550, 559, 561–563, 567, 575n
  Vol. 11: 338, 362, 367, 569, 713, 714n, 721, 722, 736
Chadwick, J.   **Vol. 1:** xlii
  Vol. 6: 330, 379
  Vol. 7: 32p
  Vol. 8: 210ff, 269, 278, 678, 704f, 755, 784, 808, 810, 812
  Vol. 9: 117, 126, 134, 152, 156, 165, 174, 185, 216, 219, 307, 321
  Vol. 10: 330, 399, 408–410, 449, 450, 465
  Vol. 11: 13, 15, 39–43, 87, 227, 228, 250, 259, 263, 267–269, 274, 275, 280, 567, 646, 649
  Vol. 12: 111, 280p, 526
chain reaction   **Vol. 9:** 67, 72, 75, 395–398, 437, 441f, 462, 465f, 679
  Vol. 11: 12, 91, 333, 363, 569, 580, 581, 599–601, 614, 615, 625, 626, 646, 650
  Vol. 12: 84
Chalmers, S.D.   **Vol. 2:** 265
Chandrasekhar, S.   **Vol. 8:** 714ff
Chang, T.S.   **Vol. 11:** 358p
Chapman, J.C.   **Vol. 1:** 412
  Vol. 2: 236
characteristic X-rays, *see* X-ray spectra; X-rays
charge (*see also* nuclear charge)   **Vol. 11:** 566, 567, 575, 577, 578, 580, 589–591, 593–595, 600, 612–614, 620–622, 624, 658, 664, 672
  Vol. 12: 43–45, 47, 51, 54, 60, 61, 65–68, 255, 259, 287, 357, 359, 365, 499, 501
  conservation   **Vol. 7:** 181, 191
  density   **Vol. 7:** 61, 77, 115, 117, 124, 135, 161, 162, 197, 203, 205, 214, 215, 242, 484
  distribution   **Vol. 7:** 12, 24, 27, 70, 72, 75, 76, 81–83, 92, 93, 130, 132, 134, 138, 139, 146, 178, 188, 198, 202, 203, 212, 235, 236, 484, 485
  elementary   **Vol. 12:** 60, 61, 65, 67
  fluctuation   **Vol. 7:** 37, 38, 43, 50, 204, 205, 209, 239–246, 457, 458, 483, 487, 490, 491, 493, 500–503
  magnetic, *see* magnetic charge
  number, *see* atomic number; effective charge
  point   **Vol. 7:** 12, 61, 70, 124, 130, 171, 178, 183, 188, 199, 216, 220, 449
  ratio in fission, *see* fission, symmetrical/asymmetrical
charge-free fields   **Vol. 7:** 64, 65, 126
charge/mass ratio   **Vol. 11:** 566
  Vol. 12: 51, 60, 65, 66
charity (*see also* love)   **Vol. 10:** 97, 160, 188, 189, 202, 203, 216
Charlier, C.V.L.   **Vol. 3:** 7, 84, 95, 110, 638
Chase, M.   **Vol. 10:** 551, 552n
chemical
  activity/reactions/processes/properties
    Vol. 5: 72–73, 278, 301f, 307, 346
    Vol. 8: 171f, 175f, 179, 325, 562, 658, 661, 759, 762, 764, 767
    Vol. 12: 61, 68, 438, 440, 446, 447
  bond, model of   **Vol. 2:** 107, 138ff, 152ff, 220, 224
  bonds   **Vol. 11:** 568, 671
  elements   **Vol. 10:** 103, 149, 176, 208, 210, 373–377, 379, 384–386, 388, 390, 393, 396, 399, 409–412, 436, 437, 510, 511, 556
  explosives   **Vol. 11:** 229, 614, 626, 692, 697
  kinetics   **Vol. 10:** 131, 137, 150, 152, 166, 484, 486, 497, 500
  relationships   **Vol. 4:** 338ff
Chemical Society (Copenhagen)
    Vol. 11: 479, 483, 488, 490
  lecture before   **Vol. 3:** 19, 20, 221–226
Chemical Society (London)   **Vol. 6:** xii, xv, 317, 371, 373, 480
    Vol. 9: 7, 91, 117
chemistry   **Vol. 10:** ix, xxvi, xxxii, 3, 4, 6, 10, 13, 14, 16, 22–24, 30, 46, 54, 55, 58, 62, 67, 68, 86, 131, 136, 148–150, 166, 176, 177, 179–181, 183, 208, 210, 211, 214, 215, 285, 291, 344–346, 351, 352, 358, 359, 364, 365, 385, 393, 407, 434, 467–469, 498, 499, 501, 510, 511, 539, 540, 542, 551, 552, 557, 559, 560
    Vol. 11: 102, 133, 270, 353, 539, 548, 565, 612, 620, 657, 662, 670, 692, 697
    Vol. 12: 42, 84, 126, 127, 257, 261, 326, 329, 382, 384, 438–441, 444

Cherwell, Lord   **Vol. 1:** 448
   Vol. 2: 133, 166, 237, 304, 305, 306,
      415, 416, 419, 421, 422, 424, 594,
      627, 629
   Vol. 5: 314
   Vol. 11: 26, 27, 36–39, 41, 235, 236,
      242, 246, 255*n*, 256, 257, 260, 262,
      269, 275, 276, 278, 280
   Vol. 12: 7*n*, 22, 26
Chew, G.   **Vol. 10:** 455*p*
Chievitz, E. (O. Chievitz' daughter)
   Vol. 8: 731*f*, 748*f*
Chievitz, I.   **Vol. 12:** 452, 457
Chievitz, J.H.   **Vol. 10:** xxviii
   Vol. 12: 126, 451, 452, 456, 457
Chievitz, O.   **Vol. 6:** 442, 444, 449, 451
   Vol. 8: 704*f*, 731*f*
   Vol. 10: 224*p*
   Vol. 11: 452, 457
   Vol. 12: 126, 448*p*, 451, 451*p*, 452,
      453, 453*p*, 454–460, 502, 503, 505, 529
Chievitz, P.   **Vol. 12:** 452, 457
China   **Vol. 10:** 157, 160, 583
Chita   **Vol. 12:** 125, 448*p*
Chladni figures   **Vol. 10:** 346, 359
Chladni, E.   **Vol. 10:** 359*n*
chlorine   **Vol. 12:** 45
choice of nature, *see* free choice
Chou, P.-Y.   **Vol. 6:** 304, 308
Christensen, A.   **Vol. 10:** 250*p*
   Vol. 11: 406*p*
Christian Bohr Memorial Grant   **Vol. 12:** 414,
      417, 418, 422–424
Christian IV   **Vol. 10:** 267*n*
Christian V   **Vol. 12:** 102
Christian VI   **Vol. 11:** 426, 430
Christian X   **Vol. 11:** 288, 341, 342, 345–347,
      353, 386, 387, 390, 391, 394, 395, 398,
      402, 407, 408, 412, 416, 417, 421–424,
      426–428, 430, 431, 470, 472, 473, 475,
      479, 482, 510*p*, 511, 520, 673*p*, 683,
      686, 693, 699, 728, 730–732
   Vol. 12: 138, 140
Christianity   **Vol. 10:** 254, 263, 555
Christiansen, C.   **Vol. 1:** xix, xx, xxvii, 3, 5,
      7, 15, 17, 94, 97, 98, 99, 109, 125, 176,
      301, 497, 498, 499, 504, 505, 506, 507
   Vol. 2: 4, 110, 332, 334, 492–497, 511,
      617, 618, 619, 624
   Vol. 3: 68, 284, 285, 294
   Vol. 5: 132*f*, 140

   Vol. 6: xx*f*
   Vol. 10: xxviii, 298, 312*p*, 313, 314,
      319, 345, 359
   Vol. 11: 353
   Vol. 12: 112, 122, 123, 131*p*, 144, 149,
      150, 153, 154, 164–166, 198, 202, 212,
      217, 306, 307, 309, 310, 316, 317,
      320, 321
Christiansen, J.   **Vol. 10:** 312*p*
Christiansen, J.A.   **Vol. 1:** xxvii
   Vol. 8: 727, 729
   Vol. 11: 354, 406*p*, 488, 490
   Vol. 12: 100, 127
Chrom, S.A.   **Vol. 10:** 74
chromosome   **Vol. 10:** 13, 168, 499, 544, 548,
      553, 554, 557
chronicles   **Vol. 10:** 254, 263
Chuang Tzu   **Vol. 10:** 479
Churchill, W.   **Vol. 1:** xlv
   Vol. 11: 15, 20–27, 30, 32–39, 41, 42,
      47, 89, 96, 233, 236, 237, 240–242,
      245, 252, 255, 256, 258–261, 263, 268,
      271, 273, 275, 277, 287, 289–292
Cini, M.   **Vol. 10:** 455*p*
civilization, challenge to   **Vol. 10:** 143, 160,
      205, 287, 292, 415
   Vol. 12: 186, 196, 278, 378
Clark–Duane experiments   **Vol. 5:** 38–40,
      121–124, 319–326
Clark, G.L.   **Vol. 5:** 38–40, 121, 123,
      319, 323*f*
Clark, R.B.   **Vol. 10:** 126
classical
   concepts/ideas/pictures   **Vol. 7:** 103, 153,
      171, 183, 197, 224, 275, 295, 351,
      457, 458
      Vol. 12: 11, 47, 344, 345, 357, 364, 494
   dispersion, *see* dispersion
   electrodynamics (*see also* electrodynamics,
      classical/ordinary)   **Vol. 7:** 64, 93,
      103, 104, 126, 146, 153, 154, 171, 178,
      183, 188, 212, 216, 226, 303, 304, 320,
      342, 344
   electromagnetism   **Vol. 7:** 24, 60, 62, 71,
      73, 77, 99, 112, 120, 123, 125, 131,
      133, 135, 150, 159, 165, 172, 181, 184,
      190, 198, 200, 202, 236, 246, 331, 343,
      400, 402, 414, 415
   electron theory, *see* electron theory
   equations of motion   **Vol. 12:** 47, 223

mechanics (*see also* mechanics, classical/
  ordinary; Newtonian mechanics)
    Vol. 7: 60, 62, 102, 104, 123, 125, 152,
    154, 171, 183, 236, 279, 290, 292, 303,
    305, 330, 331, 333, 334, 342–344, 347,
    351, 376, 400, 408, 413, 420, 504, 507
    Vol. 11: 613, 621, 622, 659, 664,
    669, 671
    Vol. 12: 4, 20, 44, 47, 51, 156, 168,
    222, 223, 233, 344, 345, 356, 500, 502
  physics **Vol. 5:** —
    Vol. 7: 197, 219, 220, 228, 249, 263,
    266, 273, 275, 285, 293, 295, 297, 311,
    318, 321, 324, 331, 332, 335, 341–343,
    488, 492
    Vol. 10: —
    Vol. 11: 530, 567, 658, 660, 661, 663,
    664, 666, 670–672
    Vol. 12: 12, 14, 43, 47, 156, 157, 168,
    169, 209, 214, 222, 223, 234, 348, 351,
    356, 364, 374, 375, 378
  resonance problem **Vol. 12:** 47
  statistical mechanics, *see* statistical
    mechanics, classical
  statistics (*see also* statistical mechanics)
    Vol. 12: 22, 25, 344, 346
Clausen, C.P. **Vol. 10:** 250p
Clausen, H.N. **Vol. 12:** 152, 164
Clausius, R. **Vol. 1:** 241, 352
    Vol. 5: 273
Clausius–Clapeyron equation **Vol. 12:** 32n
Clay, J. **Vol. 7:** 328
    Vol. 9: 597f
    Vol. 12: 38p
Clemenceau, G.B. **Vol. 11:** 250
closing of electron groups (*see also* Pauli
    principle) **Vol. 5:** 35, 59, 260, 266,
    279, 335, 424–425, 428, 430–436,
    501, 508
cloud chamber **Vol. 7:** 356, 363
    Vol. 10: 383, 404, 410–412, 435
    Vol. 11: 574, 576, 577, 579, 587, 588,
    592, 594, 598, 599
    Vol. 12: 54
  tracks (*see also* fission fragment tracks)
    Vol. 8: 155, 163, 212, 233, 235f, 245f,
    255, 315f, 318, 320f, 325, 328f,
    375f, 405f, 465, 467f, 486, 548,
    554f, 585, 587, 665, 678, 727, 729,
    734f, 749, 786, 790, 793, 800, 810

    Vol. 9: 201, 350, 353, 413, 431, 443,
    457f, 591–592
Clusius, K. **Vol. 11:** 243, 244
Cockcroft, J.D. **Vol. 7:** 32p
    Vol. 8: 269, 278, 652f, 704f
    Vol. 9: 134, 135, 165, 173, 201, 216,
    220, 612
    Vol. 10: 330, 399, 409, 449
    Vol. 11: 281, 360, 436p, 568, 646, 647,
    649, 651
    Vol. 12: 14, 84
Cohen, B. **Vol. 11:** 325–327, 329, 330
Cohen, I.B. **Vol. 12:** 15, 16
Cohen, R.S. **Vol. 7:** 55, 56
    Vol. 9: 55
coherence (*see also* phase relations)
    Vol. 5: 23, 44, 60, 62, 68, 114, 309, 356
    Vol. 9: 12, 492f, 497, 501f, 515,
    574f, 577, 614
Cohn, E. **Vol. 10:** 299n
Colby, W.F. **Vol. 5:** 29, 383, 385, 387ff
Cold War **Vol. 11:** 83
collective momentum, time derivative
    Vol. 1: 316
collective motion **Vol. 9:** 32, 38, 80, 153f,
    168, 177, 186f, 193, 197f, 211, 233,
    237, 255, 278, 293, 314, 327, 332, 498,
    529, 539, 609
collective motion of electrons **Vol. 1:** 308ff
  in case atoms are not small **Vol. 1:** 324ff
  momentum of **Vol. 1:** 310
collision
  diameter **Vol. 8:** 253, 276f, 285f, 344,
    348–349, 430–431, 440, 443–444, 449,
    454f, 462, 491, 544, 608
  problems **Vol. 1:** xxiv
    Vol. 12: 5, 6, 500, 502
  problems, absorption of α-rays
    Vol. 1: xxiv, 555, 559, 573
  time of (*see also* time of collision)
    Vol. 2: 10, 20–21, 22, 61–62, 72, 500
    Vol. 5: 70–73, 180ff, 195ff, 276, 300,
    302, 305, 346f
collisions (*see also* absorption of charged
    particles) **Vol. 2:** 5, 9, 10, 18ff, 59ff,
    96, 178, 179, 184, 628
    Vol. 5: 33, 53, 69–75, 76, 79f, 83,
    89f, 108, 114, 160, 164, 175–206, 221,
    259, 265, 275ff, 280, 300, 302, 305f,
    311, 334, 337, 341, 346–351, 360f,
    364f, 396, 415, 419, 448, 452, 454ff

Vol. 6: 34, 40, 71, 97, 105, 112, 119f, 127f, 131, 135, 138, 150f, 154ff, 163f, 166, 204, 211, 228, 242–243, 311, 322, 324, 382–383, 407, 417, 434, 449
Vol. 7: 15, 84, 86, 90, 95, 140, 141, 144, 148, 228, 230, 294, 346, 350, 454, 513
Vol. 8: —
Vol. 11: 568, 576–578, 581, 592, 594, 596, 603, 614–616, 624, 626, 627
between atoms and free electrons **Vol. 3:** 237ff, 256ff, 367, 404, 561ff
of the 1st kind **Vol. 8:** 27, 31f, 206
of the 2nd kind **Vol. 5:** 70, 179, 195, 275, 341
  Vol. 6: 383, 405
  Vol. 8: 27, 683, 685, 689, 692, 769ff, 780, 782
pseudo-classical **Vol. 8:** 290, 292, 295
radiative **Vol. 8:** 226f, 290, 292
collisions, separate **Vol. 1:** 308ff, 331, 346, 354, 371
between electrons and metal atoms **Vol. 1:** 312ff
duration of **Vol. 1:** 527
increment in momentum **Vol. 1:** 312–313
mutual among electrons **Vol. 1:** 315ff, 323
Cologne **Vol. 10:** 25, 162p, 164, 484, 488–490, 492, 494, 594
colour of ions **Vol. 4:** 240, 309, 403
Colsmann, C. **Vol. 3:** 285, 294
Columbia University **Vol. 5:** 29, 318
  Vol. 9: 57, 61, 71f, 81, 349f, 352f, 380, 550, 552f, 635, 637, 660, 662
  Vol. 11: 190p, 370, 564, 569, 711, 712, 721n
combination
law/principle (*see also* spectral laws; Ritz law) **Vol. 1:** xxvii–xxviii, 571
  Vol. 3: 71, 251, 373, 423
  Vol. 4: 206, 438, 471, 553, 613
  Vol. 5: 121ff, 157, 161, 275, 278
  Vol. 6: 153, 382f, 388f, 391, 396
  Vol. 10: 338, 339, 375, 376, 386, 388, 436, 437
lines **Vol. 3:** 275, 444, 523ff, 541ff
Comments (to Marshall) **Vol. 11:** 69, 70, 72, 160, 163, 168, 320
Committee on Atomic Energy (U.S.) **Vol. 11:** 66, 69

communication **Vol. 10:** xxxv, xxxvii, xl, xlv, 43, 71, 83, 84, 93, 95–97, 119, 123, 126, 128, 129, 141, 143, 159, 160, 168, 183–185, 189, 196, 199, 276, 277, 286, 291, 405, 428, 444, 479–481, 483, 491, 545, 549, 567, 568, 570, 572–574
commutation relation (*see also* non-commutativity) **Vol. 5:** 272, 447f, 452f, 501
  Vol. 6: 95f, 125, 153, 161, 163, 165f, 168, 179
  Vol. 7: 4–6, 8, 17, 20, 21, 24, 26, 30, 64–70, 116, 126–130, 162, 198, 201–203, 212, 306, 307, 310, 319, 332, 348, 349, 468
  Vol. 9: 104
Como Conference, 1927 **Vol. 6:** 5, 29ff, 37, 44, 98, 109f, 113, 147f, 264, 287
  Vol. 7: 349, 351, 354, 392, 493, 495
  Vol. 12: 342
Lecture, Bohr's **Vol. 5:** ix, 93f
  Vol. 6: vii, 4, 7f, 21, 26–35, 41–51, 57f, 98, 109–136, 147–158, 197, 264–268, 287–293, 362, 432–441, 478
  Vol. 7: viii$n$, 4, 230, 282, 284, 349, 351
  Vol. 10: xxiii, xxiv, 6
compassion **Vol. 10:** 97
complacency of cultures **Vol. 10:** 160, 204, 248
complementarity **Vol. 1:** xl–xlii, xlv–xlviii
  Vol. 5: ix, 93, 95, 215–216, 222
  Vol. 6: —
  Vol. 7: —
  Vol. 8: 228, 231, 253, 299, 306, 343, 348, 438, 499, 629, 631, 649, 805
  Vol. 9: 87, 104f, 108, 111, 305, 320, 641, 643
  Vol. 11: 11, 52, 134, 337, 367, 542, 550, 660, 661, 665, 671
  Vol. 12: 13, 62, 69, 70, 157, 170, 186, 196, 200, 204, 224, 534
between charity/love and justice **Vol. 10:** 160, 203, 279
  Vol. 11: 134
  Vol. 12: 13, 62, 70, 180, 186, 188, 195, 196
between contemplation and volition **Vol. 10:** 105, 152, 160, 200
between cultures **Vol. 10:** 97, 105, 188, 204, 216, 220, 247, 248, 279

between individual and community
 Vol. 10: 143, 203
between individuals **Vol. 10:** 286, 292
between instinct and reason **Vol. 10:** 105,
 245, 246, 536, 537
between mechanicism and vitalism
 Vol. 10: 4, 11, 71, 199, 279
between physics and biology **Vol. 10:** 25,
 34, 92, 105
between subjectivity and objectivity
 Vol. 12: 180, 188
between thoughts and sentiments/feelings/
 emotions **Vol. 10:** xlv, 159, 168, 245,
 279, 491
 Vol. 12: 13, 62, 70, 180, 186, 188,
 196, 224
between truth and clarity **Vol. 12:** 180,
 185, 188, 195, 463, 464
between use and meaning of words
 Vol. 10: 48, 71, 306, 308
between values and their background
 Vol. 10: 97
conditioned by introspection **Vol. 10:** 47
corresponding to change of focus
 Vol. 10: 123
in atomic physics **Vol. 10:** xxxii, xl, 30, 46,
 62, 70, 76, 77, 90, 93, 103, 105, 112, 121,
 129, 149, 166, 178, 179, 181, 183, 197,
 213, 215, 244, 278, 284, 286, 290, 291,
 389, 405–407, 445, 467, 468, 476, 495,
 496, 498, 500, 539, 541, 565, 574
in biology **Vol. 10:** ix, 11, 14, 22, 24, 61,
 131, 152, 415, 467, 468, 474, 475, 488,
 500, 544, 548, 551, 591
in children's play **Vol. 10:** 96
in classical physics **Vol. 10:** 25, 34, 432
in psychology **Vol. 10:** xlvi, 215, 286, 291,
 405, 480, 481, 519, 521, 536, 537, 591
notion of **Vol. 10:** v, vi, ix, xxiii–xxv,
 xxvii, xxxii, xlviii, 6, 16, 31, 33, 43–45,
 104, 109, 129, 131, 137, 142, 158, 189,
 200, 285, 290, 340, 444, 471, 472, 476,
 477, 484–486, 501, 503, 583, 587,
 588, 591
of social problems **Vol. 10:** 216, 405, 415
of states of mind **Vol. 10:** 93, 105, 143
replacing causality **Vol. 10:** 41, 59, 244
completeness **Vol. 7:** 427, 428
 Vol. 10: 30, 121, 213, 379
complex structure (*see also* multiplet structure)
 Vol. 3: 280*ff*, 517*ff*, 536*ff*, 546*ff*, 556

Vol. 4: 116*ff*, 157, 214*ff*, 286,
 591*ff*, 642*ff*
 Vol. 5: 37, 59, 251*f*, 282, 284, 288,
 332, 411, 413, 425, 428*f*, 501
 of alkali spectra **Vol. 4:** 596*ff*, 645*ff*
 of series lines **Vol. 4:** 570*ff*, 626*ff*
compound nucleus (*see also* nuclear
 constitution/structure)
 Vol. 10: 412, 413
compound scattering, *see* scattering, multiple
Compton
 effect/recoil (*see also* Bothe–Geiger
 experiment; Compton–Simon
 experiment) **Vol. 5:** 3–6, 23*f*, 58*f*,
 75*ff*, 93*ff*, 105*f*, 115–116, 190, 204,
 209–212, 213–216, 275, 310, 318–319,
 352, 382, 417, 421, 496, 504, 513
 Vol. 6: 17, 19*f*, 37, 39*f*, 69, 79*f*, 87*f*,
 115, 120*ff*, 139*f*, 148, 150*f*, 162–163,
 165, 173, 175, 186, 359, 394, 416*f*, 422
 Vol. 7: 172, 173, 184, 185, 192, 193,
 232, 264, 294, 314, 350, 356
 Vol. 9: 13, 109, 591
 Vol. 10: 19, 402
 Lectures **Vol. 10:** vii, 172, 182, 189, 591
 scattering (*see also* scattering of
 electrons/protons) **Vol. 7:** 4, 323
 Vol. 12: 47, 364
 wavelength **Vol. 7:** 31, 37
 Vol. 9: 12*f*, 109*f*, 126, 131*f*
Compton–Simon experiment **Vol. 5:** 69, 78,
 87, 210*f*, 215, 221, 276, 440, 443,
 484*f*, 499
 Vol. 6: 115, 139, 148
Compton, A.H. (*see also* Compton
 effect/recoil) **Vol. 3:** 571
 Vol. 5: 3–6, 23, 69, 78, 87, 89, 93,
 105*f*, 115, 207–211, 215, 221, 235,
 275*f*, 288, 306, 323, 353, 382, 440,
 443, 484*f*, 487, 496, 499, 504
 Vol. 6: 36, 37, 100, 136, 139, 158, 415
 Vol. 7: 346
 Vol. 10: 440, 441
 Vol. 11: 51, 368*p*, 639
 Vol. 12: 47, 118*n*, 495
Compton, K.T. **Vol. 5:** 6, 487
 Vol. 8: 681
 Vol. 10: 189
computer, *see* calculation device
Comrades' Assistance Fund (Denmark)
 Vol. 11: 376, 702, 703, 734

Conant, J.   **Vol. 11:** 15
conceptual frame   **Vol. 10:** xxxviii, xl,
    xlix, 25, 56, 83, 84, 86, 98, 102, 105,
    121, 123, 130, 152, 158, 174, 178, 189,
    249, 277, 280, 285, 290, 317, 322, 351,
    364, 501, 503
conditionally periodic systems   **Vol. 3:** 82*ff*,
    117*ff*, 258
  spectra of   **Vol. 3:** 93*ff*
conditions
  experimental   **Vol. 10:** 128, 158, 159, 166,
    196, 197, 212, 213, 215, 279, 284, 290,
    404–407, 428, 445, 568, 570, 574
    for description   **Vol. 10:** xxiv, xxvii, xxxiii,
      xxxviii, xl, xlix, 11, 14, 26, 188,
      279, 567
    for experience   **Vol. 10:** 428, 446
    for logical comprehension   **Vol. 10:** 92
    observational   **Vol. 10:** xxiii, xxiv, xxvii,
      xxxii, xl, xlix, 3, 6, 9–11, 14, 41, 61,
      90, 119–123, 184, 219, 444, 519, 522
Condon, E.U.   **Vol. 6:** 405
    Vol. 9: 87, 95, 230, 281, 308, 322
    Vol. 10: 409
conduction electrons   **Vol. 9:** 25*f*, 596*f*,
    600*f*, 603*f*
conduction of electricity   **Vol. 1:** 142,
    335*ff*, 553
  change in magnetic field   **Vol. 1:** 387
  conductivity   **Vol. 1:** 335
  superconductivity   **Vol. 1:** xlv
  under periodic forces   **Vol. 1:** 153
conduction of heat   **Vol. 1:** 142, 335*ff*
conduction, electric   **Vol. 6:** 144, 235, 252,
    349, 398
conductivities, table   **Vol. 1:** 145, 340
conductivity
  electrical (*see also* superconductivity)
    Vol. 2: 22, 287, 423
  thermal   **Vol. 2:** 221, 287, 423
    Vol. 8: 270, 279
configuration space   **Vol. 6:** 32, 34, 98,
    102–103, 128, 154, 174, 181, 394, 434
    Vol. 8: 219, 503
    Vol. 9: 70, 373*f*, 386*f*, 658
Confucius   **Vol. 6:** xx
conjugate variables (*see also* action–angle
    variables)   **Vol. 6:** 22, 49*f*, 80, 88,
    95*ff*, 125*f*, 153*f*, 158, 160, 167, 204,
    210*f*, 348, 420, 466
    Vol. 7: 373

connection
  with ordinary mechanics, *see* analogy
  with classical theory, *see* correspondence
    with classical physics
conscience   **Vol. 10:** 199, 280
consciousness   **Vol. 6:** 189*f*, 205, 212, 215,
    235, 253, 259, 272, 279, 298, 431
    Vol. 10: xxxi, xli, xlii, xlv, xlvii,
    xlix, 6–10, 14, 25, 35, 71, 84, 92, 93,
    105, 123, 129, 143, 152, 159, 160, 168,
    184, 185, 201, 215, 279, 280, 285, 291,
    315, 321, 482, 485, 486, 536, 537, 545,
    546, 549, 550, 566
conservation
  laws (*see also* energy conservation;
    energy–momentum conservation)
    Vol. 7: 165*n*, 225, 271, 304, 315, 333,
    351, 356, 364, 392, 405, 418, 448, 450
    Vol. 9: 6, 88, 105, 207, 230, 276, 309,
    322, 492*f*, 499, 510, 512*f*, 555, 614
    Vol. 10: 7, 41, 43, 59, 88, 89, 182, 183,
    189, 203, 215, 340, 396, 447, 451
    Vol. 12: 365
  of angular momentum   **Vol. 9:** 11, 135*f*,
    275, 493–494, 513
  of charge (*see also* charge conservation)
    Vol. 9: 8, 112, 132, 136, 570*f*
  of energy (*see also* energy
    conservation/balance)   **Vol. 9:** 4–14,
    85–89, 96*f*, 103*ff*, 107*ff*, 111*ff*, 118,
    132, 136*ff*, 145, 152, 201, 275, 280,
    307, 310, 321, 324, 417*f*, 446, 513,
    547*f*, 568*ff*, 605, 676
  of energy and momentum (*see also*
    causality; coupling between emission
    and absorption)   **Vol. 5:** 4*f*, 12,
    13–23, 27, 30*f*, 33, 38, 41, 43, 48, 65,
    67*f*, 69–75, 78, 83*f*, 87, 89*f*, 93–96,
    101, 104, 107–109, 114*ff*, 173–174,
    178*f*, 183–192, 194*f*, 198–206,
    209–216, 221, 258*f*, 265*f*, 271, 273,
    275–276, 280, 300, 302, 305, 314*f*,
    317, 334, 345–352, 390, 392–393,
    415–416, 419, 439, 440, 442, 443, 447,
    452, 455*f*, 484*f*, 490
    Vol. 8: 31*ff*, 36, 211, 246, 269, 278,
    339, 409–413, 498*f*, 658, 661, 712*f*,
    754, 807, 808–809
  of energy and momentum in the quantum
    theory   **Vol. 3:** 497*ff*

of momentum   **Vol. 7:** 82, 99, 100, 105, 117, 138, 150, 151, 155, 157, 162, 163, 180, 181, 190, 191, 276, 294–296, 314, 315, 347, 353, 355, 359

Constable, J.E.R.   **Vol. 9:** 185

constitution

   of atoms, *see* atomic constitution/structure

   of atoms and molecules, *see the element or compound in question*

   of molecules, *see* molecular structure

   of nuclei, *see* nuclear structure

constraint   **Vol. 3:** 44, 530, 558

   Vol. 4: 598, 646

"constraint", *see* "Zwang"

contact electricity   **Vol. 1:** 150

contemplation   **Vol. 10:** 143, 152, 160, 168

continuous energy transfer, *see* energy transfer, continuous

contradiction, principle of   **Vol. 10:** 508, 509

control rods   **Vol. 11:** 582, 603, 604, 615, 626

Cooke, H.L.   **Vol. 1:** 234, 347

Cooper, L.   **Vol. 12:** 9

cooperative movement   **Vol. 10:** 260, 271

coordinate space, *see* configuration space

Copenhagen

   "Irrlehre"   **Vol. 5:** 231, 461, 463, 470, 472, 478, 481, 508

   "Putsch"   **Vol. 5:** 84, 88, 351, 441, 444

Copenhagen Academy, *see* Royal Danish Academy of Sciences and Letters

Copenhagen Ball Club   **Vol. 11:** 372

Copenhagen Conference

   1929   **Vol. 6:** 304–309

   1932   **Vol. 8:** 821

      Vol. 9: 10, 11, 115–118, 570f

   1933   **Vol. 8:** 224

      Vol. 9: 570ff

   1934   **Vol. 9:** 19f, 572f, 642f

   1935   **Vol. 8:** 709f

   1936   **Vol. 8:** 807

      Vol. 9: 27, 28, 38, 582ff, 599ff, 609

   1937   **Vol. 8:** 810

   1947   **Vol. 9:** 50

   1952   **Vol. 8:** 261, 589–592

      Vol. 9: 83, 527–529

   on "Unity of Knowledge", 1936

      Vol. 9: 583, 585, 599, 601

Copenhagen interpretation   **Vol. 10:** xxiv

"Copenhagen spirit"   **Vol. 9:** 544

Copenhagen, University of (*see also* lectures, Bohr's; professorship (teaching post) at University of Copenhagen)

   Vol. 2: 4, 106, 329ff, 344, 496, 556f, 591ff, 610f

   Vol. 5: 41, 131, 138, 219, 270, 273, 318

   Vol. 6: xxiv, 110, 186, 196f, 202, 255ff, 262, 278, 283, 425, 427

   Vol. 8: 16, 235, 244, 265, 733, 735, 794

   Vol. 9: 207, 217, 221, 431, 457, 621, 639, 660

   Vol. 11: 6, 15, 63, 64, 133, 146–148, 303, 308, 318, 319, 338n, 339, 354, 357, 394, 396, 490n, 512, 513n, 521, 563, 669, 729

   Vol. 12: 5, 16, 102, 107–109, 112, 114, 115, 121–125, 127, 128, 131p, 134, 138, 140, 144, 149, 150, 172n, 198, 199, 202, 203, 209–215, 217, 248, 249, 254, 258, 265, 274, 291, 293, 300, 301, 315, 320, 327, 329n, 330, 332–334, 337, 338, 348, 350, 355, 382, 384, 452, 457, 463, 465, 518, 519, 525, 528, 529, 531, 537

biophysical laboratory   **Vol. 12:** 114, 326, 327, 329, 332, 333

institute for theoretical physics, *see* Institute for Theoretical Physics (Copenhagen)

Institute of Physical Chemistry

   Vol. 11: 512n

mathematical institute   **Vol. 11:** 456, 460, 512n, 528p, 727

   Vol. 12: 210, 215

physical chemistry institute   **Vol. 12:** 125

physical laboratory, H.C. Ørsted Institute

   Vol. 12: 382, 384

physiological laboratory   **Vol. 12:** 149, 150

Physiological Institutes   **Vol. 11:** 512n, 516, 524

Rockefeller Institute   **Vol. 12:** 534

Copernicus, N.   **Vol. 10:** 54

   Vol. 12: 371

copper, absorption in   **Vol. 2:** 37

Corbino, O.M.   **Vol. 1:** 121

core (*see also* atomic core)   **Vol. 8:** 240, 329ff, 372, 592, 609f, 617, 750

Corinaldesi, E.   **Vol. 7:** 216

coronal spectra, *see* spectra, coronal and stellar

corpuscle, *see* particle

correspondence principle   **Vol. 1:** xxvii–xxix, xxxiv–xxxvi

   Vol. 3: 4, 249ff, 351, 377, 378, 381–387, 428, 478ff

Vol. 4: 72, 75, 148*ff*, 211*ff*, 283*ff*, 450, 476, 553, 614, 660
Vol. 7: 31, 35, 36, 47, 59, 60, 63, 64, 73, 78, 104, 120, 123, 124, 126, 132, 136, 154, 165, 167–194, 209, 305, 306, 344, 348, 350, 402, 415, 447, 449, 451, 454, 464, 467, 471, 474, 477, 478
Vol. 10: 30, 34, 339, 377, 378, 389, 390, 395, 396, 401, 402, 406, 424–426, 439–441, 452, 453, 499, 519, 522, 548
Vol. 12: 44, 45, 47, 115, 348, 350, 357, 364
and atomic structure   **Vol. 3:** 566*ff*
and conservation of energy and momentum
    Vol. 3: 272*ff*
and fixing of stationary states   **Vol. 3:** 482*ff*
and nature of radiation   **Vol. 3:** 483*ff*
development of   **Vol. 3:** 4, 5
first indication of generalization of
    Vol. 3: 5, 47–52
first use of term   **Vol. 3:** 21, 247
formulation of   **Vol. 3:** 5, 6
correspondence with classical physics (*see also*
    analogy; transition from micro- to
    macro-mechanics)   **Vol. 5:** 8*ff*, 12, 21*f*,
    28, 35*ff*, 43*ff*, 55, 58, 61, 64, 67, 73, 90,
    101, 105–110, 113, 124, 145, 147, 149,
    152, 156, 165*f*, 168*f*, 171, 182, 184,
    187, 191, 193, 197, 199, 201, 204, 206,
    222, 238–239, 251*f*, 256–267, 272,
    276–280, 282–285, 288*f*, 300, 303,
    330*ff*, 340, 346, 367–369, 371, 373, 383,
    385, 404*ff*, 417, 420*f*, 424*f*, 428*ff*,
    432–436, 447*f*, 452, 457*f*, 462*f*, 484*f*,
    490, 492, 498, 506, 508, 512
    Vol. 6: 13*ff*, 59*ff*, 65, 80, 88, 94–95, 97,
    98, 111, 124*f*, 127, 133–134, 135*f*,
    152–153, 154, 156*ff*, 229, 244,
    263–264, 268, 271, 275, 286, 292, 296,
    301, 307, 314, 384, 389, 391, 393, 396,
    398, 400, 402*f*, 406, 424, 426*f*, 429*f*,
    458, 460, 462
    Vol. 8: 30, 210, 215*f*, 219–220, 355,
    379, 411*f*, 498, 502, 661, 662, 694,
    699*f*, 702*f*, 712*f*, 723
    Vol. 9: 8, 9, 87, 93, 96, 104, 107–108,
    109, 113, 125, 129–132, 260–261, 264,
    304*f*, 309, 319, 322, 375, 523, 548
cosmic rays   **Vol. 7:** 227, 321, 452, 454
    Vol. 8: 235, 265, 571, 576*f*, 597, 624,
    633, 638, 699, 765, 810*f*, 813

Vol. 9: 13, 109, 126–127, 166, 174, 425, 453, 578*f*, 597*f*, 653, 670
Vol. 11: 569, 711*n*
Vol. 12: 51, 54
cosmology   **Vol. 7:** 321, 379
    Vol. 10: 84, 275
    Vol. 11: 669
    Vol. 12: 371
cosmotron   **Vol. 11:** 569
Coster, D.   **Vol. 1:** xxxv
    Vol. 3: 601
    Vol. 4: 22, 23, 26, 30, 31, 32, 34, 36,
    40, 41, 251, 253, 324, 326, 327, 335,
    337, 339, 409, 410, 411, 461, 464, 481,
    482, 486, 490, 492, 493, 498, 507, 517,
    519, 523, 525, 530, 531, 537, 547, 581,
    588, 634, 639, 675–681, 688, 700,
    716, 729
    Vol. 5: 34, 251*f*, 278, 422
    Vol. 6: 390
    Vol. 8: 694, 797
    Vol. 10: 400, 419
    Vol. 11: 514, 522
Cosyns, M.   **Vol. 7:** 32*p*
Cotton, A.A.   **Vol. 10:** 447
    Vol. 12: 496
Coulomb energy   **Vol. 7:** 236
Coulomb force   **Vol. 8:** 50–51, 186–187, 192,
    210, 212, 215, 226, 232, 253, 289*ff*,
    297–299, 300, 304, 337*f*, 380, 385,
    396, 400, 426, 429–443, 444*f*, 449,
    450–451, 453, 485, 521, 525, 631–632,
    634*ff*, 678, 755, 813, 822
coupling   **Vol. 3:** 509, 528, 530
    between an electron and its own field
        Vol. 7: 171, 176, 184, 187
    between atom and radiation   **Vol. 7:**
        171, 183
    between electron groups   **Vol. 4:** 66, 69,
        74, 78, 147, 177, 225, 294, 646
    between electron orbits   **Vol. 5:** 59, 255,
        262, 278, 411, 413, 431*f*, 434, 436,
        454, 456
    between emission and absorption (*see also*
        causality; conservation of energy and
        momentum)   **Vol. 5:** 12*f*, 20*ff*, 41, 43,
        65, 75–90, 102, 113, 179, 190*ff*, 195,
        204*f*, 276, 280, 311, 336, 350*ff*, 362*f*,
        415*f*, 419*f*, 448*f*, 452*f*, 454*ff*,
        484*f*, 497

between modes   **Vol. 9:** 43, 46, 239, 262, 298, 316, 328–329, 332, 336, 372, 392, 421, 449, 563

between outer electrons and atomic core **Vol. 3:** 528*ff*, 558*ff*

between particles   **Vol. 9:** 34, 38, 82, 186, 189, 198, 209, 229*ff*, 236*f*, 246, 253*f*, 259, 261, 267, 269, 276, 278, 282, 293, 308*f*, 312, 314, 322*f*, 325, 327, 477, 489, 498, 505, 508, 618, 642, 644, 646*f*, 669

between space and time   **Vol. 7:** 233

between the particles of the atom **Vol. 7:** 171, 183

in collisions (*see also* reciprocity of collisions)   **Vol. 5:** 70, 179, 195

in molecules   **Vol. 5:** 346

of angular momenta   **Vol. 9:** 30, 198, 237, 261

principle   **Vol. 3:** 492*ff*

viewpoint   **Vol. 3:** 348, 356

Cour, P. la   **Vol. 10:** 267*n*

Courant, E.   **Vol. 11:** vii, 711, 712

Courant, H.   **Vol. 11:** 711, 712

Courant, N.   **Vol. 11:** 710, 712

Courant, R.   **Vol. 1:** xviii
Vol. 4: 24, 694, 696, 700
Vol. 6: 42, 436
Vol. 8: 694
Vol. 11: 327, 330, 370–372, 668*p*, 670, 710–712, 714, 734
Vol. 12: 100

covalent bond (*see also* chemical bond, model of)   **Vol. 6:** 392–393, 397
Vol. 9: 274, 305, 319

Crane, H.R.   **Vol. 5:** 95
Vol. 9: 31, 201*ff*

creation/annihilation   **Vol. 10:** 450, 451

Crick, F.H.C.   **Vol. 10:** 24, 168, 551, 555

critical deformation energy   **Vol. 9:** 70, 361, 365*f*, 369–374, 375, 377, 381*ff*, 386*ff*, 401, 404, 407*f*, 432*ff*, 438*ff*, 458*ff*, 463*ff*, 470*ff*, 477*f*, 484, 658, 661, 663, 666

Croce, B.   **Vol. 12:** 14

Crookes, W.   **Vol. 6:** 374

cross section (*see also* Ramsauer effect) **Vol. 5:** 80, 85, 309, 311

capture/loss   **Vol. 8:** 204*f*, 256, 263*f*, 362, 364, 382, 417, 426, 529, 531–539, 575, 577, 586*ff*, 591*f*, 598–603, 608–624

classical/geometrical   **Vol. 8:** 29, 296, 303, 346*f*, 351, 431, 452, 454*ff*, 463, 473, 563, 564

collision   **Vol. 8:** 29, 362*f*, 377, 379, 387, 389, 466

fission (*see also* fission cross-section) **Vol. 8:** 790

ionization   **Vol. 8:** 365, 367

pair-production   **Vol. 8:** 340, 802*f*

scattering   **Vol. 8:** 211, 254, 295*ff*, 299, 303, 347, 351, 431, 435, 436–437, 442, 446*f*, 450, 452, 454*ff*, 463, 465, 472, 480, 631

crossed electric and magnetic fields **Vol. 5:** 63, 224, 256*f*, 262, 264, 278*f*, 383, 446, 450–451, 454, 455

Crowther, B.M.   **Vol. 9:** 201

Crowther, J.A.   **Vol. 8:** 5

Crowther, J.G.   **Vol. 10:** 410
Vol. 11: 7*p*

Croxson, C.   **Vol. 2:** 469
Vol. 3: 130
Vol. 8: 672

Croze, F.   **Vol. 2:** 364

crystals   **Vol. 11:** 616, 627
Vol. 12: 31, 34, 47, 51, 60, 65, 66, 287

crystal structure   **Vol. 5:** 303*ff*, 312

cubical atoms   **Vol. 3:** 23

culture
human (*see also* human culture/society) **Vol. 7:** 336, 394, 410, 422

national (*see also* Danish culture/society) **Vol. 7:** 336

Cumming, M.   **Vol. 11:** 25*n*

Curie, I.   **Vol. 7:** 32*p*
Vol. 8: 665
Vol. 9: 125, 134, 135, 152, 166, 174, 185, 216, 219, 309, 323, 344, 365, 407*f*, 536, 591–592, 652
Vol. 10: 410, 412, 450
Vol. 11: 515, 523, 531, 569

Curie, M.   **Vol. 2:** 119, 124
Vol. 3: 361
Vol. 6: 36, 37, 318, 375, 415, 423
Vol. 7: 32*p*
Vol. 8: 213, 642, 664–665
Vol. 9: 100, 134, 592
Vol. 10: 75, 102, 385, 393, 447
Vol. 11: 515, 523, 530, 531, 565, 574, 587, 646, 649
Vol. 12: 60, 66, 255, 259, 277, 524

Curie, P. **Vol. 10:** 102
    Vol. 12: 277
Curtis, W.E. **Vol. 2:** 324, 327, 328, 374, 377,
    378, 379, 505, 506, 507
    Vol. 8: 696*f*
Curtiss, F.L. **Vol. 3:** 4, 41, 458
curvature of space **Vol. 7:** 378
cut-off distance/parameter (*see also* adiabatic
    limit) **Vol. 8:** 7*ff*, 51*ff*, 79*ff*, 94*ff*, 122,
    165, 241, 245, 365, 367, 393, 397, 485,
    632, 636, 737*f*, 772, 775
Cuthbertson, C. **Vol. 8:** 10, 643, 665–667
Cuthbertson, C. and M. **Vol. 2:** 7, 8, 31, 33,
    34, 53, 54, 74, 201, 222, 408, 614
    Vol. 3: 194, 196
    Vol. 8: 63, 65*f*, 85*f*, 100*ff*, 144, 665
cyclotron (*see also* accelerator) **Vol. 8:** 232,
    244, 246, 591, 731*f*, 734*f*
    Vol. 9: 2, 3, 217, 221, 560, 590*f*, 626,
    628*f*, 631*f*, 635, 637*f*, 640
    Vol. 10: 413, 449, 452
    Vol. 11: 243, 244, 359, 517, 518, 524,
    526, 530, 531, 557–560, 568, 569
    Vol. 12: 211, 216
Czermak, P. **Vol. 1:** 201, 321

Dagbladet (Danish newspaper) **Vol. 12:** 131*p*
Dale, H. **Vol. 11:** 25–27, 242
Dalton, J. **Vol. 3:** 582
    Vol. 4: 84, 85
    Vol. 6: 374
    Vol. 10: 55, 176, 208
    Vol. 11: 658, 663
Danish American Society **Vol. 11:** 732
Danish Cancer Committee **Vol. 11:** 373, 375,
    682, 683, 685, 686, 726, 729
    Vol. 12: 212, 217
Danish Committee for the Support of Refugee
    Intellectual Workers
    Vol. 11: 11, 409*n*
    Vol. 12: 124, 434, 435
Danish culture/society (*see also* culture,
    national) **Vol. 11:** 12, 61, 64, 137,
    140, 341, 344, 346, 350, 352, 353, 357,
    364, 382, 390, 391, 398, 399, 403–405,
    408, 409, 413, 414, 427, 430, 448–452,
    544, 546, 551, 552, 652
    Vol. 12: 125, 129, 180–182, 188, 189,
    191, 239, 244, 332, 333, 337, 338, 434,
    435, 453–455, 458–460, 469, 471,
    474, 475

Danish Liberation Council **Vol. 12:** 454, 460
Danish Medal of Merit **Vol. 11:** 510*p*, 511*n*
Danish National Museum **Vol. 10:** 250*p*,
    271*p*, 303
    Vol. 12: 106*n*
Danish national radio **Vol. 12:** 12, 55*p*, 59,
    63, 107, 119, 128, 495, 528, 529
Danish Physical Society (*see also* lectures,
    Bohr's) **Vol. 2:** 6, 114, 283, 301, 320,
    554, 555, 624, 628, 633
    Vol. 5: 76–77, 91, 300, 435
    Vol. 6: 10, 316, 479
    Vol. 7: 520
    Vol. 8: 247, 659
Danish Sports Union **Vol. 11:** 735
Danish Student Association **Vol. 11:** 736
Danish Subscription Fund for the Benefit of
    Israel (Den Danske Israelindsamling)
    Vol. 11: 699*n*
Danjelsen Trust **Vol. 12:** 241, 246
Danjelsen, C.D. **Vol. 12:** 241, 245
Danjelsen, K. **Vol. 12:** 241, 245
Dannebrog (Danish order) **Vol. 12:** 102,
    138, 140
Danysz, J. **Vol. 2:** 79, 80, 84
    Vol. 8: 111, 114, 149*f*, 154, 721
D'Arc, J. **Vol. 11:** 250
Darwin, C.G. **Vol. 1:** 114, 520, 521, 554, 555
    Vol. 2: 4, 5, 18, 19, 20, 22, 27, 29, 82,
    97, 98, 330, 428, 429, 501, 580, 594,
    615
    Vol. 3: 496, 497
    Vol. 5: 13–19, 27, 67, 81, 294, 314–318
    Vol. 6: 4, 30, 32, 90, 112, 157*f*, 304,
    305, 308, 310–316, 318, 348–349, 412,
    415, 478
    Vol. 7: 260, 261*p*, 301
    Vol. 8: 5*ff*, 50*ff*, 54, 59, 61, 77*ff*, 82,
    93*ff*, 98, 109, 113, 152, 163, 250, 434,
    436, 489, 566, 667, 672, 709*f*
    Vol. 10: 330–333, 387, 393, 396, 399,
    400, 404, 406, 419, 460
    Vol. 11: 227
    Vol. 12: 17*p*, 111, 280*p*, 286, 287
Darwin, C.R. **Vol. 8:** 5
    Vol. 11: 445
    Vol. 12: 286
Darwin, K. **Vol. 10:** 464, 465
Darwinism **Vol. 10:** 543, 547, 554
Daunt, J.G. **Vol. 12:** 38*p*

Dauvillier, A.   **Vol. 4:** 22, 24, 30, 31, 32, 252, 316, 323, 338, 463, 482, 487, 492, 493, 522, 525, 675, 676, 677, 678, 696
   Vol. 5: 260, 266, 279
Davies, A.C.   **Vol. 4:** 53
Davis, B.   **Vol. 2:** 482
   Vol. 3: 239, 257, 497
   Vol. 5: 5, 237, 294, 318–319, 323
   Vol. 8: 20f, 214, 698
   Vol. 10: 397
Davis, M.B.   **Vol. 11:** 65, 325, 329
Davis, W.   **Vol. 9:** 60
Davisson–Germer experiment (*see also* reflection of electrons)   **Vol. 5:** 210f
   Vol. 6: 4, 7, 70, 77, 84, 394
Davisson, C.J.   **Vol. 5:** 210f
   Vol. 6: 4, 7, 58, 60, 62, 64, 70, 77, 84, 394
   Vol. 10: 402, 440
   Vol. 12: 47
Davy, G.   **Vol. 7:** 328
Davy, H.   **Vol. 10:** 345, 358
   Vol. 11: 445
de Broglie wave (*see also* Broglie, L. de; phase wave)   **Vol. 6:** 23, 39, 70f, 101–102, 132, 139, 142, 156, 163f, 167, 177, 229, 245, 264f, 287f, 393, 417, 421
de Broglie wavelength, *see* wavelength, de Broglie
de Broglie's relation (*see also* Broglie, L. de)
   Vol. 7: 293, 307
Dean, G.   **Vol. 11:** 361, 562
Debierne, A.   **Vol. 8:** 752
Debye, P.   **Vol. 1:** 121, 180, 182, 188, 229, 304, 306, 309, 342, 400, 402, 403, 404, 578, 579, 580, 581
   Vol. 2: 123, 293, 331, 336, 337, 338, 351, 443, 445, 448, 449, 558ff, 603, 604
   Vol. 3: 10, 13, 59, 60, 62, 63, 69, 80, 106, 130, 133, 145, 147, 149, 150, 158, 164, 193, 199, 222, 264, 265, 267, 270, 315, 330, 342, 344, 351, 352, 430, 431, 493, 495, 505, 506, 534, 535, 607
   Vol. 4: 51, 214, 252, 323, 366, 449, 476, 712
   Vol. 5: 4, 78, 137, 353
   Vol. 6: 36, 37, 144, 318, 394, 415
   Vol. 7: 32p
   Vol. 8: 19, 685, 688
   Vol. 10: 440

   Vol. 11: 219
   Vol. 12: 376p
decay, radioactive (*see also* α-decay/particles/ rays; β-decay/particles/radiation/rays)
   Vol. 8: 4, 171, 175, 179, 329, 621, 734, 751, 764
   Vol. 12: 51, 287
Dee, P.I.   **Vol. 9:** 336
   Vol. 10: 410
   Vol. 11: 436p
definition/observation   **Vol. 10:** 45, 61, 62, 89, 200, 406, 539, 542
degeneracy   **Vol. 3:** 88ff, 119, 122, 128, 149
   Vol. 5: 51–52, 54, 64, 67, 145–147, 149–152, 303, 343, 394–395, 445f, 450f, 500
Delaunay, C.   **Vol. 3:** 639, 640
delayed neutrons   **Vol. 9:** 68, 350f, 353, 357, 369, 383–386, 407f, 566
delayed-choice experiment   **Vol. 7:** 256
Delbrück, J.   **Vol. 10:** 489
Delbrück, Manni   **Vol. 10:** 479, 481, 489
Delbrück, Max   **Vol. 9:** 10, 22–24, 28, 141, 533f, 537, 544–547, 570f
   Vol. 10: xliii, 13n, 20, 21p, 22–25, 32p, 36p, 162p, 164, 168, 460, 471n, 554, 555, 557, 558, 560–562, 565, 594
   Vol. 12: 17p
Delsasso, L.A.   **Vol. 9:** 201, 203
δ-function   **Vol. 7:** 26, 66, 68, 69, 95, 107, 127, 129, 130, 148, 156, 198, 430
δ-rays   **Vol. 2:** 238
Democritus   **Vol. 10:** 53, 55
   Vol. 12: 59, 63
Dempster, A.J.   **Vol. 5:** 112
   Vol. 9: 367, 370
Dennison, D.M.   **Vol. 5:** 231, 334, 383, 385ff, 391, 393
   Vol. 6: 145, 158, 397
   Vol. 10: 427, 443
density
   fluctuations   **Vol. 7:** 46, 451, 453, 454
   function   **Vol. 7:** 35, 173, 185, 471
   measurement   **Vol. 7:** 46, 451, 453
density of energy levels, *see* level density
Department of Scientific and Industrial Research (U.K.)   **Vol. 11:** 227n
Descartes, R.   **Vol. 10:** xliii, xliv, 3n, 7, 20, 117, 175, 546, 550
   Vol. 11: 669
   Vol. 12: 371

[67]

Dessau, B.   **Vol. 11:** 525*p*

Destouches, J.   **Vol. 7:** 261*p*, 328, 329
Vol. 9: 593*f*

Destouches, P.   **Vol. 7:** 329

detached observer   **Vol. 10:** xxxvi, xlix, 90, 92, 93, 563–565, 567–570, 572–575

detailed balancing   **Vol. 9:** 47*f*, 392, 492*f*, 499*f*, 508*f*, 510–512, 614, 617

determinism (*see also* causality)   **Vol. 7:** 388, 389, 392, 393, 399, 402, 403, 406, 413, 415, 417, 419
Vol. 10: xlv, 45, 85, 87–91, 94, 103–105, 121, 127, 129, 130, 136, 165, 178, 194–196, 201, 208, 209, 211, 213, 276, 279, 404, 432, 439, 442, 445, 453, 540, 542, 555, 568, 573
Vol. 11: 670, 671

deuterium   **Vol. 8:** 560, 571, 585, 587, 710*f*
Vol. 11: 228, 229, 365, 613, 615, 616, 623, 626, 628, 629

deuteron   **Vol. 9:** 14, 75, 135, 156, 201, 230, 257*f*, 337, 349*ff*, 353, 367, 377, 387–388, 477*ff*, 483–484, 552, 560, 588*ff*, 623*ff*, 631, 638, 640, 647*f*, 650, 652
Vol. 11: 569

Deutsch, M.   **Vol. 8:** 764

Deutsche Physikalische Gesellschaft
Vol. 5: 301

Devaux, P.   **Vol. 7:** 328

Dewar, J.   **Vol. 10:** 373

dialectics   **Vol. 10:** 169, 247, 575, 576

diamagnetism   **Vol. 1:** 393

diatomic gases, infrared spectra of
Vol. 3: 49, 190*ff*

Dibner, B.   **Vol. 10:** 362*n*

dice playing   **Vol. 7:** 281, 282, 358

Dieke, G.H.   **Vol. 6:** 37, 412, 415

dielectric constant of hydrogen, *see* hydrogen, dielectric constant of

dielectric description   **Vol. 8:** 24, 80, 95, 109, 113, 117, 216, 255, 393, 395, 398, 400, 494–497, 699*f*

Diesselhorst, H.   **Vol. 1:** 145, 222, 225, 336, 339

diffraction   **Vol. 8:** 88, 104, 207, 253*f*, 289, 296, 330, 385, 407, 439*f*, 442, 445, 447, 450, 452, 455*f*, 458
of atoms   **Vol. 6:** 132, 156, 167

of electrons (*see also* electron diffraction)
Vol. 6: 29, 101–102, 163, 190, 334*ff*, 339, 346, 360, 363–364, 367, 394, 431
Vol. 7: 293

of light   **Vol. 6:** 17*ff*, 23, 137–138, 190, 420, 431
Vol. 7: 294, 352, 353

of X-rays   **Vol. 6:** 387, 394

diffuse series   **Vol. 2:** 356, 361*f*, 366, 370, 385, 407, 509

diffusion   **Vol. 2:** 423
Vol. 8: 329, 479, 482*f*, 563
angle of   **Vol. 1:** 392
coefficient   **Vol. 1:** 327, 376
direction of   **Vol. 1:** 391
free   **Vol. 1:** 327
in magnetic field   **Vol. 1:** 391
self   **Vol. 1:** 394

dilemma, *see* paradox/dilemma

dimensional analysis   **Vol. 1:** 320
Vol. 2: 97–99, 125, 162, 268, 291, 304, 306, 594, 627
Vol. 8: 16, 241, 299, 337, 454–455, 819
Vol. 9: 234, 371

dimensions
atomic   **Vol. 8:** 6*ff*, 50, 52, 77, 79, 92*f*, 95, 130, 163, 169, 173, 177, 269, 278, 329, 366, 368, 394, 398, 457, 461, 463, 467, 492, 508, 510, 533, 543, 603, 608, 622, 632, 634*f*, 666
nuclear   **Vol. 8:** 50, 77, 92*f*, 130, 163, 485, 634

Dingle, J.   **Vol. 12:** 38*p*

dipole
electric   **Vol. 2:** 96, 98, 323, 325, 437–438, 593, 632
Vol. 5: 17, 44–45, 55, 85, 105, 145*f*, 149*ff*, 309*ff*
moment   **Vol. 9:** 45, 95, 194, 237*f*, 262, 278, 315, 327, 544, 607*f*
radiation   **Vol. 9:** 32, 44, 194, 237*f*, 278, 315, 327, 336, 544
transition   **Vol. 9:** 38, 262, 278, 327

Dirac
equation (*see also* electron theory, relativistic; quantum mechanics, relativistic)   **Vol. 7:** 36, 226, 246, 465, 479
Vol. 8: 210
Vol. 9: 12, 18, 107, 125, 549, 592*ff*
symbolism   **Vol. 7:** 177, 188

theory (*see also* electron theory; hole theory)
    Vol. 7: 31, 33–36, 48, 181, 191–193,
    203, 220–222, 241, 242, 442, 447, 449,
    464, 466, 470
    Vol. 10: 212, 339, 404, 406,
    447, 450
Dirac, P.A.M.   **Vol. 1:** xxxvii–xxxviii
    Vol. 5: 94, 216, 225, 338, $370f$, 501
    Vol. 6: $3f$, 7, 9, 12, 17, 19, 23, 36, 37,
    39, 44, 45, 52, 95, 97, 100, 103–105,
    $124f$, 138, 144, 146, 153, $155f$, 158,
    160, 162, $164ff$, 175, 191, $267f$, $291f$,
    318, 333, 339, 347, 348, 359, $393ff$,
    402, 412, 415, 417, $421ff$, $426ff$, 432,
    $439f$, 459, 461, 463
    Vol. 7: 3, 4, $32p$, 64, 66, 126, 127, 167,
    169, 177, 188, 192, 193, 226, 262, 265,
    308, 347, 363, 377, 442, $447n$
    Vol. 8: 224, 677, $715f$, 803
    Vol. 9: $6f$, 10, $11f$, 107, 110, $125f$,
    132, 134, $135ff$, 166, 174, 196, 305,
    319, 534, 547–550, $570f$, 599, $601f$,
    604, 606, 677
    Vol. 10: 339, 399–401, 403, 404, 406,
    414, $422p$, 426, 427, 441, 443, 447,
    450, $455p$, 460
    Vol. 11: 199, 201, 567
    Vol. 12: 11, $17p$, $38p$, 48, 210, 215,
    348, 351
direction of time (*see also* irreversibility)
    Vol. 6: 46, 50, 327, 438, 440, 452, 466
Dirichlet, G.P.L.   **Vol. 10:** xxix
discontinuity   **Vol. 6:** 10, 14, 17, 19, 21, 23,
    33, 52, 56, 59, 61, 63, 69, 75, $80f$, 88,
    91, $93f$, 97, 114, 124, 127, 142, 144,
    148, $151f$, 154, $160f$, 163, 166, 178,
    185, 227, 241, 261, 282, 419, 421, 423,
    433, 462
disease   **Vol. 10:** 29, 35, 67, 148, 201
disintegration (*see also* transmutation;
    transformation/transmutation)
    Vol. 8: 171, 175, 179, 186, 192, 221,
    257, $267ff$, $277f$, 286, 313–316, $468f$,
    549, 563, $576f$, 651, 652, $712f$,
    783–784, 797, $821f$
    Vol. 11: 90, 102, 123, 124, 127, 241,
    515, 523, 529, 531, 646, 649
    Vol. 12: 12, 42, 51, 52, 54, 61, 68, 79,
    81, 266, 287, 500, 502
disintegration of nucleus, *see* nuclear
    transmutations

dispersion (*see also* scattering of light)
    Vol. 1: xxxv–xxxvi, 581
    Vol. 2: 6, 7–8, 21, 45–46, 52–55, 61,
    63, 73, 74, 75, 90, 93, 106, 114, 117,
    140, 153, 194, 201–202, 222–223,
    $336ff$, 408, 448–450, 530, $564ff$, 577,
    587, 604, 624
    Vol. 3: $192ff$, 368, $495ff$
    Vol. 5: 9, 12, $17f$, 21, 53, 58, 64, 67,
    81, 90, 102, 106, $113f$, 145, 149, 166,
    169, 181, 196, $251f$, 257, $263f$, 273,
    279, 309, 317, $358f$, 396, 398–401,
    403, 417, 420, 492, 506
    Vol. 6: 3, 13, $17f$, 76, 82, 94, 124,
    138, 153, 156, 179, 359, 393, 395, 460
    Vol. 7: 36, 176, 177, 187, 306, 348,
    464, 466, 479
    Vol. 10: 401, 402, 425, 427,
    440, 441
    Vol. 12: 20, 42, 115, 348, 351, 358,
    359, 364
  nuclear (*see also* Breit–Wigner formula)
    Vol. 9: 49, 332, 344, 377, 388, 392,
    $491ff$, 497, 507, 508–510, 612, 617
  of radiation   **Vol. 8:** $8ff$, 33, 53, $63ff$, 70,
    83, $85ff$, $98ff$, 117, 131, 133, $143ff$, 160,
    171, 175, 179, 216, 256, $272ff$, $281ff$,
    371, $378f$, 396, 400, 492, $498f$, 504,
    515–516, 665, 666, $699ff$, 795, 803, 819
  optical   **Vol. 9:** 247, 249, $268f$, 278,
    312, 325
  theory, Kramers' (and Heisenberg's)
    Vol. 5: 40–47, 53, 57, 67, 86, 91, 221,
    256–257, 258, 261, 262–264, 265, 267,
    $279f$, 299, 303, $335f$, 357, 396,
    403, $484f$
displacement current   **Vol. 7:** 214
displacement law, spectroscopic   **Vol. 3:** 9,
    17, 179
    Vol. 4: 53, 107, 433, 600, 648
displacement laws, radioactive
    Vol. 2: 213–214, 418–419, 440
dissociation   **Vol. 2:** 130, 138, 142, 151, 156,
    180, 237, 459, 481, 539–540, 625
    Vol. 5: 82, 185, 199
dissymmetry, *see* symmetry
distinction between measuring tool and object
    Vol. 10: 131, 137, 517, 521
distribution law   **Vol. 7:** 177, 187
distribution of energy loss, *see* probability
    distribution of energy loss

[69]

distribution, normal velocity   **Vol. 1:** 310

disturbance of object by measurement
   Vol. 6: 40, 91, 104, 114, 148, 165, 171,
   181–182, 206, 216, 233, 249, 266,
   289, 349, 352, 418

divisibility   **Vol. 12:** 59–61, 63, 65, 68

DNA   **Vol. 10:** 13, 24, 168, 214, 485, 486,
   499, 551, 552

doctor's degree (Bohr's 1911)   **Vol. 12:** 3, 4*n*,
   5, 102, 108, 121, 131*p*, 135, 136, 138,
   140, 144, 149, 150, 209, 213

doctor's dissertation   **Vol. 1:** xix
   attempt at English publication   **Vol. 1:** 103,
      109*ff*, 114*ff*, 531, 545, 549, 551
   correspondence about   **Vol. 1:** 398*ff*
   defense of   **Vol. 1:** 97*ff*
   early reactions to   **Vol. 1:** 101
   translation of   **Vol. 1:** xxi, 291–395

Dodgson, C., *see* Carroll, L.

Dodson, R.W.   **Vol. 9:** 56, 365

Dolejsek, U.   **Vol. 4:** 493, 530

Donder, T. de   **Vol. 7:** 32*p*
   Vol. 12: 376*p*

doorway state   **Vol. 9:** 43

Doppler effect   **Vol. 5:** 4, 23, 106, 112
   Vol. 6: 18, 23, 29, 31, 37, 120*f*, 150*f*,
      165, 173, 316, 420
   Vol. 7: 84, 86, 140, 141
   Vol. 9: 298, 378*f*

Dorgelo, H.B.   **Vol. 5:** 236, 279

Dorsey, H.G.   **Vol. 1:** 62, 64, 65

Døssing, Mrs   **Vol. 11:** 318, 319

Døssing, T.   **Vol. 11:** 60, 310, 312, 314,
   316–318, 703*n*

double-scattering experiment   **Vol. 8:** 210

double-slit experiment   **Vol. 6:** 15, 365–366
   Vol. 7: 255, 293–296, 352–360,
      373, 507

double refraction   **Vol. 5:** 63

doublet
   electric, *see* dipole, electric
   spectral   **Vol. 5:** 35, 229*ff*, 235*ff*, 288–289,
      330, 339, 424*ff*, 459*f*, 467*ff*, 493,
      501, 508

doubling of spectral lines (*see also* structure of
   hydrogen lines)   **Vol. 2:** 304, 320,
   323–329, 364–367, 379–380, 469, 505,
   506–507, 591, 593, 603, 629, 632

Dræby, E.   **Vol. 11:** 358*p*

Dreyer, C.D. *or* J.L.E.   **Vol. 1:** 534, 535

Driesch, H.   **Vol. 10:** 4, 5

drop vibration   **Vol. 1:** 88

Drosophila   **Vol. 10:** 22, 23, 554

Droste, G. v.   **Vol. 9:** 387

Drude, P.   **Vol. 1:** 97, 124, 132, 133, 134,
   143, 144, 159, 172, 173, 174, 175, 224,
   225, 226, 232, 257, 258, 298, 299, 300,
   337, 338, 339, 354, 366, 408, 413, 424,
   428, 433, 446, 447, 448, 451, 453, 476
   Vol. 2: 7, 31, 33, 85, 202, 287, 449,
      577, 587
   Vol. 3: 193, 195, 196
   Vol. 8: 63, 65, 85, 100, 452, 567, 666

Druyvesteyn, M.J.   **Vol. 12:** 38*p*

dualism/duality (*see also* "Zwang")
   Vol. 5: 24, 36, 60, 66, 71, 178, 183,
      194, 198, 235, 239, 289, 406
   Vol. 6: 8, 9, 27, 59, 61, 63, 108
   Vol. 7: 212, 347
   Vol. 10: 15

duality
   in biology   **Vol. 10:** 137, 152, 169
   principle of   **Vol. 10:** 505–509

Duane, W.   **Vol. 2:** 344, 476, 485
   Vol. 4: 493, 498, 530, 531
   Vol. 5: 5, 38*ff*, 121, 123, 294,
      319–326, 496

Dubrovnik Cultural–Historical Museum
   Vol. 12: 234

Duffendack, O.S.   **Vol. 5:** 384, 386

Duke University   **Vol. 9:** 39

Dulles, J.F.   **Vol. 11:** 327, 330

Dumas, L.   **Vol. 10:** 4*n*

Dumond, J.W.M.   **Vol. 10:** 330

Duncanson, W.E.   **Vol. 9:** 185, 188

Dunning, J.R.   **Vol. 9:** 28, 57, 61*ff*, 71, 365,
   380, 385, 401, 404, 407*f*, 441, 465,
   469, 553*ff*

Dunoyer, A.   **Vol. 3:** 255

Dupreel, E.   **Vol. 7:** 328

Dürr, K.   **Vol. 7:** 328

Dyggve, E.   **Vol. 11:** 406*p*

dynamical
   concepts   **Vol. 7:** 22, 322
   quantities   **Vol. 7:** 105, 154, 392

dynamical theory of gases, *see* kinetic theory;
   statistical mechanics

dynamics (*see also* kinematics)
   Vol. 12: 223, 224

Dyson, F.J.   **Vol. 7:** 49, 50
   Vol. 10: 455*p*, 556

Easter conference (*see also* Copenhagen
    Conference)   **Vol. 7:** 446
Eastman, G.   **Vol. 11:** 281
Eberstadt, F.   **Vol. 11:** 325, 329
Eckart, C.   **Vol. 7:** 165*n*
Eckmann, B.   **Vol. 7:** 328
École Polytechnique (Paris)   **Vol. 11:** 711*n*
Edda   **Vol. 10:** 255, 265
Eddington, A.S.   **Vol. 3:** 660, 662
    Vol. 4: 686
    Vol. 5: 484, 486
    Vol. 7: 260, 261*p*, 300, 301
    Vol. 9: 6, 555
    Vol. 10: 409, 540, 542, 573
Eden, A.   **Vol. 11:** 47, 277*n*, 287
Eder, J.M.   **Vol. 4:** 702
Edinburgh Lecture, Bohr's, 1930   **Vol. 6:** 312,
    317, 351–352, 480
Edinburgh Lectures, Bohr's, 1949 (*see also*
    Gifford Lectures)
    Vol. 7: 489, 492, 502, 503
    Vol. 9: 617, 668
Edinburgh, University of, Tait Chair of Natural
    Philosophy   **Vol. 12:** 530
Edison, T.A.   **Vol. 11:** 359
Edlén, B.   **Vol. 1:** xxvii
    Vol. 6: 385, 395
    Vol. 8: 788
    Vol. 10: 379, 556
education   **Vol. 10:** 105, 169, 206, 219, 229,
    234, 246, 254, 255, 257–259, 263, 264,
    267, 268, 270, 306, 308, 352, 354, 355,
    365, 367, 369, 572, 585
    Vol. 11: 203, 205, 351, 364, 369, 416,
    417, 454, 459, 494–499, 584, 607, 617,
    630, 652
    school   **Vol. 12:** 59, 63, 102, 120, 122, 126,
    128, 155, 166, 248, 249, 399, 400,
    403, 404
    science   **Vol. 12:** 3, 94–96, 122, 182, 184,
    190, 191, 194, 212, 217, 269, 300, 302,
    317, 321, 348, 350, 429, 431, 438, 440,
    512, 513
    university   **Vol. 12:** 125, 136*n*, 327,
    330, 404*n*
effective
    charge   **Vol. 8:** 236, 238, 257, 259*f*, 309,
    320, 326, 328, 330*f*, 379, 461, 528,
    531, 540*ff*, 548, 556, 559, 591, 608*f*,
    746, 748
    nuclear charge   **Vol. 4:** 393, 501, 533

quantum number   **Vol. 4:** 501, 533, 581
Egmont Petersen Foundation
    Vol. 12: 182, 190
ego (*see also* self)   **Vol. 10:** 6, 71, 93, 215,
    286, 291, 292
Ehrenberg, W.   **Vol. 9:** 156
Ehrenfest's theorem   **Vol. 8:** 291*f*
Ehrenfest, P.   **Vol. 1:** xxxii, xxxvi
    Vol. 2: 338, 341, 436, 437, 438, 458,
    459, 537, 565, 567, 604
    Vol. 3: 7, 11, 12, 15, 16, 22, 24, 28, 29,
    30, 31, 32, 33, 34, 35, 36, 37, 38, 39,
    40, 70, 74, 75, 76, 79, 80, 83, 84, 105,
    159, 199, 258, 310, 316, 318, 333, 361,
    363, 369, 370, 372, 377, 381, 388, 392,
    393, 394, 395, 405, 406, 409, 430, 433,
    452, 469, 470, 473, 483, 484, 487, 493,
    594, 609–632, 634, 652, 660, 662, 666,
    667, 687
    Vol. 4: 14, 15, 29, 51, 136, 353, 354,
    356, 361, 362, 423, 450, 476, 556,
    616, 683
    Vol. 5: 5, 25, 26, 28, 77, 105, 116, 181,
    197, 226–230, 255, 262, 277*f*, 294,
    326–333, 442, 444, 504, 507
    Vol. 6: 36, 37–41, 44, 127, 154, 181*ff*,
    304, 305, 306*f*, 308, 389, 412,
    415–418, 426*f*, 438
    Vol. 7: 346, 353, 358, 368, 369
    Vol. 8: 22, 224, 667, 670, 675
    Vol. 9: 8, 10, 36, 100, 567*ff*, 597*f*
    Vol. 10: 32*p*, 395, 398, 439, 445, 448,
    527, 528
    Vol. 12: 17*p*, 355
Ehrenhaft, F.   **Vol. 8:** 651
Eibe, T.   **Vol. 12:** 400, 404
eigenfunction/eigenvalue, *see* proper vibration
Einstein Birthday Volume   **Vol. 7:** viii, 267,
    276, 277
Einstein weighing experiment   **Vol. 6:** 319,
    368–369
Einstein–de Haas effect   **Vol. 2:** 130, 263,
    335, 395
Einstein, A.   **Vol. 1:** xvii, xxx, xxxix, xli, 270,
    273, 376, 378, 453, 454
    Vol. 2: 123, 128, 137, 147, 165, 177,
    263, 395, 501, 532, 533, 633–635
    Vol. 3: 6, 20, 22, 45, 63, 70, 72, 73,
    147, 209, 234, 244, 256, 285, 295, 342,
    351, 362, 366, 374, 381, 402, 412, 413,
    414, 422, 423, 478, 480, 484, 489, 491,

[71]

496, 497, 498, 570, 571, 572, 586, 587, 609, 611, 613, 615, 620, 629, 632, 634, 660, 662

Vol. 4: 24, 25, 27, 28, 29, 30, 97, 344, 345, 348, 355, 423, 435, 436, 450, 470, 476, 553, 613, 659, 683, 685, 686, 697, 698

Vol. 5: 3, 5, 7, 12, 23, 24–28, 31–34, 41, 44f, 68, 76f, 79, 84, 86, 90, 98, 103ff, 107, 109, 116, 132, 139f, 183, 191, 193, 198, 205f, 220, 226f, 229f, 234, 238, 256, 260, 263, 267, 274f, 279, 303f, 314, 326–330, 350, 353, 382, 402, 415ff, 497, 504, 507

Vol. 6: xx, 4, 8, 10, 16f, 20, 21f, 29f, 32, 36, 37ff, 44, 46, 50f, 60, 62, 64f, 69, 75f, 81, 83, 91f, 94, 100, 101–103, 114f, 124, 144, 148, 153, 158, 160, 162ff, 178, 184, 191, 205, 213, 227f, 230, 233, 241, 243, 245, 250, 260, 281, 308, 318, 319, 328, 359, 363, 367ff, 380, 382, 393, 396, 402, 405, 412, 415ff, 418–421, 423, 431, 452f, 455, 467

Vol. 7: viii, 3, 17, 32p, 51, 250–258, 265, 267, 269, 270, 273, 274n, 276–281, 283p, 285, 286p, 290, 292, 295–297, 304, 305, 309, 312, 319, 329, 339, 341, 342, 344–347, 351–353, 355, 357, 358, 362–366, 368–377, 379–381, 389, 401, 416, 434, 458–462, 480, 481, 488, 491, 522

Vol. 8: 27, 88, 104, 667, 670, 672, 675, 710f

Vol. 9: 8, 93, 95, 108, 165, 174, 304f, 308, 318f, 322, 570f

Vol. 10: xxxiii, 40, 58, 59, 85, 103, 142, 158, 176, 187, 204, 209, 211, 338–340, 350, 363, 376, 377, 384, 386, 388, 391, 395, 400–402, 406, 409, 418, 425, 432, 433, 437–441, 445, 446, 448, 449, 477, 503, 504, 558, 564, 568, 570, 573, 591

Vol. 11: 6, 276n, 372, 530, 567, 568, 576, 591, 670

Vol. 12: 43, 44, 47, 99, 118, 119, 149, 150, 155, 156, 167, 168, 209, 214, 221, 282, 356, 358, 363, 364, 371, 372, 374, 375, 376p, 378, 494–496, 529, 531

Einstein, Podolsky and Rosen paper
Vol. 7: viii, 14, 17, 281, 371–374, 425–430, 461, 462, 503, 507, 510, 511, 521, 524

Eisenbud, L. **Vol. 9:** 508

Eisenhower, D.D. **Vol. 11:** 83, 327n, 360, 366, 367, 368p, 369, 619n, 639, 646, 650, 710n, 717

"Ekliptica" discussion circle **Vol. 2:** 632
Vol. 6: xxiv
Vol. 10: 299–301

Ekman, V.W. **Vol. 5:** 383, 385

elastic scattering, *see* scattering, elastic

Eldridge, J.A. **Vol. 5:** 145, 147, 148, 153

electric
conduction **Vol. 12:** 7, 8, 39
current, as adiabatic change **Vol. 12:** 39
dipole, *see* dipole, electric
discharge in gases **Vol. 12:** 42, 60, 66, 255, 259, 316, 321, 340, 341
doublet, (*see also* dipole, electric)
Vol. 1: 390
fields **Vol. 12:** 24, 28, 60, 66, 357, 368
moment, *see* dipole, electric
oscillations **Vol. 12:** 106, 236
power transmission **Vol. 12:** 106, 236
resistance **Vol. 12:** 4, 5, 7–9, 22–24, 26–28, 30, 33, 39, 499, 501

electrical
charge, *see* charge
engineering **Vol. 11:** 530, 557, 559
field (*see also* field, electromagnetic)
Vol. 7: 9, 42, 44, 60, 76, 81, 124, 134, 138, 202, 205, 213, 214, 233, 330
oscillator, *see* oscillator
polarization, *see* polarization

electricity **Vol. 10:** 56, 67, 120, 342, 344–347, 351, 357, 358, 360, 364, 498, 499
Vol. 11: 364, 366–368, 658, 663, 670
Vol. 12: 20, 60, 65, 84, 220, 356, 438, 440

electrochemistry **Vol. 10:** 68, 345, 358, 359
Vol. 12: 438, 440

electrodynamics **Vol. 10:** 177, 195, 210, 338, 376, 379, 386, 388, 389, 402, 406, 436, 453
Vol. 11: 657, 659, 663, 664
Vol. 12: 43–45, 47, 222, 233, 281, 344, 345, 356, 357

electrodynamics, classical/ordinary (*see also* mechanics, classical/ordinary; classical electrodynamics)

Vol. 2: 113, 126, 128, 162*ff*, 172–175, 179, 180, 182, 241–247, 254–255, 258, 263, 280, 287*ff*, 308, 312*f*, 337, 351, 353, 358*ff*, 393*f*, 404, 433, 435, 443*ff*, 460, 473*f*, 552*ff*, 584, 588

electrolysis **Vol. 10:** 56, 345, 359
    Vol. 11: 612, 616, 620, 628
    Vol. 12: 60, 65, 66

electrolytic dissociation, theory of
    Vol. 12: 125, 438, 441

electromagnetic
  field (*see also* field, electromagnetic)
    Vol. 11: 670
    Vol. 12: 6, 221, 359, 365
  forces **Vol. 11:** 657, 663
  mass **Vol. 12:** 20
  moment **Vol. 12:** 365
  radiation **Vol. 11:** 671
    Vol. 12: 6*n*, 221
  theory (*see also* Maxwell theory)
    Vol. 12: 4, 20, 220, 221, 223, 255, 259
  waves **Vol. 12:** 43, 356

electromagnetic field, measurability of
    Vol. 10: 407, 531, 532

electromagnetic radiation
  from atoms **Vol. 1:** xxx–xxxvi
  measurements of electromagnetic fields
    Vol. 1: xli–xlii
  virtual oscillators **Vol. 1:** xxxv–xxxvi

electromagnetism **Vol. 10:** 29, 30, 57, 68, 85, 87, 127, 148, 149, 175, 178, 209, 210, 342–344, 346, 349, 351, 354, 357, 358, 360, 362–364, 368, 389, 401, 402, 407, 432
    Vol. 11: 479, 483
    Vol. 12: 107, 122*n*, 156, 168, 222, 306, 307, 309, 310, 317, 322, 356, 371, 428, 431
  lectures on **Vol. 2:** 331, 515, 517

electromotive force **Vol. 1:** 348

electron
  annihilation/creation, *see* pair creation
  binding **Vol. 11:** 511, 512, 514, 515, 520–523, 566, 568, 613, 621, 623
    Vol. 12: 44–47, 51, 52, 209, 214, 215, 222, 356, 364, 381, 383
  capture (*see also* capture/loss of electrons)
    Vol. 6: 376, 458
    Vol. 11: 568
    Vol. 12: 44, 45

capture by α-particles **Vol. 5:** 71, 82, 95, 181, 184–185, 192, 197, 199–200, 205–206, 334, 337–338
configuration symmetry of **Vol. 12:** 45
constitution **Vol. 12:** 508
core, *see* core
coupling **Vol. 12:** 39
definition of **Vol. 12:** 255, 259
diameter **Vol. 7:** 31, 33, 170–172, 175, 176, 179, 181, 183, 184, 186, 187, 189, 190, 192, 193, 219–221, 226, 463, 466, 470
    Vol. 9: 5, 8, 12, 93, 104*f*, 108*ff*, 117, 126, 131*f*, 548, 605*f*
diameter, classical **Vol. 6:** 401*ff*, 445*ff*
diffraction (*see also* diffraction of electrons)
    Vol. 5: 95, 215
discovery of **Vol. 11:** 530, 566, 574, 587, 658, 663
    Vol. 12: 43, 45, 51, 108, 221, 252, 356
free **Vol. 11:** 566
    Vol. 12: 7, 39, 44, 52, 364
free mobility in metals of **Vol. 12:** 6, 22–25, 27, 28, 499, 501
gas **Vol. 12:** 22, 25, 26, 39
groups **Vol. 6:** 387–391, 402, 458
interaction **Vol. 12:** 30*n*, 39, 47
interference (*see also* reflection of electrons)
    Vol. 6: 105, 270, 295, 365–366
lattice **Vol. 12:** 5, 7, 9, 23, 27, 30, 31, 33, 499, 501
loss, *see* capture/loss of electrons
mass **Vol. 7:** 180, 190, 216, 219, 221, 222, 227, 303, 332
    Vol. 12: 357
motion **Vol. 12:** 39, 45, 357
orbit/trajectory (*see also* orbital velocity)
    Vol. 5: 30, 35*f*, 59, 87–88, 122, 157–171, 188*f*, 202*f*, 223, 235, 239, 243, 247, 251*f*, 272, 276*f*, 279, 288–289, 331, 335, 383, 385, 406, 411, 413*f*, 423, 425–432, 434, 436, 440, 443, 446, 450–451, 467–482, 490, 503, 506
    Vol. 6: 13, 52, 104, 112, 155, 161*ff*, 172*ff*, 311, 339*ff*, 347*f*, 366, 379*ff*, 384, 388, 392*ff*, 422, 460
    Vol. 7: 104, 153, 180, 190, 260, 305, 347–349, 352, 353, 356–358, 361, 378, 390, 402, 416, 447, 449

Vol. 8: 34, 51, 56, 60f, 67, 77f, 80, 89, 93f, 96, 104, 117, 133, 163, 169, 172, 177, 184, 189, 255, 309, 382, 394, 398, 494–495, 531, 723, 752, 772, 775f, 818f
    Vol. 12: 43–48
pairing **Vol. 12:** 9
particle/wave **Vol. 7:** 219, 307, 315, 332, 350
polarization **Vol. 6:** 309, 347, 349
positron pairs **Vol. 7:** 37, 44, 171, 174, 184, 185, 197, 215, 216, 221, 226, 242–244, 315, 321
radius **Vol. 8:** 638, 798
reflection, *see* reflection of electrons
scattering (*see also* reflection of electrons; scattering of electrons)
    Vol. 6: 105, 309
secondary, *see* secondary electrons
shells **Vol. 7:** 279
    Vol. 12: 45, 364
spin (*see also* magnetic moment of electron; spin of electron; spin) **Vol. 6:** 3, 7, 52, 158, 305, 307, 310f, 337–349, 391f, 394, 398, 404, 422, 457
    Vol. 12: 11, 48, 51
system, change of state **Vol. 12:** 30, 33
theory (*see also* Dirac theory) **Vol. 1:** 503
    Vol. 2: 4, 61, 71, 325, 337, 362, 554, 556
    Vol. 8: 4f, 141, 169, 172, 177, 496, 799
    Vol. 10: 337–340, 407, 437, 452
theory, classical **Vol. 6:** 136, 157f, 268, 292, 348, 359, 395, 401
    Vol. 12: 3–5, 20, 22, 25, 43, 135, 136, 138, 140, 144, 149, 150, 209, 213, 340, 341, 356–358
theory, relativistic (*see also* Dirac equation; quantum mechanics, relativistic; relativistic quantum mechanics)
    Vol. 6: 4, 7, 268, 292, 330, 333, 339, 347–348, 356, 395, 402, 426f
    Vol. 9: 4, 8, 13, 125f, 132, 166, 174, 305, 319, 547, 549, 578f, 668
    Vol. 12: 365
wave **Vol. 12:** 23, 27
electron distribution in atoms
revision of **Vol. 4:** 40ff, 662
table of **Vol. 4:** 330, 456, 479, 488, 521, 582, 635

electron theory of metals **Vol. 1:** 575, 579
book by P. Suter **Vol. 1:** 119
dissertation **Vol. 1:** 291–395
history **Vol. 1:** 297ff
lecture notes **Vol. 1:** 113, 194, 446–471
lecture on **Vol. 1:** 411–419, 581
electronic orbits, classification of
    Vol. 4: 418, 419, 456, 479
electronics **Vol. 11:** 574, 587
electrons
mobility of **Vol. 2:** 93
motion in metal **Vol. 1:** 132ff
secondary, *see* β-particles, secondary
stopping of, *see* stopping of charged particles
electrons, free **Vol. 1:** 300, 304
speed **Vol. 1:** 309
velocity distribution **Vol. 1:** 307
electrostatic forces **Vol. 12:** 51
elementary
particle (*see also* particle) **Vol. 6:** 136, 146, 157–158, 209, 231f, 247f, 265, 267f, 288ff, 401–403, 407
    Vol. 9: 87f, 103, 105, 107f, 110, 117, 125f, 140f, 166, 174, 186, 207, 216, 219, 230, 258, 274, 279, 374, 376, 415, 445, 490, 506–507, 605
    Vol. 10: 12, 34, 87, 210, 452, 454, 500, 547, 551, 561, 562
    Vol. 11: 569, 613, 622, 646
process, *see* individual/elementary process
elements
element 72 **Vol. 4:** 22, 23, 26, 30ff, 316, 338, 405, 463, 482
of first period (H, He) (*see also* hydrogen and helium)
    Vol. 4: 222, 287ff
of second period (Li–Ne) **Vol. 4:** 53, 65ff, 140ff, 223ff, 291ff, 389ff, 459, 479
of third period (Na–A) **Vol. 4:** 153ff, 229ff, 247ff, 394ff, 459, 479
of fourth period (K–Kr) **Vol. 4:** 162ff, 233ff, 302ff, 399ff, 459, 480
of fifth period (Rb–Xe) **Vol. 4:** 171, 241, 310ff, 404, 459, 480
of sixth period (Cs–Rn) **Vol. 4:** 172, 242, 311ff, 404, 459, 480
periodicity of **Vol. 4:** 658
physical and chemical properties of
    Vol. 4: 20ff, 181ff
relationships among **Vol. 4:** 441, 472ff
Elisabeth II **Vol. 11:** 735

Elisabeth, Queen of Belgium   **Vol. 12:** 376
elliptical orbit (*see also* Keplerian laws/orbit)
       Vol. 2: 116, 174, 181, 244–251, 290,
       328–329, 357*ff*, 363*ff*, 374, 380, 397,
       401–403, 405, 406–407, 465*ff*, 629
Ellis, C.D.   **Vol. 4:** 261
       Vol. 7: 32*p*
       Vol. 8: 651, 785*f*
       Vol. 9: 100, 134, 137*f*
       Vol. 10: 399, 411, 412, 451
Ellis, E.   **Vol. 10:** 23
Elsasser, W.M.   **Vol. 5:** 80
       Vol. 6: 138
       Vol. 8: 30
       Vol. 9: 156
       Vol. 10: 14, 461, 480, 482, 485,
       486, 497*n*
Elster, J.   **Vol. 8:** 668
embryology   **Vol. 10:** 55, 61, 166, 559, 560
emission
   of radiation (*see also* coupling between
       emission and absorption)   **Vol. 5:** 7*ff*,
       24, 32*f*, 43*ff* 58–59, 60, 65, 102*ff*,
       112*ff*, 117, 122*ff*, 145, 149, 179, 195,
       275*ff*, 299, 309, 317, 398*f*, 415,
       419, 492
       Vol. 6: 40, 52, 127, 133, 140, 154, 156,
       173, 177, 226, 240, 360, 380, 382*ff*,
       388, 402, 405, 417, 445, 447
       Vol. 8: 21, 30*ff*, 117, 264, 598, 605,
       610, 619, 621*f*, 637*f*, 675, 680, 687,
       689, 691, 802, 812
   process   **Vol. 7:** 23, 200, 345, 348
   spontaneous   **Vol. 7:** 201, 345
emotion/feeling   **Vol. 10:** xxix, xli, xlv, 47,
       62, 160, 245, 285, 291, 483, 491, 511,
       512, 568
Encyclopædia Britannica   **Vol. 12:** v*n*, 10–12,
       494, 508, 526
energy
   conservation/balance (*see also* conservation
       laws)   **Vol. 7:** 8, 36, 75, 82, 134, 138,
       180, 181, 190, 191, 297, 314, 315, 347,
       353, 355, 371, 468
       Vol. 10: 7, 68, 411, 412
       Vol. 11: 671
       Vol. 12: 6, 221
   density   **Vol. 7:** 201, 222
   dissipation   **Vol. 10:** 376, 386, 436
   equipartition   **Vol. 10:** 433

   equivalence with mass, *see* mass,
       equivalence with energy
   exchange/transfer   **Vol. 7:** 84, 140, 297,
       298, 304, 314, 333, 346, 350, 354, 355,
       361, 362, 364, 365, 369;
       Vol. 10: 136, 150, 178, 387,
       397, 410
   fluctuation   **Vol. 7:** 238, 245, 487, 491,
       493, 494
   negative   **Vol. 7:** 242, 468
   of formation of hydrogen, *see* hydrogen,
       heat of combination/dissociation of
   of photon, *see* photon energy
   sources, *see* power sources
   transfer, continuous   **Vol. 8:** 17, 22, 32,
       118, 121, 125, 274, 283, 395, 399*f*,
       498, 500
   transfer, quantized   **Vol. 8:** 17, 20, 22*ff*, 28,
       112, 114, 184, 189, 198, 200, 241, 246,
       256, 381, 395*f*, 399*f*, 498, 672
energy–frequency relation   **Vol. 12:** 47
energy–momentum
   conservation (*see also* conservation laws)
       Vol. 10: 131, 137, 158, 183, 196, 197,
       208, 213, 277, 402, 406, 425,
       439, 451
   exchange/transfer   **Vol. 10:** 104, 158, 183,
       197, 211, 213, 278, 410
       Vol. 11: 102, 671
       Vol. 12: 23, 27, 39, 44, 54, 157, 170,
       209, 214, 282, 364
ensemble (*see also* identical systems)
       Vol. 6: 101, 320, 325, 469*f*
Enskog, D.   **Vol. 1:** 102, 476
entelechy (*see also* life force)   **Vol. 10:** 6
entropy (*see also* irreversibility; second law of
       thermodynamics)   **Vol. 6:** 320*ff*, 326*ff*,
       400, 438, 440, 449*ff*, 467*ff*
       Vol. 9: 259
       Vol. 10: 10, 119, 136, 150, 151, 166,
       208, 432, 434, 493
       Vol. 11: 298, 670
       Vol. 12: 221
enzymes   **Vol. 10:** 136, 151, 214, 499,
       544, 548
epistemology   **Vol. 7:** 51, 263, 265–267, 277,
       312, 316, 339–381, 383, 384, 403, 417
       Vol. 10: xxiii–xxviii, xxx, xxxii, xxxiii,
       xxxv, xxxix, 6, 11, 13–15, 18, 25, 39,
       41, 43, 45, 47, 58, 60, 64, 96, 112, 141,
       157, 167, 179, 241, 243, 245, 249, 275,

279, 298, 314, 317, 320, 322, 404, 428, 444, 445, 448, 454, 477*n*, 480, 482, 483, 485, 486, 489, 490, 500, 501, 503, 518, 519, 521, 522, 527, 528, 546, 550, 559, 560, 567, 569, 571, 573, 585, 591
Vol. 11: 81, 134, 350, 371, 712
Vol. 12: v, 86, 87, 90, 117, 158, 171, 211, 217, 224, 366
attitude to religion **Vol. 1:** xx–xxi
complementarity **Vol. 1:** xl–xlii, xlvi–xlviii
early thoughts **Vol. 1:** xx, 499, 503
indeterminacy relations
Vol. 1: xxxviii–xxxix
measurements of physical quantities
Vol. 1: xxxix–xl, xli–xlii
EPR, *see* Einstein, Podolsky and Rosen paper
Epstein, P.S. **Vol. 2:** 474, 632
Vol. 3: 7, 13, 43, 50, 51, 52, 54, 62, 69, 84, 87, 95, 106, 121, 136, 139, 140, 142, 143, 144, 150, 165, 192, 196, 197, 222, 262, 264, 267, 314, 315, 320, 328, 330, 335, 372, 383, 409, 430, 463, 467, 505, 506, 534, 536, 598, 637–642, 686
Vol. 4: 29, 37, 39, 213, 358, 368, 369, 449, 476, 688, 689
Vol. 5: 277, 318
Vol. 8: 673
Vol. 10: 395
equations of motion **Vol. 7:** 306, 447, 449
equilibrium, laws of **Vol. 11:** 669
equilibrium, mechanical heat **Vol. 1:** 300
equipartition of energy **Vol. 8:** 17
Vol. 9: 54, 422, 450
equipartition principle, *see* energy equipartition
equipartition law **Vol. 2:** 179, 450*ff*
equivalence between matrix and wave
mechanics **Vol. 6:** 3, 8*f*, 96, 126, 422
equivalence principle, *see* relativity, equivalence in theory of
ergodic hypothesis **Vol. 6:** 320
Eriksen, B. **Vol. 11:** 358*p*
Eriksen, E. **Vol. 11:** 452*n*
Errera, I. **Vol. 7:** 32*p*
"Ersatzoszillator" **Vol. 5:** 21
Eskimoes **Vol. 12:** 128, 478, 479
Esmarch, L. **Vol. 10:** 347, 361
Establier, M. **Vol. 7:** 261*p*, 302
ether waves **Vol. 12:** 252
ethics **Vol. 10:** 141, 160, 202, 279, 286, 292, 298, 317, 318, 322, 345, 358
Vol. 12: 158, 159, 171

ethnography/ethnology (*see also* anthropology)
Vol. 10: 187, 203, 240, 245, 248
Eton College **Vol. 12:** 285
Eucken, A. **Vol. 2:** 458, 459
Vol. 6: 397
Euclid **Vol. 11:** 669, 670
Euclidean geometry **Vol. 10:** 141, 157
Eudoxus **Vol. 11:** 669
Euler's constant **Vol. 8:** 9, 57, 492
Euler, H. **Vol. 2:** 473
Vol. 7: 46, 451, 453
Vol. 8: 803
Vol. 9: 28, 582*ff*, 618
Euler, L. **Vol. 5:** 271, 273, 277
Evans, E.J. **Vol. 1:** 570, 571, 572, 573, 585
Vol. 2: 121, 274, 299, 333, 334, 385, 386, 387, 388, 399, 469, 494, 495, 501, 509, 516, 518, 558, 560, 561, 563, 583, 596, 597, 603, 615, 617, 619
Vol. 3: 130, 216, 589
Vol. 6: 385
Vol. 8: 672, 695, 697
Vol. 10: 378, 391, 392, 396, 418
Evans, G.R. **Vol. 8:** 813
evaporation **Vol. 9:** 27, 29*f*, 39, 209*f*, 216, 220, 241*ff*, 252, 255, 262, 268*f*, 276*f*, 298, 313*f*, 326*f*, 335*f*, 342, 344, 349, 350, 353, 378, 382, 384, 386, 388, 419, 423–428, 432*f*, 448, 451–454, 458*f*, 586*f*
Eve, A.S. **Vol. 10:** 413, 420
Vol. 12: 526
even/odd nuclei **Vol. 9:** 36, 38, 65, 94, 117, 197, 231, 236, 274, 280, 307, 321, 345, 367*f*, 380*f*, 384, 387, 440, 465, 478, 583*f*, 609, 623, 625
Everdingen, E. van **Vol. 1:** 159
Evershed, J. **Vol. 3:** 164
evolution **Vol. 10:** 97, 131, 137, 285, 291, 545, 546, 549, 550, 552–554, 557, 559–562, 565, 571, 574
Vol. 11: 537, 547
Ewald, P.P. **Vol. 3:** 691
excessive screening, *see* screening
exchange effect **Vol. 8:** 211, 225, 437, 442, 449, 502, 709*ff*, 762
exchange forces (*see also* spin exchange)
Vol. 9: 34, 36, 38, 79, 193, 197, 236, 524, 549, 582*ff*, 610, 618, 649, 651
exchange relation, *see* commutation relation

excitation
    of atoms/spectra   **Vol. 6:** 94, 97f, 123f,
        133, 138, 143, 152, 157, 180, 385
    of nuclei   **Vol. 6:** 405
    of radiation   **Vol. 3:** 254ff
    residual   **Vol. 8:** 599, 611, 613–614,
        619f, 790
excited state, *see* stationary state
exclusion of stationary states
        Vol. 4: 59ff, 143
exclusion principle (*see also* closing of
        electron groups; Pauli principle)
    Vol. 6: 24, 136, 144, 146, 158, 391,
        394–395, 397–398, 404, 425
    Vol. 7: 203, 242, 348, 377
    Vol. 8: 210, 269, 278, 306
    Vol. 9: 20, 27, 34, 81, 94, 106, 111f,
        117, 193, 197f, 231, 236, 274, 279f,
        304, 307, 319, 321, 523, 549, 580ff,
        584, 649, 651
    Vol. 10: 339, 379, 394, 403, 427, 438,
        442, 450
    Vol. 12: 11, 22, 25, 39, 48, 51, 364
existence and permanence of quantum numbers
        Vol. 3: 473, 508, 538
existence, meaning of   **Vol. 10:** 160
Exner, F.   **Vol. 5:** 29f, 490
        Vol. 6: 192
expectation value   **Vol. 7:** 43, 243
experience   **Vol. 7:** 259, 390, 399, 413, 427,
        504, 508
experiments on radioactivity, *see* radioactive
        methods, Bohr's experiments on
explanation
    concept of   **Vol. 10:** xlix, 7, 35, 539, 541
    modes of   **Vol. 12:** 185, 195
    physical   **Vol. 10:** 33, 34, 90, 175, 178,
        179, 377, 391
expression, limitations of   **Vol. 12:** 154, 166

Faber, K.   **Vol. 3:** 286, 295, 296
Fabre, J.H.   **Vol. 7:** 20
        Vol. 10: 185
Fabricius, K.   **Vol. 11:** 406p, 428, 431
Fahlenbrach, H.   **Vol. 9:** 185, 188
Fahrenheit, D.G.   **Vol. 12:** 428, 430, 431
Faisal I   **Vol. 11:** 692, 692p, 697, 698p
Fajans, K.   **Vol. 1:** xxiii
        Vol. 2: 119, 214, 236, 419, 428, 594
        Vol. 4: 13, 691
        Vol. 6: 378

        Vol. 10: 385, 417
        Vol. 12: 287
Fano, U.   **Vol. 8:** 521, 523, 567
Faraday
    Lecture, Bohr's, 1930   **Vol. 5:** 216
        Vol. 6: 316f, 319, 328, 330, 371–408,
            454, 480
        Vol. 7: 4, 463, 465, 466
        Vol. 9: 7, 13, 91–97, 103, 117, 156,
            306, 320, 676
    Medal   **Vol. 6:** 371, 480
Faraday effect   **Vol. 5:** 50, 340
Faraday, M.   **Vol. 1:** xvii
        Vol. 3: 45, 583
        Vol. 4: 84, 429, 434, 467, 469
        Vol. 5: vii, 129, 136, 273
        Vol. 6: 360, 373f, 407f
        Vol. 10: 54, 56, 148, 209, 337, 349, 363
        Vol. 11: 445, 530, 658, 663, 670
        Vol. 12: 220, 233, 371
Faust parody   **Vol. 7:** 463, 465
Faust, C.   **Vol. 7:** 411
Favrholdt, D.   **Vol. 6:** xxii
        Vol. 10: 302n
Faxén, H.   **Vol. 5:** 240
        Vol. 8: 304, 457, 567, 631, 634
Faye, J.   **Vol. 10:** 300n
Feather, N.   **Vol. 9:** 384
        Vol. 10: 410
Fechner, G.T.   **Vol. 10:** xliii, 7
Federation of American Scientists
        Vol. 11: 51
Feenberg, E.   **Vol. 7:** 489, 492
        Vol. 9: 70, 79ff, 369f, 372, 646ff, 650
Feld, B.T.   **Vol. 6:** 330
Fermi
    distribution   **Vol. 9:** 34, 385f, 582ff, 597f
        Vol. 12: 23, 24, 27, 28
    gas   **Vol. 9:** 30, 259, 539, 649f
    sea   **Vol. 9:** 27, 603f
Fermi–Dirac statistics (*see also* identical
        particles)   **Vol. 6:** 50, 144f, 349, 397,
        398, 404f, 466
        Vol. 8: 211, 437, 442, 755
        Vol. 9: 25, 94f, 117
        Vol. 10: 403, 444
        Vol. 12: 23, 27
Fermi, E.   **Vol. 1:** xlii
        Vol. 5: 71, 185–186, 200, 216, 305, 334
        Vol. 6: 142–144, 318
        Vol. 7: 32p

Vol. 8: 33*f*, 215, 225, 256, 496, 499, 518, 567, 572, 660*f*, 737*f*, 765
Vol. 9: 13*f*, 17, 20, 56, 60–63, 66*f*, 100, 134, 145, 152, 154*ff*, 166*f*, 170, 175, 178, 275, 280, 307, 309, 321, 323, 344, 365, 379, 384, 386, 407*f*, 428, 455, 534, 540*f*, 550–555, 572*f*, 580*f*, 634, 636, 652
Vol. 10: 102, 412, 443, 446, 451
Vol. 11: 569, 581, 601, 646, 650
Vol. 12: 39
fermion   **Vol. 10:** 443, 452
fermion field quantization   **Vol. 7:** 4
Ferretti, B.   **Vol. 12:** 38*p*
ferromagnetism   **Vol. 6:** 439*f*
   Vol. 10: 427, 446, 447
   Vol. 12: 51
Feshbach, H.   **Vol. 9:** 508
Festskrift, Copenhagen University, 1929
   Vol. 6: 110, 196–197, 202, 255–273, 278, 283, 425–426, 427
Feynman, R.P.   **Vol. 7:** 49
   Vol. 10: 455*p*, 478, 483
Fiala, F.   **Vol. 7:** 328
Fichte, J.G.   **Vol. 10:** xxxi*n*
field theory   **Vol. 6:** 23, 136, 157, 421
field, electromagnetic (*see also* classical electromagnetism; electrical field; magnetic field)
   and matter   **Vol. 7:** 64, 87, 93, 99, 126, 142, 146, 150, 178, 179, 188, 199, 203, 212, 215
   and photons   **Vol. 7:** 33, 71, 131
   average/mean   **Vol. 7:** 29, 41, 62, 70, 73, 79, 91, 113, 125, 130, 131, 133, 137, 145, 160
   concept   **Vol. 7:** 19, 33, 34, 93, 99, 146, 150, 172, 184, 199, 203, 215, 221, 222, 319, 457, 458, 469, 470, 474, 478
   critical   **Vol. 7:** 73, 74, 78, 112, 133, 136, 159, 206–208
   equations   **Vol. 7:** 24, 30, 182, 191, 201
   fluctuation   **Vol. 7:** 4, 13, 22–24, 26, 29, 30, 34, 37, 38, 41, 43–46, 48, 71–75, 101, 113, 131–134, 152, 160, 179, 189, 200, 201, 203, 204, 213, 214, 220, 451, 453, 457, 458, 468, 470, 487, 490, 499
Fierz, M.E.   **Vol. 5:** 220, 238
   Vol. 7: 329
   Vol. 9: 78, 542*f*
   Vol. 10: 418, 555, 572

Vol. 11: 243, 244
Vol. 12: 38*p*
finalism   **Vol. 10:** xxxi, xxxii, 11, 184, 544, 548
fine structure (*see also* complex structure; multiplet structure)   **Vol. 5:** 158, 162, 238, 251*f*, 277, 282, 284, 288–289, 387*f*, 391, 393, 406, 445*f*, 448, 450–452, 467–476, 478, 480*f*, 484*f*, 497–498
   Vol. 6: 349, 389, 396, 457
   Vol. 8: 184, 189, 672*f*
   Vol. 9: 110, 197, 236–237, 281, 309, 322, 621, 634, 637
   Vol. 10: 395, 412, 434, 447, 451
   Vol. 12: 344, 345, 348, 351, 381, 383
   constant   **Vol. 6:** 314, 402, 428*f*
   Vol. 7: 37, 78, 136, 471, 472
   Vol. 9: 106*ff*, 113, 130*f*, 578*f*
   effect of magnetic field upon   **Vol. 3:** 158
   of hydrogen lines (*see also* doubling of spectral lines)   **Vol. 2:** 343, 469, 474, 595
   Vol. 3: 129*ff*, 259
   spectrum of hydrogen   **Vol. 7:** 48
finite mass of nucleus, *see* reduced mass
Finland   **Vol. 11:** 341, 355, 387*n*, 503–507, 728
Finsen Institute   **Vol. 11:** 516, 524, 673*p*, 686*n*
   Vol. 12: 327, 330
Finsen, N.   **Vol. 10:** 35, 75, 259, 269
   Vol. 11: 516*n*, 544, 551
   Vol. 12: 330*n*, 454, 459
first law of thermodynamics (*see also* conservation of energy and momentum)   **Vol. 5:** 33, 273, 415, 419
Fischer, H.   **Vol. 7:** 328
fission   **Vol. 2:** 10
   Vol. 11: 12, 102, 103, 229, 340, 363, 564, 569, 579–583, 598–600, 602–605, 614, 615, 624–626, 646, 649, 650, 718
   Vol. 12: 12, 84, 211, 216
   barrier   **Vol. 9:** 66*f*, 70–71, 370*ff*, 375*ff*, 380, 382*f*, 386*ff*, 658
   cross section (*see also* cross section, fission)   **Vol. 9:** 58, 65, 71, 74*f*, 342, 344*f*, 349, 351*f*, 365*f*, 377*ff*, 388*f*, 397*f*, 477*ff*, 484, 559, 589

deuteron-induced **Vol. 9:** 75, 349, 351, 367, 387–388, 477, 479, 483–484, 552, 588*ff*, 626, 628*f*, 631, 638, 640

fragments **Vol. 9:** 53, 55*f*, 66, 70, 72*f*, 344, 350*ff*, 357, 361, 365*f*, 371, 373, 378*f*, 384*ff*, 389, 407*f*, 413, 431*f*, 436, 441, 443, 457*f*, 461, 465, 467–473, 550*f*, 553, 557*ff*, 564, 566, 626, 628, 634, 637, 662*f*, 679

fragment tracks (*see also* branching of tracks) **Vol. 8:** 233, 235*ff*, 245*f*, 254, 311, 315*f*, 318, 320*f*, 325, 328*ff*, 371, 375–376, 405*f*, 467*f*, 486, 548, 575*f*, 585, 587, 727, 729, 731*f*, 734*f*, 749, 789*f*

mass ratio in **Vol. 8:** 326, 331, 376

probability **Vol. 9:** 71, 73, 344*f*, 349*ff*, 361, 366, 370, 373*ff*, 380*ff*, 387*f*, 398, 433*f*, 437*ff*, 459*f*, 462*ff*, 469, 472, 477*f*, 484, 553, 561*ff*, 662*f*, 666

symmetrical/asymmetrical (*see also* statistical distribution of fission fragments) **Vol. 9:** 72–74, 82, 387, 467–473, 559, 662–663, 665, 669–670

threshold, *see* critical deformation energy

track **Vol. 9:** 413, 431*f*, 443, 457*f*

width **Vol. 9:** 375*ff*, 380*ff*, 659

yield **Vol. 9:** 54, 74, 345, 349, 351, 375, 377, 379, 384, 386*f*, 401, 404, 438*f*, 441, 463*ff*, 469, 472*f*, 479, 484, 626*f*

Flamm, L. **Vol. 2:** 60, 77
Vol. 3: 12, 344, 352, 686
Vol. 8: 130, 147, 372, 519, 567, 668, 779, 781

Flammersfeld, A. **Vol. 8:** 567

Fleck, A. **Vol. 2:** 119
Vol. 10: 385

Fleming, W.A. **Vol. 8:** 787

Fletscher (*recte* Fletcher) **Vol. 1:** 520, 521

Flint, H.T. **Vol. 7:** 172, 184

Florance, D.C.H. **Vol. 2:** 501
Vol. 8: 672

fluctuations (*see also* black-body energy fluctuations; field, electromagnetic, fluctuation; straggling; vacuum fluctuations) **Vol. 5:** 29, 30–31, 32, 56, 273, 344, 390, 392, 415, 419
Vol. 8: 14, 263, 355, 417, 466, 470, 547, 551, 553, 591*f*, 599*ff*, 617, 668, 683

Vol. 9: 108, 342, 367, 418, 433, 435, 447, 459, 461, 477, 556, 558, 563, 617, 649, 651

fluorescence **Vol. 8:** 662, 681, 685, 688, 692
Vol. 12: 503, 504

fluorescent light, polarization of **Vol. 5:** 34, 48–65, 66, 143–154, 156, 303, 335, 344, 358*f*, 394–397, 431*f*, 495

fluorescent radiation **Vol. 5:** 114, 122, 416, 420, 512

fluorine **Vol. 11:** 577, 593

Flügge, S. **Vol. 9:** 387
Vol. 11: 243, 244

Fock, A.V. **Vol. 7:** 166*n*, 230
Vol. 11: 7*p*, 59*p*, 316, 317

Fog, M. **Vol. 11:** 56, 301, 302, 332, 703*n*

Fokker, A.D. **Vol. 2:** 9, 339, 498–501, 618, 620
Vol. 5: 226
Vol. 8: 22*ff*, 643, 667–676, 712*f*
Vol. 12: 38*p*

Foldy, L.L. **Vol. 9:** 82, 669

folk high school **Vol. 10:** 257, 260, 267, 271

Folse, H. **Vol. 7:** 283*n*

Foote, P.D. **Vol. 3:** 38, 448, 468, 592
Vol. 5: 40, 145, 148, 295, 320, 326, 512

force
electrical **Vol. 7:** 30, 103, 153, 201, 203, 204, 224
mechanical **Vol. 7:** 117, 163
short-range **Vol. 7:** 227

forces inversely proportional to *n*th power of distance **Vol. 1:** 320

forces, interatomic, *see* interatomic forces

Forck (*recte* Forch, C.) **Vol. 1:** 62

Ford Foundation **Vol. 11:** 371, 715*n*

Ford Motor Company **Vol. 11:** 367

Ford, E. **Vol. 11:** 367, 716

Ford, Henry, I **Vol. 11:** 367, 716

Ford, Henry, II **Vol. 11:** 720

Ford, W.C. **Vol. 11:** 367, 639

foreword to German translation of early papers **Vol. 3:** 27, 303
German text **Vol. 3:** 308–323
translation **Vol. 3:** 325–337

formalism, mathematical **Vol. 7:** 41, 227, 368, 369, 372, 377

Försterling, K. **Vol. 5:** 112

Foster, J.S. **Vol. 3:** 599

Foundation for the Benefit of the Country (Fondet til Fædrelandets Vel) **Vol. 11:** 730

Fourier analysis   **Vol. 8:** 34, 381, 411, 413, 499, 634, 637f, 798, 801f
Vol. 11: 670
Fourier, J.B.J.   **Vol. 11:** 670
Fowler lines/series   **Vol. 2:** 116, 121–122, 123, 170–171, 200, 275, 285, 298–300, 326–328, 352, 385–388, 399, 503ff, 532f, 558f, 561f, 582f
Vol. 10: 378, 390, 392
Fowler, A.   **Vol. 1:** xxix, 570, 571, 572, 573
Vol. 2: 8, 121, 122, 170, 171, 200, 274, 275, 276, 285, 298, 300, 324, 325, 326, 327, 328, 333, 352, 385, 386, 387, 395, 399, 405, 406, 502–509, 533, 559, 560, 562, 563, 582, 583, 588, 593, 615, 616, 617
Vol. 3: 168, 170, 178, 215, 216, 516, 545, 579, 589, 590, 591
Vol. 4: 54, 66, 206, 279, 333, 334, 340, 442, 445, 473, 474, 564, 565, 576, 622, 631
Vol. 5: 278
Vol. 6: 385f
Vol. 8: 14, 643, 695ff
Vol. 10: 390–392, 418
Fowler, E. (R.H. Fowler's wife)
Vol. 8: 677, 788
Fowler, R.D.   **Vol. 9:** 56, 365
Fowler, R.H.   **Vol. 3:** 458
Vol. 5: 7, 9, 56f, 68f, 71, 81ff, 181, 184, 196f, 199, 225, 231, 295, 300, 308, 334–340, 350, 362f, 370f, 487ff, 492
Vol. 6: 8–9, 14, 36, 37, 412, 415, 421–424, 457f
Vol. 7: 260, 301
Vol. 8: 29, 32ff, 203f, 206, 210ff, 256, 372, 498, 529, 567, 643, 657, 660, 677–680, 709f, 788
Vol. 9: 6, 10, 534, 555–556
Vol. 10: 378, 399, 400, 410
Vol. 11: 217, 220
Vol. 12: 17p
Fowler, W.A.   **Vol. 9:** 201, 203
Fox, M.   **Vol. 11:** 563
frame of reference   **Vol. 7:** 85, 141, 199, 265, 296, 504, 507
Francis, R.F.   **Vol. 2:** 619
Vol. 4: 32, 678
Francis, W.   **Vol. 1:** 111, 112
Franck–Hertz
collisions, *see* collisions of the 1st kind

experiment   **Vol. 2:** 86–87, 92, 117, 331, 335, 408–409, 559, 563
Vol. 5: 27, 70, 179, 194–195, 275
Vol. 6: 10, 71, 80, 88, 94, 123, 138, 166, 228, 242, 382
Vol. 8: 17–22, 24, 27, 117, 156f, 184f, 189f, 198, 199–200, 269, 278, 686
Vol. 9: 304, 318
Vol. 10: 377, 397, 439
Franck, J.   **Vol. 1:** xxxi
Vol. 2: 86, 87, 92, 117, 201, 202, 236, 331, 335, 408, 409, 559, 563
Vol. 3: 20, 22, 23, 25, 180, 237, 238, 239, 257, 291, 299, 322, 336, 368, 404, 468, 469, 494, 496, 591, 592, 600, 618, 627, 631, 644, 645, 650
Vol. 4: 20, 24, 29, 53, 64, 65, 94, 132, 219, 288, 353, 375, 385, 446, 475, 593, 643, 659, 672, 675, 676, 693–700, 721, 740
Vol. 5: 27, 31, 50, 52–56, 70, 72ff, 81, 83ff, 108, 145, 147f, 153f, 179, 188f, 193f, 203, 206, 229, 239, 275, 295, 299ff, 305ff, 311, 340–352, 362ff, 369f, 396f, 402, 407
Vol. 6: 71, 80, 88, 138, 228, 242, 382f, 395
Vol. 7: 344
Vol. 8: 18ff, 27, 29f, 33f, 156f, 184, 189, 643, 656, 658, 660f, 663, 680–695, 713, 715, 759f, 765, 780, 782, 786
Vol. 9: 28, 31, 304, 318, 578f
Vol. 10: 58, 134, 377, 397, 400, 471, 472
Vol. 11: 217, 220, 328, 331, 372, 567
Vol. 12: 44
Frank, I.   **Vol. 8:** 497, 567
Frank, P.   **Vol. 6:** 452
Vol. 7: 267
Vol. 10: 16, 19, 36p
Frankfurter, F.   **Vol. 11:** 20–22, 26, 27, 30–33, 36–42, 45, 46, 48, 71, 98, 240, 241, 247–251, 253, 257, 258, 260–262, 266–269, 272–280, 284, 285, 290–293, 324, 327, 329, 330
Frankfurter, M.   **Vol. 11:** 247, 250, 282
Franklin, B.   **Vol. 10:** 55, 344, 347, 355, 358, 361, 368
Vol. 11: 530, 531, 568
Franklin, R.   **Vol. 10:** 24
Fraser, R.   **Vol. 10:** 169, 474, 475
Fraunhofer, J.   **Vol. 3:** 289, 297

Frederik II   **Vol. 10:** 267*n*
   Vol. 11: 544, 551
Frederik VI   **Vol. 12:** 152, 163
Frederik VIII   **Vol. 11:** 387*n*, 682, 685
Frederik IX   **Vol. 10:** 74
   Vol. 11: 346, 347, 353, 354, 422, 424,
   426–432, 434, 435, 478–482, 484, 485,
   488–491, 732
   Vol. 12: 240, 244
Fredholm integral equation   **Vol. 1:** 332, 362
Fredholm, I.   **Vol. 2:** 473
   Vol. 3: 653, 654
   Vol. 5: 312
free
   choice   **Vol. 6:** 45*f*, 104*ff*, 205, 212, 228,
      243, 261, 267, 272, 282, 291, 297, 370,
      382, 400, 471
   energy   **Vol. 6:** 326*f*, 449, 453
      Vol. 10: 91, 104, 119, 151, 166,
      544, 548
   will   **Vol. 6:** 206, 215–217, 220*f*, 234,
      250–251, 275–276, 302
Free French Forces   **Vol. 11:** 250*n*
freedom
   for the individual   **Vol. 10:** 206
   minimal for permitting life   **Vol. 10:** 11, 34
   of experimentation   **Vol. 10:** 89, 104, 406,
      573, 574
   of will/choice   **Vol. 7:** 275, 336, 409, 422
      Vol. 10: xxv, xxviii, xxx, xxxi, xlii, 6,
      8, 10, 17, 19, 20, 35, 45, 47, 60, 62, 71,
      94, 105, 160, 185, 201, 202, 216, 279,
      280, 286, 292, 318, 322, 485, 486, 517,
      519, 520, 522, 540, 542, 588
      Vol. 11: 134
      Vol. 12: 224
Fremming, A.   **Vol. 10:** 224*p*
Frenkel, Y.I.   **Vol. 9:** 27, 29, 70, 209,
   241–242, 313, 326, 369, 660, 666
   Vol. 11: 7*p*
frequency
   concept   **Vol. 7:** 343, 350
   condition, Bohr's (*see also* postulates,
      Bohr's)   **Vol. 2:** 111, 167*ff*, 193, 222,
      233, 293, 325, 335, 351, 357, 364, 394,
      443*ff*, 468, 559, 562
      Vol. 5: 44, 104, 107, 111, 121, 258,
      265, 272, 275, 277, 280, 315, 447,
      449, 452*f*
      Vol. 8: 20, 184, 189
   distribution   **Vol. 7:** 88, 142

high   **Vol. 7:** 221, 238, 349
relation   **Vol. 3:** 71, 242, 245, 310, 326,
   341, 350, 366, 401, 422, 476, 560, 570
   Vol. 4: 191, 264, 345, 351, 437, 471,
   552, 613, 659
Fresnel, A.J.   **Vol. 8:** 727, 729
   Vol. 10: 209
Freuchen, M.   **Vol. 12:** 129*p*, 506, 507
Freuchen, P.   **Vol. 12:** 128, 129*p*, 478, 479,
   505–507
Freud, S.   **Vol. 10:** 566
Fricke, H.   **Vol. 1:** 114, 573, 574, 575
   Vol. 2: 617
   Vol. 3: 200
   Vol. 4: 17, 37, 133, 493, 500, 530, 532
Friedmann, F.   **Vol. 2:** 78
   Vol. 8: 110, 114, 148, 668
   Vol. 12: 15
friendship   **Vol. 10:** 202
Friis, A.   **Vol. 10:** 221
   Vol. 11: 338*n*, 399, 400, 406*p*,
   409, 410
   Vol. 12: 124, 125, 434, 435
Frisch, O.R.   **Vol. 1:** xliv
   Vol. 9: 16*f*, 20, 28, 52–54, 55*ff*, 62*ff*,
   66*ff*, 156, 209, 248, 342, 344*f*, 361,
   365, 369, 374, 407*f*, 431*f*, 457*f*, 469,
   473, 535, 550*ff*, 556–566, 588*ff*, 622*ff*,
   636, 659, 662
   Vol. 10: 412
   Vol. 11: 12, 358*p*, 569, 579, 598,
   646, 649
   Vol. 12: 12
Frivold, O.E.   **Vol. 4:** 670
Frobenius, G.   **Vol. 5:** 312
From, F.   **Vol. 10:** 36*p*
Fröhlich, H.   **Vol. 12:** 8, 38*p*, 39, 530
Füchtbauer, C.   **Vol. 5:** 114
Fues, E.   **Vol. 4:** 405, 587, 600, 639, 648, 717
   Vol. 6: 304, 462*f*
   Vol. 12: 17*p*
fundamental postulates of the quantum theory
   Vol. 3: 39*ff*, 401*ff*, 422, 455–500
fundamental assumptions   **Vol. 3:** 569*ff*
Fürtbauer (*recte* Füchtbauer, C.)   **Vol. 4:** 695
fusion   **Vol. 9:** 419–420, 448
   Vol. 12: 84
Fysisk Tidsskrift (Danish journal)
   Vol. 11: 353
   Vol. 12: 114, 123, 307, 310, 337, 338

[81]

Gaerttner, E.R.   **Vol. 5:** 95
Gagnebin, S.   **Vol. 7:** 328
Gale, A.J.V.   **Vol. 9:** 557f
Galilei, G.   **Vol. 1:** xvii
　　Vol. 3: 45
　　Vol. 5: 129, 136, 273f
　　Vol. 7: 388, 399, 413
　　Vol. 10: xxxvii, 53, 67, 85, 117, 127,
　　174, 194, 208, 405
　　Vol. 11: 657, 662, 669
　　Vol. 12: 272, 371
Galvani Congress 1937   **Vol. 12:** 105, 110,
　　111, 274, 525
Galvani, L.   **Vol. 10:** xxxviii$n$, 9n, 51p, 52,
　　54, 55, 68, 344, 358, 415
　　Vol. 12: 99, 105, 272
galvanomagnetic phenomena   **Vol. 1:** 383ff
γ-ray
　emission/spectrum   **Vol. 9:** 6, 19, 32, 36,
　　38, 54, 80, 95, 97, 113, 131, 137, 146,
　　152ff, 167f, 175, 177, 201ff, 208f,
　　237, 262, 276, 278, 281, 298, 310f,
　　315, 323f, 327, 336f, 384, 388, 423,
　　427, 451, 454, 549, 555, 567, 593f,
　　609, 626f, 677
　microscope, *see* microscope
　scattering (*see also* scattering of γ-rays)
　　Vol. 8: 220, 228, 704f, 794f, 798,
　　804, 807
　spectra   **Vol. 8:** 704f
γ-rays   **Vol. 2:** 120, 211, 214, 238, 374,
　　424–425, 428, 441, 581, 584, 586,
　　589ff, 597
　　Vol. 5: 94f, 215
　　Vol. 10: 22, 393, 398, 409, 410, 451
　　Vol. 11: 574, 587
　　Vol. 12: 60, 66, 287
gamete   **Vol. 10:** 13, 544, 548, 559, 560
Gammelholm School   **Vol. 12:** 128, 135, 136,
　　138, 140, 144, 209, 213, 452, 456
Gamow barrier   **Vol. 7:** 442
Gamow, G.   **Vol. 6:** 25, 304, 308, 405
　　Vol. 7: 18, 19, 32p, 260, 261p
　　Vol. 8: 210, 269, 278, 344, 349, 653,
　　678, 704f, 754f
　　Vol. 9: 6, 8, 13, 19f, 25, 27, 29, 31, 32,
　　36ff, 87, 95, 127, 134, 137, 141, 147, 153,
　　165, 173, 186f, 189, 230, 250, 254, 256,
　　281, 308f, 322, 367, 374, 384, 388, 535,
　　540, 547f, 555, 567–573, 578f, 600f,
　　603ff, 609, 642, 644, 651

　　Vol. 10: 166, 330, 409, 412, 414,
　　449–451, 489
　　Vol. 11: 7–9, 216, 568
　　Vol. 12: 13, 14, 78p, 79–82, 518, 519
Gans, R.   **Vol. 1:** 121, 159, 278, 284, 285,
　　383, 389, 390, 417
　　Vol. 2: 128
　　Vol. 3: 497
Gant, D.H.T.   **Vol. 9:** 479
Garbasso, A.   **Vol. 2:** 321, 323, 360,
　　510–511, 616
　　Vol. 3: 58, 158, 439
Gardie, M. De la   **Vol. 12:** 482, 483
gas, rarefied   **Vol. 11:** 565
gases
　kinetic theory of, *see* thermodynamics
　"Knudsen"   **Vol. 12:** 112
　paramagnetic   **Vol. 12:** 386, 387
　properties of   **Vol. 12:** 290–293, 300, 301,
　　306, 309, 316, 317, 321, 322
Gaulle, C. de   **Vol. 11:** 250n
Gaunt's stopping formula, *see* stopping
　formula, Gaunt's
Gaunt, J.A.   **Vol. 8:** 208f, 213, 215f, 225,
　　241, 256, 371, 499, 567, 791
Gauss
　error function, *see* normal distribution
　integral/theorem   **Vol. 7:** 42, 43
Gauss, C.F.   **Vol. 12:** 372
Gaussian distribution   **Vol. 8:** 15, 111, 114,
　　134ff, 147, 170f, 173f, 177, 179, 255,
　　263, 440, 473ff, 477ff, 482, 484, 520,
　　551, 553, 561f, 564, 601ff, 633
Gauthier, *see* Goucher, F.S.
Gaviola, E.   **Vol. 5:** 145, 148
Gehrke, E.   **Vol. 2:** 321, 355, 360
Geiger counter   **Vol. 11:** 567, 574, 578,
　　587, 596
　　Vol. 12: 84
Geiger–Bothe experiment, *see* Bothe–Geiger
　experiment
Geiger–Nuttall relation   **Vol. 9:** 27, 95, 141,
　　308, 322, 603f
Geiger's rule/law   **Vol. 8:** 61, 83, 98, 331, 550
Geiger, H. (*see also* Bothe–Geiger experiment)
　　Vol. 2: 29, 78, 87, 88, 89, 96, 97, 98,
　　99, 161, 189, 238, 343, 589, 596, 615
　　Vol. 5: 34, 68f, 75ff, 82f, 87, 89f, 178,
　　190, 194, 204, 215, 221, 276, 295, 303f,
　　351, 352–354, 440, 443, 484f, 499
　　Vol. 6: 139, 185, 379, 405, 442, 444

Vol. 8: 4, 6, 14, 16, 61, 75, 77, 83f, 91f, 98f, 110, 114, 148, 157ff, 163, 170, 173f, 177f, 632f, 708
Vol. 9: 127
Vol. 10: 384, 398, 399, 404, 409
Vol. 11: 567
Vol. 12: 6n, 265
Geitel, H.   **Vol. 8:** 668
Geitler, J.R.   **Vol. 3:** 17, 688
Gell-Mann, M.   **Vol. 10:** 455p
gene, *see* genetics
general relativity, *see* relativity, general
general spectral laws
   for arc spectra   **Vol. 4:** 564, 622
   for elements of higher atomic number
     Vol. 4: 563, 621
   for series spectra   **Vol. 4:** 564, 621
   for spark spectra   **Vol. 4:** 564, 622
generalization   **Vol. 6:** 46, 59, 61, 63, 78, 80, 85, 88, 105, 116, 125, 135, 149, 152, 157f, 164–165, 168, 203, 208, 229, 244, 261, 271, 282, 297, 359, 379, 469
genetics   **Vol. 10:** 13, 22–25, 94, 105, 130, 136, 137, 151, 164, 165, 168, 180, 214, 246, 408, 468, 469, 471, 472, 485, 486, 487n, 489, 490, 492, 552, 554n, 557, 590, 594
Geneva Conference on Peaceful Uses of Atomic Energy, 1955   **Vol. 11:** 83, 365–367, 617, 630, 631, 721, 722
genotype   **Vol. 10:** 13, 246, 544, 549, 554
Gentner, W.   **Vol. 8:** 567
   Vol. 9: 43f, 50, 262, 298f, 315, 328, 389, 393, 543
geodesy   **Vol. 10:** 157
   Vol. 11: 400, 410, 657, 662, 669
geography   **Vol. 10:** 241, 300
geometry (*see also* multi-dimensional geometry)   **Vol. 10:** 42, 303, 346, 360, 511–513
   Vol. 11: 669, 670
  non-Euclidean   **Vol. 11:** 670
   Vol. 12: 371
George VI   **Vol. 12:** 84
Gerdien, H.   **Vol. 5:** 352
Gerlach, W. (*see also* Stern–Gerlach experiment)   **Vol. 3:** 484
   Vol. 4: 367, 449, 476, 596, 645, 712
   Vol. 5: 255, 262, 278, 504
   Vol. 6: 318, 333f
   Vol. 10: 339

German Cultural Institute (Copenhagen)
   Vol. 11: 13
Germanist Congress, Second International
   Vol. 10: 140, 141, 143
Germer, L.H.   **Vol. 5:** 210f
   Vol. 6: 4, 7, 77, 84, 394
   Vol. 10: 402, 440
   Vol. 12: 47
Gerritsen, A.N.   **Vol. 12:** 38p
Gertner, I.V.   **Vol. 10:** 342p
Gex, M.   **Vol. 7:** 328
g-factor, *see* gyromagnetic factor
ghost waves   **Vol. 7:** 346
giant dipole   **Vol. 9:** 50
Gibbs, J.W.   **Vol. 1:** 182, 183, 306
   Vol. 5: 273
   Vol. 6: 320, 324–325, 400, 470
   Vol. 12: 372
Gieseler, H.   **Vol. 4:** 606, 652
Gifford Lectures (*see also* Edinburgh Lectures, Bohr's, 1949)   **Vol. 10:** 172–174, 179, 474n, 587
   Vol. 11: 78, 327, 330
Gijsman, H.M.   **Vol. 12:** 38p
Gilbert, C.W.   **Vol. 9:** 337
Gilbert, W.   **Vol. 10:** 343, 357
Gill, E.   **Vol. 10:** 414
Gilson, E.   **Vol. 7:** 411
   Vol. 10: 82
Gjelsvik, A.   **Vol. 12:** 17p
Glaisher, J.W.L.   **Vol. 2:** 74
   Vol. 8: 144
Glasgow, University of   **Vol. 11:** 347, 436p, 438
glasnost   **Vol. 11:** 83
Glasoe, G.N.   **Vol. 9:** 61, 365, 380
Glass, B.   **Vol. 10:** 483n
Glasson, J.L.   **Vol. 2:** 89
   Vol. 8: 159
Glazebrook, R.I.   **Vol. 8:** 787
Gmelin, P.   **Vol. 2:** 183
God (*see also* providence)
   Vol. 10: xxxix, 555
God's Eye View   **Vol. 10:** xxxix
Goeppert-Mayer, M., *see* Mayer, M.G.
Goethe, J.W. von   **Vol. 5:** 35, 411
   Vol. 6: vi, xxi
   Vol. 10: 318, 322, 557n
   Vol. 12: 159, 172
gold   **Vol. 11:** 97, 105, 139, 143

[83]

gold medal, *see* Royal Danish Academy of
    Sciences and Letters
gold, absorption in **Vol. 2:** 30, 35, 36
Goldberger, M.L. **Vol. 10:** 455p
Goldhaber, M. **Vol. 9:** 204, 256
Goldman, H. **Vol. 7:** 328
Goldschmidt, J. **Vol. 11:** 714
Goldschmidt, R.B. **Vol. 10:** 554
Goldschmidt, V.B. **Vol. 3:** 286, 295
Gonseth, F. **Vol. 7:** 328, 329
Goodeve, C.F. **Vol. 9:** 19
Gordon, V. **Vol. 11:** 7p
Gordon, W. **Vol. 8:** 210, 272, 281,
    437, 567
    Vol. 10: 447
Gorter, C.J. **Vol. 10:** 455p
    Vol. 12: 38p
Göttingen **Vol. 2:** 126, 331, 558, 560,
    561, 563
    Vol. 5: 23f, 46, 53, 88f, 92, 145,
    147f, 153, 220, 223, 228f, 235, 239,
    299, 301f, 308, 311, 312, 334, 345,
    351, 355f, 361, 363, 364ff, 370, 377f,
    396, 407, 411, 413, 441, 443, 445, 447,
    449, 452, 495, 503, 507
    Vol. 6: 9, 30, 42, 325, 436
    Vol. 10: 378, 400, 409, 414, 424–427,
    474, 475, 558, 560
Göttingen, seven lectures in (*see also* "Bohr
    Festspiele") **Vol. 3:** 40
    Vol. 4: 23ff, 341–419
Göttingen, University of **Vol. 11:** 371
    Vol. 12: 84, 326, 328, 329
Götze, R. **Vol. 3:** 510, 539
    Vol. 4: 24, 401, 402, 404, 576, 631, 696
Goucher, F.S. **Vol. 2:** 482
    Vol. 3: 239, 257, 592
    Vol. 8: 20f
    Vol. 10: 397
Goudsmit, S.A. **Vol. 1:** xxxvi
    Vol. 5: 28, 80, 83, 228–231, 233ff,
    237f, 279, 283, 285, 287–289,
    329–333, 339f, 371, 373ff, 406,
    459–466, 470, 472, 474, 478, 481,
    484f, 497f, 501, 507f
    Vol. 6: 3, 37, 52, 136, 158, 304, 305,
    392, 412, 415, 457f
    Vol. 7: 301
    Vol. 8: 633, 698
    Vol. 9: 100, 304, 319
    Vol. 10: 339, 418

Gowing, M. **Vol. 11:** 14n
Graae, F. **Vol. 10:** 250p
Graaff, R. van de **Vol. 9:** 216, 220
Graetz, L. **Vol. 8:** 708
Gram, C. **Vol. 12:** 134
Gram, H. **Vol. 11:** 399, 409
gravitation **Vol. 6:** 157f, 260, 281, 368–369
    Vol. 8: 667, 670, 675
    Vol. 9: 8, 132, 568ff, 668
    Vol. 12: 60, 65, 220, 221, 371, 378
gravitational
    constant **Vol. 7:** 366, 476, 479
    effects **Vol. 7:** 182, 191, 366
    field **Vol. 7:** 320, 366
    force **Vol. 7:** 182, 191, 367
    mass **Vol. 7:** 319, 366
    theory, *see* relativity, general
    wave **Vol. 7:** 476, 479
Gray, J. **Vol. 10:** 23
Great Terror of 1937–38 **Vol. 11:** 10
Greek philosophers **Vol. 7:** 465, 503
Green, G.K. **Vol. 9:** 365
Greenland Foundation **Vol. 11:** 736
Grimm, H. **Vol. 3:** 601
    Vol. 4: 405
Groenewold, H.J. **Vol. 12:** 38p
Gromyko, A. **Vol. 11:** 68
Grønbech, V. **Vol. 1:** xx
    Vol. 10: 298
Groot, H. **Vol. 3:** 496
Groot, S.R. de **Vol. 10:** 493
    Vol. 12: 38p
Grosse, A. v. **Vol. 9:** 401, 404, 441, 465
Grossmann, M. **Vol. 8:** 667
Grotrian, W. **Vol. 8:** 681, 686
ground state, *see* normal state; permanent state
group
    structure of electron configuration
    Vol. 6: 387–391, 402, 458
    velocity **Vol. 6:** 23, 29, 31, 76ff, 84f,
    117f, 149, 421
Groves, L. **Vol. 11:** 16, 17, 30, 39, 40, 42,
    56, 86, 87, 245, 250, 259, 267–269,
    272, 274, 275, 286, 324, 329
Grundtvig, N.F.S. **Vol. 10:** 256–258, 265,
    267, 269, 366p
    Vol. 12: 128
Gruner, P. **Vol. 1:** 119, 138, 139, 142, 153,
    198, 199, 200, 204, 218, 228, 237, 285,
    318, 319, 322, 333, 341, 348, 370
Grünmach **Vol. 1:** 62, 65

Guericke, O. von   **Vol. 10:** 343, 357
Guertler, W.   **Vol. 1:** 147, 231, 343
Guldberg Medal   **Vol. 12:** 139, 141
Guldberg, C.M.   **Vol. 12:** 139*n*
Gullion, E.A.   **Vol. 11:** 321
Gurney, R.W.   **Vol. 5:** 184, 199
   Vol. 6: 405
   Vol. 9: 87, 95, 185, 230, 281, 308, 322
   Vol. 10: 409, 412
Gustafson, T.V.   **Vol. 10:** 559, 561
Guthrie lecture, seventh   **Vol. 3:** 37, 38,
   415–446
Guye, C.E.   **Vol. 7:** 32*p*
g-value   **Vol. 5:** 289, 333, 412, 414, 423, 427
Gyldendal (Danish publishing house)
   Vol. 12: 518, 519
   Quantum Books   **Vol. 12:** 94, 96, 534
gyromagnetic factor   **Vol. 6:** 310, 349

Haakon, King of Norway   **Vol. 12:** 138, 140
Haar, D. ter   **Vol. 7:** 7*n*, 229
Haas, A.E.   **Vol. 2:** 127, 166, 512–513, 616
   Vol. 3: 40, 647
   Vol. 4: 436, 470
   Vol. 5: 37
   Vol. 10: 386, 417, 434
Haas, W.J. de   **Vol. 2:** 263, 395
   Vol. 3: 361, 363
   Vol. 4: 97
   Vol. 10: 339
   Vol. 12: 24, 28, 38*p*
Haber, F.   **Vol. 2:** 166, 237
   Vol. 5: 24, 26, 352, 402
Haber, Prof.   **Vol. 12:** 533
Hadamard, J.   **Vol. 12:** 118*n*, 495
Hadorn, E.   **Vol. 7:** 328
hafnium   **Vol. 1:** xxxv
   Vol. 4: 26, 339, 464, 482
   Vol. 6: 390–391
   Vol. 11: 514, 522
   Vol. 12: 210, 215
  discovery of   **Vol. 4:** 30*ff*
Hafstad, L.R.   **Vol. 9:** 31, 56, 256, 365,
   384, 389
Hagemann, G.A.   **Vol. 11:** 483*n*
   Vol. 12: 294*n*
Hagen, E.   **Vol. 1:** 147, 150, 156, 250, 257,
   258, 360, 366, 408, 447, 464
Hahn, L.   **Vol. 9:** 638, 640
   Vol. 11: 358*p*

Hahn, O.   **Vol. 1:** xliv
   Vol. 4: 85
   Vol. 8: 405
   Vol. 9: 28, 42, 52, 56, 59*f*, 62*f*, 66*f*,
   285, 317, 329, 342, 344*f*, 360, 365*f*,
   378*f*, 382, 407*f*, 431, 438, 457, 462,
   469, 552*ff*, 563*ff*, 622*ff*, 626, 628, 634,
   636, 659
   Vol. 10: 22, 412
   Vol. 11: 12, 29, 569, 579, 597, 598,
   646, 649
   Vol. 12: 14, 84
Halban, H. von   **Vol. 9:** 384, 386, 397
Haldane, J.S.   **Vol. 10:** 5, 6, 19
half-life, *see* lifetime
Halifax, Lord   **Vol. 11:** 16, 17, 19, 20, 22, 30,
   34, 35, 38, 42, 43, 46, 47, 86, 88, 231,
   232*n*, 233, 240, 241, 245, 251–256,
   258, 260, 261, 263, 267, 270, 272–274,
   278–280, 283–288, 322
Hall effect   **Vol. 1:** 159, 160, 161, 187, 284,
   285, 387, 388, 389, 417, 418
Hall, C.C.   **Vol. 12:** 152, 164
Hall, E.H.   **Vol. 10:** 440
Hamburg   **Vol. 5:** 228*f*, 341, 343, 410, 412,
   414, 418, 422, 426, 430, 435, 437, 439,
   441, 442, 444, 445, 449, 453, 459*f*,
   462*f*, 464, 465*f*, 470, 472, 474, 477,
   480, 507
   Vol. 6: 32, 35, 41*ff*, 326*f*, 432, 435*ff*,
   449, 452
Hamilton's
  analogy between mechanics and optics
   Vol. 6: 111, 150, 204, 210, 264, 287
  equations   **Vol. 6:** 95, 125, 153, 174*f*
  principle   **Vol. 2:** 354, 400, 435,
   441–442, 465
   Vol. 8: 675
Hamilton, C.F.H.   **Vol. 12:** 128, 482, 482*p*,
   483, 483*p*
Hamilton, W.R.   **Vol. 3:** 85, 95
   Vol. 5: 243, 247, 273, 277
   Vol. 6: 204, 210, 264, 287
   Vol. 7: 347
   Vol. 10: 402, 424, 426, 441
   Vol. 12: 47
Hamiltonian equation/function
   Vol. 5: 374*ff*, 447*f*, 452, 464*f*
   Vol. 6: 128, 154, 162, 348
   Vol. 7: 231, 262, 271, 308, 447, 449
   Vol. 10: 401–403, 406

Hammach, K.C. **Vol. 7:** 489
  Vol. 9: 79
Hammarskjöld, D. **Vol. 11:** 82, 190*p*, 192
Hammerich, L.L. **Vol. 11:** 399, 406*p*, 409
Handbuch der Physik **Vol. 7:** 458, 467
Handel, J. van den **Vol. 12:** 38*p*
Hanle, W. **Vol. 5:** 50–56, 62, 145, 147*f*,
  153*f*, 302, 341–344, 357, 396*f*, 495
Hansen, A. **Vol. 11:** 563
Hansen, H.C. **Vol. 11:** 363, 376
  Vol. 12: 107
Hansen, H.M. **Vol. 1:** xxvii, 127, 576,
  577, 586
  Vol. 2: 9, 110, 324, 507, 514–518, 554,
  555, 593, 596, 610, 611, 617, 618,
  619, 627
  Vol. 3: 9, 25, 182, 183, 287, 296, 524,
  553, 655, 656, 657, 658, 690
  Vol. 4: 32, 370, 378, 604, 650, 688, 741
  Vol. 5: 41, 50, 87, 137, *317*, 418, 421,
  426, 430, 441*f*, 444, 487
  Vol. 8: *13*, 14, 235, 643, 687, 695–697
  Vol. 10: 74, 332, 367*n*
  Vol. 11: 465*n*, 513, 521, 525*p*, 673*p*
  Vol. 12: 114, 118, 210, 214, 326, 326*p*,
  327–330, 332, 332*p*, 333, 333*p*, 334,
  336*p*, 337, 338, 368, 511, 531
Hansen, H.P.E. **Vol. 10:** xxxiv, 461, 501*n*
Hansen, K. **Vol. 8:** 235
Hansen, K.G. **Vol. 9:** 639
Hansteen, C. **Vol. 10:** 366*p*
  Vol. 11: 354
Hanstein, H.B. **Vol. 9:** 384, 386
Harding, M.C. **Vol. 5:** 47, 127
Hardy, G.H. **Vol. 1:** xviii, 520, 521, 524,
  525, 536, 537, 542, 543, 544, 545, 546,
  547, 562, 563, 580, 581, 584
  Vol. 9: 258
  Vol. 12: 180, 188
Harkins, W.D. **Vol. 9:** 253, 280, 307, 321
harmonic
  analysis, *see* Fourier analysis
  oscillator (*see also* virtual
    oscillator/radiation) **Vol. 5:** 38, 42*ff*,
    58, 105, 110, 145*f*, 148*ff*, 166, 169,
    256, 263, 276*f*, 401, 417, 420, 446,
    451, 490–491, 505
    Vol. 7: 4, 306
    Vol. 8: 8*f*, 11, 34, 66, 80, 83, 96, 99,
    109, 113, 131, 209, 216, 394, 398, 490,
    505, 699*f*, 723
    Vol. 9: 276, 498, 501*f*, 574*f*, 615
    Vol. 10: 376, 386, 389, 390, 395, 437
  oscillator in quantum mechanics
    Vol. 6: 134, 157, 173
  oscillator model of atom
    Vol. 2: 5, 47
  vibrations **Vol. 12:** 47
  vibrator, *see* Planck's theory
harmony **Vol. 11:** 129, 134, 146, 147, 344,
  404, 414, 542, 550, 670
  Vol. 12: 75, 116, 128, 158, 159, 161,
  170, 172, 174, 180, 186, 188, 196, 221,
  371, 378, 413, 417, 421, 423, 463, 464,
  470, 472, 474, 475
  between content and form **Vol. 12:** 224
Harms, F. **Vol. 8:** 209, 719, 722
Harnack, A. von **Vol. 10:** 478
Harris, P. **Vol. 10:** 74
Harris, W.T. **Vol. 9:** 385
Harrison, G.R. **Vol. 5:** 496
Harteck, P. **Vol. 11:** 243, 244
Hartley, W.N. **Vol. 10:** 373
Hartline, H.K. **Vol. 10:** 134
Hartmann, E. von **Vol. 10:** 566
Hartmann, J. **Vol. 3:** 164
Hartmann, M. **Vol. 10:** 20, 466, 467,
  469, 470
Hartree, D.R. **Vol. 5:** 168, 171
  Vol. 6: 395
  Vol. 8: 214, 644, 698
  Vol. 10: 399, 400
Harvard **Vol. 6:** 452
Harvard University **Vol. 5:** 5*ff*, 65, 68, 319*ff*,
  324, 381, 404, 495, 497*f*, 502
  Vol. 9: 39, 539, 572*f*, 583, 585
Harvey, W. **Vol. 10:** 13, 54, 67, 117
Harwell, *see* Atomic Energy Research
  Establishment (U.K.)
Hasenöhrl, F. **Vol. 4:** 436, 470
Hassel, O. **Vol. 4:** 670
Hatley, C.C. **Vol. 5:** 318
Hatt, G. **Vol. 10:** 250*p*, 299*n*
Havas, P. **Vol. 8:** 329, 405, 556, 566
  Vol. 9: 666
Haxel, O. **Vol. 9:** 78, 185*f*, 253
H.C. Ørsted Centennial Foundation
  Vol. 12: 291, 293, 300, 302, 307, 310
H.C. Ørsted Medal **Vol. 11:** 353, 354, 470,
  471, 473, 474, 478–482, 484, 485,
  488–491, 729, 733

Vol. 12: 127, 291, 293, 307, 310, 317, 322

H.C. Ørsted Museum **Vol. 11:** 354, 478, 480, 482, 483
    Vol. 12: 300, 302, 307, 310, 317, 322

He$^+$ spectrum **Vol. 1:** xxix, 567, 571, 573

heat of combination (*see also* hydrogen, heat of combination/dissociation of)
    Vol. 2: 137, 141–142, 148, 156, 230, 416

heat radiation (*see also* black-body radiation)
    Vol. 11: 530, 613, 623, 646, 648, 659, 664, 670
    Vol. 12: 20, 43, 222

heat radiation law for long periods of vibration
    Vol. 1: 377*ff*

heat rays **Vol. 1:** 357*ff*
  absorption by metals **Vol. 1:** 359*ff*
  absorption coefficient **Vol. 1:** 360*ff*, 527
  emission by metals **Vol. 1:** 367*ff*
  reflection coefficient **Vol. 1:** 366

heat transfer **Vol. 1:** 331*ff*

heavy water, *see* deuterium

Hebrew University of Jerusalem
    Vol. 11: 695*n*

Hecht, H. **Vol. 4:** 670

Hecke, E. **Vol. 6:** 194, 445, 447

Hedgran, A. **Vol. 8:** 764

Hedtoft, H. **Vol. 11:** 80, 331, 360, 363, 373, 452*n*, 573, 586

Heegaard, P. **Vol. 1:** 96, 97, 98, 99
    Vol. 12: 131*p*

Hegel, G.W.F. **Vol. 6:** xx

Heiberg, P.A. **Vol. 12:** 152, 163

Heilbron, J.L. **Vol. 2:** 105, 114
    Vol. 8: 4*f*

Hein, P. **Vol. 10:** 36*p*
    Vol. 12: 17*p*

Heisenberg resonance **Vol. 6:** 136, 158, 395, 397

Heisenberg, E. **Vol. 6:** 25

Heisenberg, W. **Vol. 1:** xxxii, xxxvii, xxxviii, xxxix
    Vol. 3: 41, 42, 183, 486, 528, 556, 649, 650
    Vol. 4: 25, 38, 39, 391, 574, 593, 599, 602, 629, 643, 647, 649, 660, 669, 672, 673, 700
    Vol. 5: ix, 23, 31, 35*f*, 46*f*, 51, 53, 55–65, 67, 75, 79*f*, 87*ff*, 92, 95, 146*f*, 150, 152, 156*f*, 168, 171, 191, 204,

215, 220–226, 229–232, 235, 237*f*, 240, 251*f*, 256–267, 272*f*, 279, 280, 283*f*, 288*f*, 293, 295, 297, 299, 303*f*, 306, 308, 312–313, 329, 334*ff*, 338, 341*ff*, 348, 354–378, 394, 396, 405, 407–408, 409, 411*ff*, 418, 421*f*, 426, 430–433, 435, 437–444, 446*ff*, 451*f*, 454–465, 467*ff*, 471*ff*, 478*ff*, 484*f*, 495*ff*, 501*ff*, 507*f*, 512
    Vol. 6: v*f*, xv, xx, 3*f*, 7*ff*, 10–12, 15–20, 21*ff*, 29, 32*f*, 36, 37, 39, 44, 45, 47, 51*f*, 60, 62, 65, 71, 75, 80, 82, 88, 93*ff*, 100, 103, 105–106, 111, 116, 120, 123*ff*, 132*ff*, 136*f*, 140, 141–142, 144, 145, 149*f*, 152*f*, 155*ff*, 159–186, 204, 211, 229, 232, 244, 248, 263*ff*, 285*f*, 289, 314*f*, 318, 324–326, 359, 364, 393, 395, 397, 399, 402, 412, 415*ff*, 424–425, 429*f*, 433, 439*f*, 446*ff*, 457, 463
    Vol. 7: 3–7, 9*n*, 11, 13*n*, 14, 32*p*, 35, 37, 38, 43, 45–47, 59, 64, 69, 82, 121, 123, 126, 129, 139, 165, 166*n*, 203, 204, 216, 222, 228, 239–246, 250, 252, 262, 265, 267, 285, 306, 308, 310, 321, 329, 347–349, 363, 405, 418, 434, 437*n*, 444, 446*n*, 448, 450, 453, 457, 458, 472–474, 477, 487, 488, 491, 493, 495, 496, 501, 502
    Vol. 8: 207, 215*f*, 219*f*, 224, 247, 269*f*, 278*f*, 498, 590, 644, 651*f*, 654, 656, 660, 662*f*, 677*f*, 699–705, 712*f*, 787, 803
    Vol. 9: 10, 12*f*, 17, 19, 21, 25, 28, 34, 35, 36, 45, 76, 87, 100, 104, 125, 127, 134, 138*ff*, 156, 165–166, 174, 193, 197, 231, 274*f*, 280, 305, 307, 319*ff*, 490, 507, 528, 533, 535, 570*f*, 573–587, 600*ff*, 604, 617
    Vol. 10: xxiv, 21*p*, 58, 88, 212, 340, 400, 401, 403, 406, 419, 422*p*, 424–428, 441–444, 447, 451, 453, 455*p*, 495, 531, 533, 547, 551, 556, 593
    Vol. 11: 13, 29, 243, 244, 264, 265, 361, 513*n*, 514, 523, 563, 567, 671
    Vol. 12: 5, 8, 10, 17*p*, 38*p*, 44, 47, 100, 115, 116, 210, 215, 223, 348, 351, 358, 500, 502

Heitler, W. **Vol. 6:** 304, 305, 397, 439*f*
    Vol. 7: 13*n*, 236
    Vol. 8: 224, 638, 709*f*, 811

[87]

Vol. 10: 403, 427, 443, 447, 455*p*
Vol. 12: 17*p*
helium **Vol. 4:** 6, 52, 63*ff*, 94, 122–139, 218*ff*
 Vol. 10: 378, 379, 387, 390, 392, 395,
  397, 401, 403, 427, 437, 442, 443, 449
 Vol. 11: 57, 298, 574, 575, 577,
  588–590, 593, 594, 613, 614, 623, 624
 Vol. 12: 12, 44, 51, 54, 357
 absorption in **Vol. 2:** 6, 32–33, 68, 73, 75,
  90, 92, 500, 581
 coplanar orbits in **Vol. 4:** 126, 381*ff*
 crossed orbits in **Vol. 4:** 128*ff*
 dispersion in **Vol. 2:** 33, 53, 222–223, 336,
  408, 449, 564–567, 587
 energies of stationary states of
  Vol. 4: 130*ff*
 ion, spectrum of, *see* helium, spectrum of
 ionization in **Vol. 2:** 86–88, 92, 171
 ionization potential of **Vol. 4:** 133, 221, 290
  Vol. 6: 392, 395
 ionized **Vol. 4:** 52, 194, 268, 349, 442,
  473, 560, 619
 metastable state of **Vol. 4:** 8, 53, 64, 94,
  132*ff*, 219, 288, 385
  Vol. 6: 395
 model **Vol. 5:** 57, 304, 503*f*
 molecule, instability of **Vol. 2:** 140,
  141–142, 154, 156, 228–229
 normal state of **Vol. 4:** 132, 136, 221, 222,
  290, 385
 ortho- and para- **Vol. 6:** 145, 395, 397
 spectra of **Vol. 3:** 276*ff*
 spectrum of **Vol. 2:** 116, 120–122, 123,
  127, 170–171, 273–276, 297–300, 325,
  326, 333–334, 335, 343–344, 352, 356,
  361–362, 370, 385–389, 399–400,
  404–408, 469, 494, 495, 503, 508,
  532*f*, 558*f*, 561*f*, 573, 574, 582*f*,
  588, 593, 595*f*, 603, 626, 629, 633
  Vol. 5: 288, 447, 451, 470, 472, 478,
  481, 493, 501, 503
  Vol. 6: 52, 145, 385*f*, 392, 395, 398
  . **Vol. 12:** 47, 357
 spectrum of ionized **Vol. 3:** 214*ff*, 588*ff*
 stopping in **Vol. 5:** 181*f*, 184, 197,
  199, 347
 the problem of **Vol. 4:** 36*ff*
helium atom
 mechanical instability of **Vol. 2:** 46,
  49–51, 104–105, 106
 perturbation calculation on **Vol. 2:** 8, 49–51

structure of **Vol. 2:** 6, 7, 8, 33, 39, 53, 68,
  106, 140, 154, 170, 200–202, 222, 264,
  290, 297, 325, 404–405, 531, 577, 587
  Vol. 3: 3, 180
Hellmann, S. **Vol. 7:** 522
  Vol. 8: 336, 823
  Vol. 9: 364, 564
  Vol. 10: 110, 588, 591–593
  Vol. 11: 358*p*, 734
  Vol. 12: 531, 532
Helmholtz, H. von **Vol. 1:** 486
  Vol. 3: 583
  Vol. 6: 374, 408
Helsinki conference, 1936 **Vol. 9:** 39,
  159–178, 609
  Vol. 11: 355–356, 501–507
Hempel, C.G. **Vol. 10:** xli, 16, 36*p*
Henderson's stopping formula, *see* stopping
  formula, Henderson's
Henderson, G.H. **Vol. 2:** 409
  Vol. 5: 181*f*, 197
  Vol. 8: 28*f*, 33, 35, 209, 241, 256, 339,
  355, 371, 395, 399, 498, 512, 529, 567,
  644, 657, 660, 706, 737*f*, 783*ff*
  Vol. 10: 412
Hendriksen, F. **Vol. 12:** 470, 472
Hendriksen, Halfdan **Vol. 12:** 127, 128,
  469–472, 531
Hendriksen, Holger **Vol. 12:** 453, 458
Henningsen, E. **Vol. 10:** 366*p*
Henriksen, K. **Vol. 10:** 299*n*
Henriksen, R. **Vol. 11:** 360, 364, 647, 651
Henriot, E. **Vol. 7:** 32*p*
Henriques, V. **Vol. 1:** xviii, 584
  Vol. 2: 332, 519–524, 618
  Vol. 12: 121, 525
heredity **Vol. 10:** 70, 91, 104, 122, 169, 198,
  246, 545, 549, 553, 554
Herglotz, G. **Vol. 3:** 637, 638
Hermann, A. **Vol. 5:** 23, 220
  Vol. 6: 10, 41
  Vol. 7: 10*n*
Hermite, C. **Vol. 5:** 259, 266, 280
Herschel, J. **Vol. 11:** 445
Hershey, A.D. **Vol. 10:** 23, 551, 552*n*
Hertz, G. **Vol. 1:** xxxi
  Vol. 2: 86, 87, 92, 117, 201, 331, 335,
  408, 409, 559, 563
  Vol. 3: 20, 237, 238, 239, 257, 322,
  336, 368, 591, 631
  Vol. 4: 353, 375, 446, 475, 659, 672

Vol. 5: 27, 70, 179, 194, 275
Vol. 6: 71, 138, 228, 242, 382
Vol. 7: 344
Vol. 8: 18*ff*, 27, 156*f*, 184, 189, 682, 693
Vol. 9: 304, 318
Vol. 10: 58, 363, 377, 397, 400
Vol. 11: 530, 567
Vol. 12: 44, 221, 371
Hertz, H.   **Vol. 4:** 434, 469
    Vol. 5: 129, 137
    Vol. 10: 350
Hertz, P.   **Vol. 3:** 9
    Vol. 5: 398
    Vol. 8: 688, 690, 692
Hertz, S.   **Vol. 10:** 224*p*
Hertzsprung, E.   **Vol. 3:** 29, 610, 660, 662
    Vol. 5: *484*, 486
Herzen, E.   **Vol. 3:** 362
    Vol. 7: 32*p*
Herzfeld, K.F.   **Vol. 1:** 448
    Vol. 2: 59, 64, 65, 78, 326, 364, 378, 446, 525–526, 616
    Vol. 3: 142, 637
    Vol. 8: 111, 114, 129, 134*f*, 148, 372
Hess, E.   **Vol. 11:** 563
heteropolar compounds   **Vol. 4:** 295
Hettner, G.   **Vol. 3:** 190
    Vol. 4: 355
Heurlinger, T.   **Vol. 4:** 453, 478
    Vol. 10: 379
Heuse, W.   **Vol. 2:** 458, 459
Hevesy, G. de   **Vol. 1:** xxiii, xxxv
    Vol. 2: 109, 111, 118, 119, 122, 123, 214, 423, 527–533, 551, 589, 613, 614, 615, 616, 617, 618
    Vol. 3: 25, 291, 292, 300, 601, 664
    Vol. 4: 22, 23, 26, 32, 339, 464, 482, 691, 700, 734
    Vol. 5: 48, 296, 308, 379, 487, 492
    Vol. 6: 378, 390, 458
    Vol. 8: 24, 246, 644, 680, 693*f*, 706–708, 731*ff*, 735, 750, 774, 777, 787
    Vol. 9: 3, 4, 28, 53, 140, 156, 216, 220, 589, 599*f*, 623*ff*, 627, 633*ff*, 640
    Vol. 10: 36*p*, 76, 167, 330, 377, 385, 393, 400, 417, 594
    Vol. 11: 217, 220, 358*p*, 373, 460*n*, 513, 514, 516, 517, 522–524, 531, 567
    Vol. 12: 210, 215, 216*n*, 265, 502, 504, 534

Hey, F.   **Vol. 3:** 286, 295
Heydenburg, N.P.   **Vol. 9:** 31, 256
Heydweiller, A.   **Vol. 2:** 237
Heyn, F.   **Vol. 9:** 244
high-frequency spectra, *see* X-ray spectra
high-voltage generator/laboratory/plant (*see also* accelerator)   **Vol. 8:** 235, 315, 731*f*, 749
    Vol. 10: 75, 398, 410, 449
    Vol. 11: 516, 518, 524, 526, 528*p*, 568, 727, 728
Hilbert, D.   **Vol. 1:** 566, 567
    Vol. 2: 126
    Vol. 4: 740
    Vol. 5: 312
    Vol. 11: 670
Hilbert, Mrs   **Vol. 8:** 692
Hill, A.V.   **Vol. 11:** 445
Hill, D.L.   **Vol. 1:** 520, 521, 536, 537
    Vol. 9: 74, 82*f*, 667, 669*f*, 680
    Vol. 11: 327, 328, 330, 331
Hinshelwood, C.   **Vol. 12:** 139*n*
Hippel, A.R. von   **Vol. 11:** 518, 526
Hippocrates   **Vol. 10:** 67
Hiroshima   **Vol. 11:** 5, 50
Hirsch, G.   **Vol. 7:** 328
history   **Vol. 12:** 366
    of civilization   **Vol. 11:** 26, 35, 156, 242, 348, 445, 537, 548, 569, 611, 619, 690, 694
    of Denmark   **Vol. 11:** 422, 423, 428, 432
    of Nordic countries   **Vol. 11:** 408*n*, 504, 506
    of science   **Vol. 11:** 399, 409, 494–499, 515, 523, 545, 551, 567
    Vol. 12: 93, 95, 103, 105, 118, 122, 128, 199, 203, 277, 428, 430
Hitler, A.   **Vol. 11:** 281, 338
    Vol. 12: 124
Hjalmar, E.   **Vol. 4:** 493, 499, 530, 532
Hjelmslev, J.   **Vol. 5:** 219
    Vol. 10: 303, 510, 511
    Vol. 11: 345*n*, 380, 382
Hobbes, T.   **Vol. 10:** 3*n*
Hodson, T.C.   **Vol. 12:** 10, 508
Høffding, G.   **Vol. 10:** 302
Høffding, H.   **Vol. 1:** xx, 7, 8, 16
    Vol. 3: 286, 295
    Vol. 5: 81
    Vol. 6: xx–xxi, xxii, xxiii, xxiv, 24, 27, 424
    Vol. 10: ix, xxviii, xliv, 7, 250*p*, 257,

266, 271*p*, 296*p*, 297–303, 306–309,
312*p*, 313–322, 324, 325, 461,
505–513, 582, 585
Vol. 12: 100, 123, 124, 136*n*, 152–154,
155*p*, 162, 164–166, 169*p*, 175, 198,
202, 212, 217
Høffer-Jensen, S.   **Vol. 11:** 358*p*
Hoffmann, F.   **Vol. 1:** 120
Højgaard Jensen, H.   **Vol. 9:** 630, 639
Vol. 12: 16, 94, 96
Holbach, P.   **Vol. 10:** 3*n*
Holberg, L.   **Vol. 10:** 255, 264
hole theory (*see also* Dirac theory; positron)
Vol. 9: 549
holism (*see also* wholeness)   **Vol. 10:** 11, 60
Holland, Aid to, Foundation
(Hollandshjælpen)   **Vol. 11:** 730
Holm, E.   **Vol. 2:** 458, 459
Holm, E.A.   **Vol. 3:** 653, 655
Holmes, O.W.   **Vol. 10:** 202
Vol. 11: 249
Holst, H.   **Vol. 3:** 659, 661
Vol. 5: 35, *41*
Vol. 6: 308
Vol. 8: *734*, 736
Vol. 12: 502, 504
Holst-Weber, S.T.   **Vol. 1:** 500, 501, 534, 535
Vol. 2: 344, 609–611, 620
Vol. 3: 29, 610
Vol. 8: *676*, 733, 735
Holter, H.   **Vol. 11:** 297
Holton, G.   **Vol. 7:** 411
Holtsmark, J.P.   **Vol. 4:** 712
Vol. 8: 224, 262, 304, 457, 567, 631,
634, 688
Homer   **Vol. 12:** 152, 164
homopolar bond   **Vol. 6:** 392–393, 397
homopolar compounds   **Vol. 4:** 295
Honda, K.   **Vol. 2:** 237
Hönigschmid, O.   **Vol. 2:** 594, 597
Hönl, H.   **Vol. 5:** 279
Honner, J.   **Vol. 6:** 195
Hooker, J.D.   **Vol. 11:** 445
Hooker, W.J.   **Vol. 11:** 445
hope   **Vol. 10:** 95, 160, 185, 199, 202
Hopf, H.   **Vol. 7:** 328
Hopf, L.   **Vol. 1:** 273, 378
Hopkins, F.G.   **Vol. 11:** 445
Horsley, L.H.   **Vol. 8:** 709*f*
Horton, F.   **Vol. 4:** 53

House of Culture (Leningrad)
Vol. 11: 198, 201
Houston, University of   **Vol. 12:** 532
Houstoun, R.A.   **Vol. 2:** 139, 152
Houtappel, R.M.F.   **Vol. 12:** 38*p*
Houtermans, F.G.   **Vol. 9:** 542, 607*f*
Vol. 10: 409
Vol. 11: 243, 244
Hove, L. van   **Vol. 10:** 455*p*
Hoyer, U.   **Vol. 2:** 5, 7, 47, 49, 107, 115,
130, 131
Vol. 5: 69
Vol. 8: 9
Vol. 10: 332*n*
Hoyle, F.   **Vol. 9:** 385
Hoyt, F.C.   **Vol. 3:** 45
Vol. 4: 467
Vol. 5: 124, 320, 324, 403, 502, 512
Vol. 7: 165*n*
H-theorem (*see also* entropy; irreversibility;
second law of thermodynamics)
Vol. 6: 320*ff*, 438, 440
Huber, P.   **Vol. 10:** 572
Huby, R.   **Vol. 9:** 618
Hückel, E.   **Vol. 6:** 144, 304
Huggins, W.   **Vol. 3:** 164
Vol. 11: 445
Hughes, A.L.   **Vol. 2:** 237
Vol. 4: 346
Hull, A.W.   **Vol. 2:** 597
human
culture/society (*see also* culture, human)
Vol. 10: xl, 92, 97, 98, 102, 105, 112,
129, 141, 143, 157, 160, 202–206, 216,
220, 240, 241, 245–248, 253–255,
257–264, 266, 268–270, 272, 278, 279,
286, 287, 292, 293, 355, 369, 545, 549
Vol. 11: 11, 50–53, 57, 61, 65, 79, 123,
124, 127, 134, 137, 141, 146, 148, 150,
152, 184, 185, 192, 203, 205, 337, 344,
352, 355, 366, 404, 414, 448, 450, 464,
467, 503, 504, 506, 530, 531, 537, 548,
617, 631
Vol. 12: 62, 70, 86, 87, 93, 95, 111,
119, 181, 189, 201, 205, 277, 371, 372
experience, *see* experience
knowledge   **Vol. 7:** 340, 379, 395–423,
523, 524
language, *see* language
life/existence   **Vol. 7:** 408, 420
Vol. 11: 12, 146, 148, 151, 153, 154,

185, 323, 343, 356, 404, 414, 428, 432, 449, 451, 455, 459, 506, 507, 536, 537, 542, 543, 547, 548, 550
Vol. 12: 13, 93, 95, 153, 157, 158, 165, 170, 171, 201, 205, 316, 320, 349, 352, 386, 388, 400, 403, 417, 423, 463, 464, 474, 475, 478, 479
limitations **Vol. 12:** 157, 170
mind **Vol. 10:** 93, 94, 105, 123, 159
Vol. 11: 434, 435
Vol. 12: 93, 95
motives **Vol. 10:** 71, 93, 105, 194, 200
problems **Vol. 7:** 38
rights **Vol. 10:** 255, 264, 279
Vol. 11: 54, 71, 124, 140, 144, 162, 183, 192, 296
understanding **Vol. 12:** 184, 186, 194, 196
values **Vol. 10:** 188, 202, 203, 254, 260, 263, 271, 276
vs. animal **Vol. 10:** 245, 246
humanities **Vol. 10:** 143, 241, 249, 300, 415, 511, 562, 563, 583
Humboldt, W. von **Vol. 6:** vi
Hume-Rothery, W. **Vol. 1:** 121, 122
Hund, F. **Vol. 3:** 508, 537
Vol. 4: 698, 699, 700
Vol. 5: 108, 193, 206, 251f, 256, 262, 279, 301, 313, 499f
Vol. 6: 9, 145, 154, 158, 397, 423, 463
Vol. 8: 30, 652, 654, 660, 662, 693f
Hupka, E. **Vol. 2:** 199
Husimi, K. **Vol. 9:** 239
Huus, T. **Vol. 8:** 235
Huxley, A. **Vol. 10:** 584
Huxley, J. **Vol. 10:** 82
Huxley, T.H. **Vol. 11:** 445
Huygens, C. **Vol. 3:** 229
Vol. 5: 129, 137, 273, 310
Vol. 6: 360
Vol. 10: 209, 426, 432
Vol. 12: 371
Huygens' principle **Vol. 8:** 271, 280, 343, 348
Hvidberg, F. **Vol. 11:** 350, 351, 375, 376, 466n
Hyde Park (Roosevelt's summer residence) **Vol. 11:** 37–39, 41, 259, 263
hydrogen **Vol. 4:** 191ff, 265ff
Vol. 7: 447, 449
Vol. 10: 157, 192, 331, 374, 377, 378, 387–392, 394–396, 402, 403, 409, 426,

427, 437, 443, 449, 463, 464
Vol. 11: 531, 569, 575–578, 580, 590, 591, 593, 596, 599, 612, 613, 615, 616, 620, 622, 623, 627, 628
Vol. 12: 42, 44, 48, 59, 60, 64, 66, 67, 255, 259, 386, 387
absorption in **Vol. 2:** 6, 30–32, 68, 73–75, 78, 85–86, 90, 92, 170, 176, 500, 581
and X-ray spectra **Vol. 4:** 267ff
bomb **Vol. 11:** 77, 615, 627
Vol. 12: 84
dielectric constant of **Vol. 5:** 63, 357ff
dispersion in **Vol. 2:** 31, 32, 53–55, 114, 222, 336, 337, 449, 568, 570, 587
fine structure of **Vol. 4:** 111ff, 118–120, 197, 214ff, 269ff, 360, 362, 447ff, 475ff, 620
heat of combination/dissociation of **Vol. 2:** 105, 130–131, 141, 156, 221, 230, 481, 539–540
ionization in **Vol. 2:** 87–88, 92
ionization potential of **Vol. 6:** 393
Stark effect of **Vol. 4:** 213, 368, 451, 477
stopping in **Vol. 5:** 181f, 196f, 337f, 347
Zeeman effect of **Vol. 4:** 214, 370
hydrogen atom
perturbation calculation on **Vol. 2:** 8, 47–48
structure of **Vol. 2:** 5, 6, 7, 32, 39, 74, 140, 154, 168, 199–200, 290, 308, 352, 474, 476–480, 528, 531, 577, 629, 633
hydrogen chloride, structure and spectrum of **Vol. 2:** 117, 230–231, 447, 457
hydrogen ion **Vol. 2:** 200, 479–480, 486
hydrogen isotope, spectrum of **Vol. 2:** 277–278, 627
hydrogen lines
effect of electric and magnetic fields on **Vol. 3:** 419–446, 597ff
hydrogen molecule **Vol. 3:** 187ff
Vol. 4: 50–52, 214
Vol. 6: 145, 392–393, 397–398
formation of **Vol. 2:** 141, 156, 226–228, 481, 485
ionized **Vol. 2:** 481ff
structure of **Vol. 1:** 581
Vol. 2: 6, 8, 31–32, 53, 68, 74, 105, 131, 138ff, 166ff, 221–225, 259, 480–488, 539, 587, 590

[91]

triatomic **Vol. 2:** 278, 344, 483–488, 630

hydrogen spectrum (*see also* Balmer formula/series/spectrum) **Vol. 2:** 110*ff*, 163, 168*ff*, 176, 199, 273–276, 277–278, 281–301, 308, 313, 315, 319, 322*ff*, 327–329, 333*ff*, 352, 356*ff*, 370, 377–380, 385–388, 399*ff*, 444*ff*, 452, 467*ff*, 474, 494, 495, 503*ff*, 559, 562, 582*ff*, 588, 593, 600, 626, 629, 634
    Vol. 3: 4, 103*ff*, 155, 156, 207, 246*ff*, 424*ff*
    Vol. 4: 49, 92, 439*ff*, 471*ff*, 558*ff*, 617*ff*, 660
    Vol. 5: 91, 157*f*, 161, 163, 237, 239, 243, 247, 251*f*, 276*ff*, 280, 282, 284, 288, 347, 362*f*, 370, 406, 407, 445–451, 454*f*, 497–498, 499, 501
    Vol. 6: 3, 52, 138, 379*f*, 383*ff*, 389, 392
    Vol. 12: 43, 44, 344, 345, 357

continuous **Vol. 3:** 164*ff*

effect of electric and magnetic fields on **Vol. 3:** 419–446

effect of electric and magnetic fields making an angle with each other **Vol. 3:** 159*ff*, 506, 536

effect of external electric field on **Vol. 3:** 135*ff*, 261*ff*, 434*ff*

effect of magnetic field on **Vol. 3:** 145*ff*, 264*ff*, 431*ff*, 593*ff*

effect of magnetic field on fine structure **Vol. 3:** 158

fine structure of **Vol. 3:** 129*ff*

relativity effect **Vol. 3:** 268*ff*

simple theory of **Vol. 3:** 103*ff*

simultaneous effect of relativity and electric field **Vol. 3:** 144

Hylleraas, E.A. **Vol. 6:** 395
    Vol. 7: 261*p*, 301
    Vol. 9: 10, 412
    Vol. 12: 17*p*

Hylling-Christensen, V. **Vol. 9:** 639*f*

hyperfine structure **Vol. 6:** 392, 404
    Vol. 9: 94, 231, 621
    Vol. 12: 342

hypnosis **Vol. 10:** 575, 576

hypotheses, Bohr's fundamental, *see* assumptions, Bohr's principal

Iacobæus, H. **Vol. 11:** 206, 208, 209

iatro-chemistry **Vol. 10:** 54

iatro-physics **Vol. 10:** 54, 61

Ibsen, H. **Vol. 10:** 303

ideal gas **Vol. 6:** 47, 50, 143*f*, 438, 440, 464, 466

idealism **Vol. 10:** 95

identical
    atoms/nuclei **Vol. 6:** 145, 397–398, 406–407
        Vol. 9: 96*f*, 111, 113*f*
    particles (*see also* Bose–Einstein statistics; Fermi–Dirac statistics) **Vol. 6:** 102, 136, 144*f*, 158, 193, 394*f*, 441, 443
        Vol. 8: 211, 437, 442, 449
    systems (*see also* ensemble)
        Vol. 6: 13, 460

"identical" elements, *see* isotopes

identity hypothesis **Vol. 10:** xliii

ideology **Vol. 10:** 546, 550

immortality **Vol. 10:** 478, 483, 485, 486

impact parameter **Vol. 8:** 6*ff*, 53, 131, 141, 152, 164, 208, 218, 241, 292, 311, 365*f*, 377, 380, 393, 397, 430, 433*ff*, 439*ff*, 444, 448*f*, 451, 453, 489, 491*f*, 499, 506, 510, 534, 544, 608, 630*f*, 635*ff*, 802

impact phenomena, *see* collisions

impact-parameter method **Vol. 8:** 208, 218–219, 225, 381, 503, 630, 636–637, 793

improvisation **Vol. 10:** 95, 96

independent-particle model (*see also* shell model) **Vol. 9:** 25, 27, 34, 38, 76, 79*ff*, 186, 193, 231, 246, 275, 278, 281, 308*f*, 322*f*, 336, 428, 455, 523–524, 582, 584, 609, 649, 651, 667, 669*f*

indeterminacy principle/relations (*see also* uncertainty principle/relations) **Vol. 10:** xxxix, 42, 43, 88, 104, 406, 407, 428, 444, 448
    Vol. 11: 660, 665, 671
    Vol. 12: 223

India **Vol. 10:** 157, 160, 487

indicator method **Vol. 11:** 373, 531, 567, 569

indicator, radioactive, *see* tracer

individual wave, *see* phase wave

individual/elementary process **Vol. 6:** 41, 75*f*, 81, 83, 94*f*, 101*f*, 116, 123–124, 134*f*, 139, 149, 152*f*, 156, 157, 228, 242*f*, 261, 263, 282, 285*f*, 327, 359*f*, 381*f*, 384, 388, 393*f*, 400, 405, 453

Vol. 7: 213, 229–234, 293, 331, 343, 345, 408, 421, 475

individuality **Vol. 10:** 5, 8, 13, 33, 41, 45, 57–60, 123, 149, 165, 242, 388, 402, 441–443

individuality (in quantum theory) **Vol. 12:** 47

individuality of particles **Vol. 9:** 88, 94, 118, 605

indivisibility **Vol. 11:** 512, 521, 670, 671
  Vol. 12: 185, 194, 233

indivisibility/closedness of atomic processes
  Vol. 10: 30, 89, 103, 122, 176, 177, 351, 364, 445, 476, 568, 570, 573

Industrial Council (Denmark)
  Vol. 11: 363, 365

industrial plants, automatic control of
  Vol. 10: 130, 135

industry **Vol. 11:** 90, 92–94, 105, 106, 108, 133, 177, 363, 531, 573, 574, 583, 586, 587, 606, 616, 627, 635, 636, 646, 648, 690, 695

inelastic scattering, *see* scattering, inelastic

inert gases **Vol. 4:** 10, 19, 79, 96, 246, 315, 595*ff*, 644*ff*
  number of electrons in n_k-orbits in
  Vol. 4: 246, 315

inertial mass **Vol. 7:** 366

infrared spectra, *see* spectra, infrared

Ingolf Expedition (1895–1896)
  Vol. 12: 316, 321

inhibition **Vol. 10:** 575, 576

inner electron groups, development of
  Vol. 4: 235, 305, 310, 313, 606, 652

Innsbruck Meeting, 1924 **Vol. 5:** 31*f*, 36, 51, 54, 341, 343, 397, 415, 419, 507

instability, *see* stability; stability of atoms and molecules

instinct **Vol. 10:** 47, 92, 105, 172, 185–187, 204, 245, 246, 285, 291, 536, 537, 545, 549

Institute for Advanced Study (Princeton)
  Vol. 7: 371
  Vol. 11: 78, 103, 320, 361

Institute for Theoretical Physics (Copenhagen)
  Vol. 10: 32*p*, 109, 110, 157, 192, 224*p*, 400, 410, 414, 425, 427, 472, 473, 484, 486, 487, 492, 495, 520*n*, 587
  Vol. 11: 7, 11, 12, 29, 32, 57, 61, 83, 103, 203, 204, 332, 333, 338, 340, 354, 356, 357, 359–361, 367, 371, 373, 375, 456, 460, 495, 499, 511–514, 516–518,

520–526, 528*p*, 529, 531, 556–560, 562, 563, 565, 579, 599, 641, 646, 649, 669, 715*n*, 716, 717, 720, 726–729, 733
  Vol. 12: viii, 5, 6, 8, 13, 24, 28, 84, 103, 108, 109, 115–117, 119, 124, 125, 138, 140, 144, 145, 149, 150, 160, 161, 172*n*, 173, 174, 192*n*, 199, 200, 203, 204, 210, 211, 214–217, 252, 277, 326, 329, 332, 333, 348–351, 356, 368, 381, 383, 386–388, 415, 422, 516

dedication of **Vol. 3:** 24*ff*
speech at dedication **Vol. 3:** 283–293
translation of speech **Vol. 3:** 293–301

Institute of Nuclear Science (Rehovot)
  Vol. 12: 487

insulators **Vol. 12:** 31, 34

intensities **Vol. 3:** 11, 102, 123, 133, 143, 151

intensity **Vol. 5:** 18, 39*f*, 85, 86, 105, 107, 109, 112, 114, 123*f*, 165*f*, 168*f*, 251*f*, 258, 264, 272, 274*f*, 277, 279, 283*f*, 311, 317, 320, 322, 325, 362*f*, 387*f*, 400, 448, 452, 472, 474, 483, 485, 506
  of spectral lines **Vol. 2:** 170, 345, 356, 379, 400*ff*

interaction
  between cells **Vol. 10:** 545, 549
  between light and matter **Vol. 3:** 20, 227–240
  between object and measuring instrument
  Vol. 9: 104–105, 139, 164, 173, 305, 320
  Vol. 10: 30, 31, 34, 41, 43, 59, 69–71, 76, 88, 90, 103, 112, 128, 129, 142, 149, 152, 158, 165, 178, 183, 189, 196, 203, 212, 243, 248, 277, 284, 290, 340, 405, 428, 444, 445, 448, 453, 476, 477, 517, 521, 539, 541, 542, 573
  Vol. 11: 134, 541, 549, 671
  Vol. 12: 62, 69, 185, 194, 200, 204, 223
  between radiation and matter **Vol. 10:** 29, 432, 435, 438
  Vol. 12: 42

interatomic forces **Vol. 2:** 61*f*, 72

interference **Vol. 5:** 3, 6, 9, 12, 15, 38*f*, 53, 68, 79, 87, 101, 103, 110*f*, 309, 353–354, 396*f*, 440*f*, 443, 448, 452, 487, 491, 493
  Vol. 8: 88, 104, 269, 272*ff*, 277, 278, 281*ff*, 286, 292*f*, 298, 304, 306, 343, 348, 388, 390, 395, 399, 536, 629, 638

of electrons (*see also* reflection of electrons)
  Vol. 6: 105, 270, 295, 365–366
  Vol. 7: 36, 255, 264, 294, 314, 315,
  347, 356–359, 475, 479
of light   **Vol. 6:** 21f, 60, 62, 64, 78, 85–86,
  92, 115, 117f, 148, 150
  Vol. 7: 356–359, 362
of wave functions   **Vol. 6:** 60, 62, 64, 78,
  85–86, 96ff, 104–105, 106, 117f, 126,
  129, 141, 149f, 153, 155
Interim Committee (U.S.)   **Vol. 11:** 47, 69
internal conversion   **Vol. 6:** 405
  Vol. 8: 679f, 802
  Vol. 9: 95, 194, 237, 239, 315,
  327, 607f
international
  control/regulation   **Vol. 11:** 5, 17, 18, 21,
  24, 27, 28, 33, 36, 37, 44, 48, 50, 63,
  65–70, 73, 80, 83, 88, 93, 100, 106,
  107, 109, 111, 114–116, 118–120, 124,
  127, 157, 161, 166, 177–179, 182, 189,
  234–236, 246, 259, 294, 295, 308, 309
  Vol. 12: 116, 359
  cooperation   **Vol. 10:** 102, 106, 141, 192,
  205, 206, 257, 259, 260, 267, 270, 272,
  287, 292, 293, 428, 536n
  Vol. 11: 4, 6, 8–10, 15, 20, 23, 26, 31,
  33, 42, 48–52, 54, 57–59, 63, 70, 71,
  73–75, 79, 80, 83, 91, 93–96, 99, 102,
  104, 107–109, 115–119, 123, 128, 133,
  138, 139, 142, 143, 150–153, 156, 157,
  160–166, 168, 175, 176, 178–181,
  183–185, 188, 189, 192, 197, 200, 203,
  205, 215, 216, 219, 224, 226, 228, 238,
  247, 251, 268, 271, 273, 275, 290, 291,
  294–298, 300, 303, 311–315, 317, 318,
  323, 324, 331, 339, 341, 343, 345, 349,
  355–357, 360–362, 365, 368, 369, 371,
  403, 405, 413, 414, 428, 432, 440, 441,
  445, 448–451, 503, 505, 513, 514, 519,
  522, 526, 530, 531, 546, 552, 562, 565,
  569, 570, 573, 585, 611, 613, 615–617,
  619, 622, 627–629, 631, 643, 644, 646,
  647, 650, 651, 693, 699, 702, 703,
  715n, 717, 722, 727, 729
  Vol. 12: 75, 101, 106, 111, 159, 172,
  183, 191, 193, 201, 205, 211, 216,
  228–230, 234, 277, 282, 291, 294, 300,
  302, 316, 318, 321–323, 332, 334, 348,
  351, 359, 371, 372

International Association of Academies
  Vol. 11: 445
International Astronomical Union
  Vol. 11: 732
International Catalogue of Scientific Literature
  Vol. 11: 445
International Congress on Light Therapy
  Vol. 7: 376, 465n
International Council of Scientific Unions
  Vol. 11: 445, 732
  Vol. 12: 302n
International Education Board   **Vol. 5:** 313,
  483, 485
  Vol. 11: 7, 514, 522
  Vol. 12: 125, 192n
International Institute of Intellectual
  Cooperation   **Vol. 7:** 260, 377, 521
International Institute of Physics, *see* Solvay
  Institute
International Labour Organization
  Vol. 11: 327n
International Research Council
  Vol. 12: 302n, 318, 323
International Union for Pure and Applied
  Physics (IUPAP)   **Vol. 7:** 260
  Vol. 11: 6
  Vol. 12: 349, 351, 359
International Union of Geodesy and
  Geophysics   **Vol. 12:** 302n
intra-atomic force, *see* binding force
intra-nuclear electrons   **Vol. 6:** 319, 378,
  403f, 406
  Vol. 9: 11f, 88f, 93f, 96, 111–114,
  166, 174, 307, 321, 414–415, 444
introspection (*see also* consciousness;
  psychology)   **Vol. 6:** 275, 301
  Vol. 10: xli, xlii, xlvii, 7, 47, 62, 71,
  92, 94, 105, 245
inverse collisions   **Vol. 6:** 383, 405
inverse cube law of force (*see also* nth power
  law)   **Vol. 8:** 230, 453, 469, 534, 563,
  634–635, 811
Ioffe, A.F.   **Vol. 6:** 307
  Vol. 7: 32p
  Vol. 11: 197, 201, 217, 230
  Vol. 12: 376p
ion   **Vol. 11:** 566, 574, 576, 577, 588,
  591, 594
ion lattice   **Vol. 12:** 22, 23, 26, 27, 30, 31, 33,
  34, 39

ionization (*see also* hydrogen ion)   **Vol. 2:** 4, 9, 10, 19, 29, 60, 81, 83, 84–90, 92, 171, 177, 238, 239, 298, 482, 485, 515, 517, 633
  Vol. 5: 71–72, 82, 83, 178, 184–187, 192, 194, 199–202, 205–206, 278, 322, 349, 384
  Vol. 8: —
  Vol. 11: 578, 595
 energy   **Vol. 5:** 123, 159, 163, 167, 170, 182, 184, 186f, 192, 197, 199, 201, 206, 276, 305, 503
 potential (*see also* binding energy)
  Vol. 2: 87, 92, 126, 165, 200, 201, 204, 335, 388, 399, 408f, 478, 482, 588
  Vol. 6: 392f, 395
  Vol. 8: 12, 18ff, 22f, 28, 45f, 118, 122, 132, 157, 197, 199, 276, 285, 355, 381, 395, 399, 488f, 498, 505, 521f, 532, 534, 542, 602, 605, 619, 622, 636, 638, 672, 681, 792, 800
 secondary   **Vol. 8:** 16, 45f, 155ff, 160, 497, 522, 757f
ionizing power   **Vol. 8:** 77, 93, 244, 258, 276, 285, 555, 584, 598, 625
ions, mobility of   **Vol. 2:** 118, 139, 152, 423, 631
Iowa State University   **Vol. 12:** 532
Ipsen, C.   **Vol. 12:** 452, 457
Ipsen, J.   **Vol. 12:** 452, 457
iron group atoms, structure of
  Vol. 2: 209–210
iron metals   **Vol. 4:** 81, 168, 234, 399, 459, 480
irrationality   **Vol. 6:** 96, 127, 148, 154, 158, 263, 265, 271, 285, 288, 297
irreversibility (*see also* direction of time; entropy; *H*-theorem; reversibility; second law of thermodynamics)
  Vol. 6: 319, 326–329, 382, 400, 449, 452–454, 467–472
  Vol. 10: xxxvi, xlvii, 12, 104, 119, 128, 131, 137, 151, 152, 165, 166, 169, 212, 492, 493, 545, 549, 561, 562, 568
Isenstein, Harald   **Vol. 10:** 36p
Isenstein, Hildegard   **Vol. 10:** 36p
Ishino, M.   **Vol. 2:** 597
Ishiwara, J.   **Vol. 1:** 102
  Vol. 2: 451, 452
isomeric nuclei   **Vol. 9:** 13, 42, 67, 127, 239, 291–296, 317, 329, 653–657, 678

isotope separation   **Vol. 9:** 72, 398, 441, 465
isotopes (*see also* radioactive isotopes/substances)   **Vol. 1:** xxii, 581
  Vol. 2: 119–120, 213, 415–425, 590, 594, 597, 627, 629, 630
  Vol. 6: 377–378, 404, 458
  Vol. 8: 171f, 175f, 179, 562, 709, 787, 811
  Vol. 11: 12, 91, 229, 566, 567, 569, 575–583, 590, 591, 593, 595–597, 599, 603, 604, 613–616, 621–627
  Vol. 12: 42, 61, 67
 difference between series spectra of
  Vol. 3: 38, 453
isotopic
 elements   **Vol. 4:** 86, 431, 468
 indicators   **Vol. 10:** 69, 76, 104, 167, 385, 393, 400
isotropy   **Vol. 1:** 300
Israel (Danish journal)   **Vol. 11:** 376
Israel   **Vol. 11:** ii$p$, 372, 375, 376, 690–699, 734
Israel Institute of Technology   **Vol. 11:** 372
Itterbeek, A.F. van   **Vol. 12:** 38p
Ivanenko, D.D.   **Vol. 11:** 7p
Iversen, C.   **Vol. 10:** 288p
Iversen, R.   **Vol. 11:** 399, 408
Izvestia   **Vol. 11:** 8, 9, 196, 206, 207, 209, 211, 217

Jacobi, C.G.J.   **Vol. 3:** 85, 86, 95, 638
  Vol. 5: 243, 247, 277
  Vol. 10: 424
Jacobsen, C.   **Vol. 10:** 302n
Jacobsen, J.C. (brewer)   **Vol. 10:** 302n, 356, 369, 594
  Vol. 11: 339, 343, 401, 403, 411, 413, 448, 451, 456, 460, 464, 466, 471, 474
  Vol. 12: 152, 162–164, 175, 182, 190, 198, 202, 533
Jacobsen, J.C.G.   **Vol. 3:** 183
  Vol. 5: 95, 213–215
  Vol. 6: 44, 304, 438, 458
  Vol. 8: 25, 204ff, 214, 224, 231, 239, 245f, 256, 261, 530, 533, 536, 564–565, 567, 584f, 587, 649, 683, 698, 727, 729, 731f, 734ff, 783f, 786f, 795, 807, 809
  Vol. 9: 2, 13, 28, 62, 68, 140, 272, 479, 484, 536, 552, 560, 574f, 588, 590–591, 626, 628ff, 635, 637f, 640, 678

[95]

Vol. 11: 217, 220, 358p, 373, 513, 518, 521, 526, 545, 551, 653p
Vol. 12: 210, 215, 320n
Jacobsen, L. **Vol. 10:** 299n
Jäger, G. **Vol. 1:** 225, 236, 339, 348
Jaeger, W. **Vol. 1:** 145, 222, 225, 336, 339
James, W. **Vol. 6:** 27
    Vol. 9: 641, 643
    Vol. 10: xxviii, xlvii, 297, 540, 542
Jammer, M. **Vol. 5:** 3, 7, 81
    Vol. 6: xxii, 27
    Vol. 10: 297n, 300n
Japan Lectures, Bohr's, 1937 **Vol. 9:** 41
Jeanneret, R.L. **Vol. 7:** 328
Jeans, J.H. **Vol. 1:** xxx, 102, 133, 155, 174, 191, 192, 201, 203, 204, 206, 208, 232, 247, 255, 256, 258, 260, 263, 268, 271, 272, 273, 274, 299, 300, 313, 321, 322, 323, 345, 358, 364, 365, 366, 369, 371, 374, 375, 378, 379, 419, 422, 425, 426, 427, 430, 431, 474, 479, 528, 529, 530, 531
    Vol. 2: 24, 124, 125, 288, 293, 393, 588
    Vol. 3: 343, 352, 362
    Vol. 5: 315
    Vol. 8: 56
    Vol. 10: 393, 433
Jeffreys, J.G. **Vol. 12:** 286
Jensen, A. **Vol. 9:** *630*, 631
    Vol. 11: 358p, 406p, 518, 525
Jensen, B.W. **Vol. 11:** 358p
Jensen, H. **Vol. 11:** 243, 244, 264, 265
Jensen, J.H.D. **Vol. 9:** 28, 77–78, 79f, 619
Jensen, J.V. **Vol. 12:** 103
Jensen, P. **Vol. 8:** 567
Jentschke, W. **Vol. 8:** 405, 555, 567
    Vol. 9: 365
Jespersen, O. **Vol. 3:** 24
    Vol. 12: 528
Jessen, B. **Vol. 11:** 406p
Jessen, K. **Vol. 11:** 406p
jet
    distribution of velocity **Vol. 1:** 87
    effective mean velocity **Vol. 1:** 86ff
    photography of **Vol. 1:** 56
    production of **Vol. 1:** 47ff
    velocity of **Vol. 1:** 50
    vibration theory **Vol. 1:** 30ff, 88
Jette, E.R. **Vol. 5:** 29, 296, 379–380
Johannsen, O.F. **Vol. 1:** 4
Johannsen, W. **Vol. 3:** 286, 295

John Harling Fellowship **Vol. 12:** 285, 287
John Jay Hopkins Laboratory for Pure and Applied Science **Vol. 11:** 736
Johns Hopkins University **Vol. 9:** 39, 57
Joint Committee of Atomic Energy (U.S.) **Vol. 11:** 327n
Joint Danish Committees for Aiding Emigrants **Vol. 12:** 434, 435
Joliot-Curie, F. **Vol. 7:** 32p, 448, 450
    Vol. 8: 405, 468f, 564, 567, 665, 704f
    Vol. 9: 11f, 14, 67, 125, 134, 135ff, 152, 166, 174, 185, 216, 219, 309, 323, 365, 384, 386, 397, 407f, 536, 591–592, 652
    Vol. 10: 410, 412, 450
    Vol. 11: 243, 244, 515, 523, 531, 569, 646, 650
    Vol. 12: 111
Joliot-Curie, I., *see* Curie, I.
Jönsson, E. **Vol. 4:** 493, 530
Joos, G. **Vol. 5:** 114, 145, 148
Jordan, H. **Vol. 12:** 131p
Jordan, P. **Vol. 5:** 84, 223, 225, 229, 232, 258f, 265f, 280, 303, 308, 312f, 338, 362f, 478, 507
    Vol. 6: 3f, 8, 26, 28f, 31, 52, 69, 71, 95, 97, 124f, 138, 153, 155, 158, 160, 164ff, 171, 175, 184, 393, 423
    Vol. 7: 4, 8, 64, 126, 166n, 230, 262, 308, 347, 449
    Vol. 8: 207, 662f
    Vol. 9: 28, 305, 319, 567
    Vol. 10: 16–21, 36p, 401, 403, 426, 427, 441, 459, 461, 465, 466, 514n, 529n, 540, 542
Jørgensen, A.V. **Vol. 10:** 221, 224p
Jørgensen, J. **Vol. 10:** 15n, 16, 36p
Joseph Pilsudski University **Vol. 7:** 260
Jost, R. **Vol. 7:** 50, 216, 485, 500
Joule heat **Vol. 1:** 350
Joule, J.P. **Vol. 1:** 152, 238, 242, 290, 353, 395, 460, 490
    Vol. 11: 445
Juel Henningsen, E. **Vol. 9:** 639
Juel, C.S. **Vol. 5:** *219*
Juennson **Vol. 3:** 653, 654
Julius, W.H. **Vol. 3:** 496
    Vol. 4: 34, 702
Juliussen, J.C. **Vol. 12:** 241, 245
Jung, C.G. **Vol. 10:** 566, 567

Jupiter   **Vol. 12:** 428, 431
justice   **Vol. 10:** 97, 160, 188, 189, 202, 203, 216, 279, 286, 292
Juul, F.   **Vol. 9:** 630

Kalähne, A.   **Vol. 1:** 62, 65
Kalckar, F.   **Vol. 1:** xliii
   Vol. 8: 224, 226f, *228*, 229f, 458, 567, 644, 709–711, 795, 798ff, 803, 806, 808, 810f, 822
   Vol. 9: 10, *18*, 19f, 22, 25, 27, 28, 29f, 31, 32, 33, 34, 39, 48f, 144, 156, 180, 184, 192, 196, 200, 211, 223–264, 272f, 284, 299, 306, 320, 369, 413, 443, 489, 493, 499, 505, 510, 539f, 542, 544, 572f, 580ff, 592, 594ff, 600f, 609f, 676ff
   Vol. 10: 420, 583
   Vol. 12: 17p, 119, 386–388, 526
Kalckar, H.   **Vol. 10:** 488
Kalckar, J.   **Vol. 10:** 330, 460, 531n, 592, 593
   Vol. 12: vii, 533, 534
Källén, G.   **Vol. 10:** 455p, 556
   Vol. 12: 38p
Kallmann, H.   **Vol. 8:** 688
Kaluza–Klein theory   **Vol. 6:** 14, 112, 157–158, 422, 439, 441, 463
Kaluza, T.   **Vol. 6:** 158
Kamerlingh Onnes Laboratory (Leiden)
   Vol. 12: 355
Kamerlingh Onnes, H.   **Vol. 1:** 447, 448
   Vol. 3: 361, 363, 610, 612, 613, 615
   Vol. 5: 440, 443
   Vol. 10: 434, 440
   Vol. 12: 5, 8, 22, 24, 26, 28, 39
Kampmann, V.   **Vol. 11:** 363, 647, 651, 652n
Kang-Fuh-Hu   **Vol. 4:** 493, 530
Kanner, M.H.   **Vol. 9:** 382, 385, 397, 404
Kant, I.   **Vol. 10:** 4, 298, 345, 358, 359, 566
Kanto earthquake   **Vol. 12:** 503n
Kapitza, A.   **Vol. 11:** 53, 57, 58, 212, 213, 219, 220, 230, 231, 295, 296, 299, 302, 303, 308, 316, 317
Kapitza, P. (P. Kapitza's son)   **Vol. 11:** 230
Kapitza, P.   **Vol. 6:** 318, 348, 349
   Vol. 8: 679
   Vol. 9: 548, 612
   Vol. 10: 399, 414, 447
   Vol. 11: 9–11, 24, 32, 33, 38, 40, 41, 53, 54, 56–59, 59p, 60, 212–216, 218–222, 231, 238, 247, 251, 257, 261,

263, 273, 275, 286, 293, 295, 296, 299–302, 304, 305, 307, 310, 312, 315–319, 325, 329, 332, 333
κ (Bohr's constant)   **Vol. 8:** 219, 253f, 330, 441–442, 501
Kapur–Peierls method   **Vol. 9:** 49, 493, 508, 614f
Kapur, P.L.   **Vol. 9:** 28, 49, 493, 508
Karpinsky, A.P.   **Vol. 11:** 217, 224
Katcoff, S.   **Vol. 8:** 562
Katzir-Katchalsky, A. (letter has "Catalski")
   Vol. 10: 492
Kauffmann, H.   **Vol. 11:** 80n, 249, 282, 288, 290, 327, 330, 361, 362p, 563
Kay, W.   **Vol. 2:** 597
   Vol. 10: 398
Kayser, H.G.J.   **Vol. 2:** 230
   Vol. 4: 34, 35, 550, 576, 611, 631, 673, 702, 703, 704
   Vol. 5: 399
   Vol. 10: 375
Kayser "Gratulationsheft"   **Vol. 4:** 34ff
k (Bohr's constant)   **Vol. 8:** 9, 35, 59, 82, 97, 132, 492, 501, 832
k (constant appearing in papers on absorption)
   Vol. 2: 6, 27, 41–42, 62, 624
Keesom, W.H.   **Vol. 1:** 448
   Vol. 2: 618
Kelvin, Lord   **Vol. 1:** 67, 153, 174, 241, 242, 243, 244, 245, 290, 299, 352, 353, 354, 355, 356, 393, 415, 416, 442, 443, 447, 448, 460, 464, 483, 486, 488, 490
   Vol. 3: 625
   Vol. 11: 214, 438, 445
   Vol. 12: 25n
Kemble, E.C.   **Vol. 3:** 46, 82
   Vol. 4: 37, 39, 136, 221, 290, 349, 386
   Vol. 5: 5ff, 65, 296, 381–382, 503
Kendall, J.   **Vol. 5:** 380
Kennan, G.F.   **Vol. 11:** 327, 330
Kennard, E.H.   **Vol. 6:** 112, 157
Kepler, J.   **Vol. 12:** 371
Keplerian
  laws/orbit (*see also* electron orbit)
   Vol. 5: 157–171, 239, 243, 247, 276f, 406, 503
   Vol. 6: 379, 384, 388
   Vol. 10: 57, 174, 194, 377, 388–390, 395, 437
   Vol. 12: 44, 45

motion   **Vol. 8:** 269, 270, 278, 279, 343f, 347f, 430, 439

Kerr, C.   **Vol. 7:** 411

Kharkov, University of   **Vol. 11:** 7p, 203, 204, 212

Khruschev, N.   **Vol. 11:** 82

Kierkegaard, S.   **Vol. 6:** xxii, 27
  Vol. 10: 256, 265, 266n, 297, 298, 303
  Vol. 12: 14, 163n
  "Stages on Life's Road"   **Vol. 1:** 501, 503

Kikuchi, S. **Vol. 9:** 239

Killian, J.R.   **Vol. 11:** 367, 368p, 716, 721n
  Vol. 12: 15

kinematical
  concepts   **Vol. 7:** 322
  principles   **Vol. 7:** 320
  quantities   **Vol. 7:** 105, 154
  variables   **Vol. 7:** 392

kinematics (*see also* dynamics)   **Vol. 12:** 223, 224, 378

kinetic theory (*see also* statistical mechanics, thermodynamics)   **Vol. 5:** 8, 42, 129, 136, 188, 193, 202, 206, 273f
  Vol. 8: 76, 79, 91, 95, 269–270, 279, 463, 473, 564, 693
  of gases (*see also* statistical mechanics, classical)   **Vol. 7:** 252, 481

King's College (London)   **Vol. 12:** 4

King, W.L. Mackenzie   **Vol. 11:** 325, 329

Kingsford, R.J.L.   **Vol. 12:** 15, 509, 510

Kinsey, B.B.   **Vol. 9:** 201

Kirchhoff's law   **Vol. 2:** 175, 286–287
  Vol. 5: 32, 112
  Vol. 6: 382

Kirchhoff, G.R.   **Vol. 1:** xxvi, 246, 357, 447
  Vol. 2: 286, 287
  Vol. 3: 208, 232, 570, 586, 591
  Vol. 4: 446, 475
  Vol. 5: 132, 139
  Vol. 9: 303, 318

Kirkby, R.M.   **Vol. 2:** 138

Kirschbaum, H.   **Vol. 2:** 370

Kleeman, R.   **Vol. 2:** 28, 29
  Vol. 8: 60f

Klein paradox   **Vol. 7:** 37, 463, 466
  Vol. 9: 4, 110

Klein–Kaluza theory   **Vol. 6:** 14, 112, 157–158, 422, 439, 441, 463

Klein–Nishina formula/theory   **Vol. 7:** 4, 192, 193, 463, 466
  Vol. 9: 109

Klein–Rosseland collisions, *see* collisions of the 2nd kind

Klein, F.   **Vol. 5:** 312

Klein, G. (O. Klein's brother)   **Vol. 5:** 384f, 391, 394

Klein, M.J.   **Vol. 5:** vii, 24

Klein, O.   **Vol. 1:** xxxii
  Vol. 3: 24, 404, 469, 592, 631, 664
  Vol. 4: 7, 693
  Vol. 5: 14, 28–29, 70, 179, 195, 256, 262, 275, 278, 296, 316, 382–394
  Vol. 6: 14, 16, 24, 26, 28, 35, 43ff, 58, 60, 62, 97, 127, 155ff, 191, 194, 256, 273, 304, 305, 307, 308, 383, 422f, 425ff, 430, 432, 435, 437ff, 446f, 462, 478ff
  Vol. 7: 4, 37, 175, 186, 255n, 260, 261p, 269, 301, 434, 449, 459, 460, 474, 477, 480, 521
  Vol. 8: 26f, 215, 219ff, 224, 247, 262, 268, 270, 278, 644, 655f, 683, 692, 701, 703, 711–717, 769ff, 780, 782, 821
  Vol. 9: 18f, 86, 88, 110, 536, 547, 592–596, 676
  Vol. 10: xxx–xxxii, 110, 173, 275n, 400, 403, 426, 447, 455p, 461, 469, 470, 495, 518n, 525, 526, 533n–535n, 543, 547, 582
  Vol. 11: 354, 513, 522
  Vol. 12: 210, 215

Klibansky, R.   **Vol. 7:** 385, 386
  Vol. 10: xxxiv$n$, 480n

Klint, K.   **Vol. 12:** 106, 241, 246

Kluyver, J.C.   **Vol. 2:** 537
  Vol. 3: 652

Knauer, F.   **Vol. 6:** 344, 347

Kneser, H.O.   **Vol. 3:** 460

Knight, F.H.   **Vol. 10:** 82

Knipp, J.K.   **Vol. 8:** 240, 252, 407, 537, 540f, 556, 566f, 750, 765, 789

Knipping, P. **Vol. 3:** 180
  Vol. 4: 53, 64, 94, 132
  Vol. 8: 681

knowledge (*see also* unity of knowledge)
  Vol. 10: 40, 97, 152, 157, 179, 212, 229, 234, 254, 263, 276, 287, 292, 293, 302, 314, 320, 356, 369, 404, 510, 511, 568, 586, 588–591
  Vol. 11: 52, 53, 64, 146, 147, 151, 153, 185, 339, 343, 361, 369, 404, 414, 448, 450, 535, 538, 540, 543, 545, 547–551, 570, 641, 643, 691, 695, 698

Vol. 12: 59, 62, 63, 70, 75, 84, 93, 95, 152, 157, 158, 164, 170, 171, 184, 194, 234, 462, 464, 487, 529–533

Knudsen, E.   **Vol. 12:** 316, 321

Knudsen, M.   **Vol. 1:** xix, 109, 127, 528, 529
Vol. 2: 106, 319, 332, 535, 554, 555, 577, 589, 610, 613, 618, 619
Vol. 3: 285, 294, 361, 620, 629, 666, 688
Vol. 5: 47, 127*f*, 129, *136*, 137
Vol. 6: 36, 37, 415
Vol. 8: *13*, 14, 644, 717–718, 751
Vol. 9: 423*f*, 451
Vol. 10: 354, 368, 435
Vol. 11: 345, 353, 383*n*, 406*p*, 480, 483*n*, 484, 525*p*
Vol. 12: 112, 113, 290, 290*p*, 291–294, 296, 297, 300, 300*p*, 301, 302, 305–311, 314*p*, 315–324, 527–529

Knudsen, O.   **Vol. 12:** 4*n*

Kobylinski, H.   **Vol. 11:** 328, 331

Koch, H. von   **Vol. 5:** *384*, 386

Koch, H.H.   **Vol. 11:** 327, 328, 330, 331

Koch, J.   **Vol. 9:** 28, 621, *638–639*, 640
Vol. 11: 358*p*

Koch, J. van   **Vol. 3:** 175, 653, 654

Koch, P.P.   **Vol. 6:** 41, 435

Kockel, B.   **Vol. 7:** 46, 451*n*, 453

Koenigsberger, J.   **Vol. 1:** 145, 146, 227, 228, 230, 340, 341, 343
Vol. 2: 237

Kofoed, H.A.   **Vol. 10:** 360*n*

Kofoed-Hansen, O   **Vol. 8:** 590
Vol. 9: 528

Kohlrausch, F.   **Vol. 1:** 171, 223, 297, 336, 458

Kohlrausch, F.K.W.   **Vol. 8:** 209*f*, 644, 719–726

Kohlschmidt, W.   **Vol. 10:** 140

Kohn, W.   **Vol. 8:** 590
Vol. 9: 529

Kolowrat-Tscherwincki   **Vol. 1:** 64, 65

Komarov, V.L.   **Vol. 11:** 224*n*

Konen, H.   **Vol. 2:** 615
Vol. 4: 34, 35, 36, 550, 612, 702–705

König, H.   **Vol. 7:** 328

Kopfermann, H.   **Vol. 8:** 224, 714, 716

Korringa, J.   **Vol. 12:** 38*p*

Kossel, W.   **Vol. 1:** xxxi
Vol. 2: 88, 89, 335, 336, 411, 412, 413, 516, 517

Vol. 3: 9, 17, 22, 23, 178, 179, 189, 200, 322, 337, 393, 591, 688, 689, 691
Vol. 4: 4, 5, 13, 14, 48, 53, 55, 57, 80, 96, 104, 107, 170, 201, 202, 203, 227, 239, 249, 251, 253, 274, 275, 276, 309, 320, 322, 323, 388, 393, 398, 402, 403, 408, 410, 423, 457, 460, 479, 481, 490, 500, 523, 532, 600, 648, 661, 707, 708, 710, 711, 713
Vol. 5: 278
Vol. 6: 387*f*, 392
Vol. 8: 158*f*, 756
Vol. 10: 394, 418
Vol. 12: 45

Kovarik, A.F.   **Vol. 2:** 238
Vol. 5: 486

Kowarski, L.   **Vol. 9:** 384, 386, 397

K-radiation, *see* X-ray spectra; X-rays

Kramers, A.   **Vol. 12:** 360

Kramers, H.A. (*see also* Bohr–Kramers–Slater theory)   **Vol. 1:** xxxii, xxxiv, xxxv, xxxvii
Vol. 2: 342, 343, 344, 536–537, 573, 574, 595, 610, 611, 619, 620
Vol. 3: 3, 7, 8, 9, 10, 11, 12, 13, 15, 16, 17, 18, 21, 24, 25, 28, 29, 30, 31, 37, 41, 42, 44, 46, 62, 95, 124, 135, 144, 145, 151, 171, 180, 182, 184, 189, 190, 192, 194, 196, 198, 199, 200, 225, 264, 270, 279, 292, 300, 301, 316, 320, 331, 335, 394, 438, 440, 454, 467, 471, 481, 488, 505, 506, 516, 535, 545, 598, 615, 650, 664, 670, 671, 674, 675, 683, 689, 690
Vol. 4: 5, 6, 9, 11, 15, 16, 25, 36, 37, 38, 39, 41, 48, 50, 52, 63, 64, 94, 125, 136, 137, 142, 213, 219, 221, 222, 288, 290, 291, 358, 369, 371, 379, 386, 451, 452, 477, 593, 595, 607, 643, 644, 652–662, 669, 671, 673, 678, 679, 680, 688, 693, 699, 700, 719, 720, 723, 726, 731, 737, 738
Vol. 5: 9–13, 19*f*, 22, 26*ff*, 31*f*, 34*f*, 40–47, 50–54, 57, 62, 66*ff*, 75, 79, 84–88, 91, 94–95, 97*ff*, 102, 113, 146*f*, 150, 152, 156*f*, 165, 191, 204, 219*ff*, 223*f*, 226, 232*f*, 235*ff*, 240, 250*f*, 254*ff*, 260, 263, 265, 267, 270, 277, 279*f*, 281–285, 293, 296*f*, 303*f*, 306, 308, 310, 328*f*, 333*ff*, 340*ff*, 348, 355*ff*, 360*f*, 365*f*, 369, 371*ff*, 383*ff*, 389–397,

400f, 405, 407, 409f, 412ff, 418,
421f, 425f, 429f, 434–444, 448f,
452ff, 457–463, 465ff, 469f, 472, 474f,
477, 479–480, 481–482, 484f, 487,
492–493, 494ff, 498, 502, 507, 512f
  Vol. 6: 3, 35, 36, 37, 39, 44, 80, 88, 94,
  100, 124, 139–140, 153, 196, 197, 256,
  273, 304, 305, 308, 318, 393, 412, 415,
  417, 422, 425–430, 438ff, 457, 478
  Vol. 7: 32p, 48, 255n, 260, 300, 301,
  306, 348, 434, 461, 462
  Vol. 8: 30f, 36, 205f, 245f, 256f,
  363f, 372, 383, 385ff, 496, 530, 536f,
  566f, 645, 659f, 662, 673, 676, 694,
  726–739, 786f, 789, 821
  Vol. 9: 7, 10, 13, 25–27, 28, 134, 305,
  319, 536, 596–605
  Vol. 10: 396, 400, 401, 419, 425, 427,
  440, 441, 454
  Vol. 11: 325, 329, 513, 521, 628n
  Vol. 12: 6n, 17p, 115, 116, 210, 215,
  348, 348p, 349, 350, 350p, 351, 352,
  355–360, 364, 502, 504, 530
Kramers, S. (H.A. Kramers' daughter)
  Vol. 8: 728, 730f, 733
Krause, R.C.   **Vol. 12:** 240, 245
Krebs, C.   **Vol. 10:** 74
Krishnan, K.S.   **Vol. 5:** 60
Kristensen, M.   **Vol. 11:** 406p
Kristensen, P.   **Vol. 8:** 590
  Vol. 9: 528
Kristiania, University of, *see* Oslo, University of
Kroeber, A.L.   **Vol. 10:** 82
Krogh, A.   **Vol. 9:** 602, 604
  Vol. 11: 348, 349, 516, 524
Kronig, R. de L.   **Vol. 5:** 58, 60, 80ff, 91,
  220, 225, 229, 232–237, 279, 296, 336,
  362ff, 397, 449, 453, 460f, 508
  Vol. 6: 16, 304, 305, 404, 439f
  Vol. 7: 301
  Vol. 8: 655f, 714ff
  Vol. 9: 10, 94
  Vol. 12: 7, 8, 17p, 30, 33, 38p
Krüger, F.   **Vol. 1:** 355
  Vol. 3: 191
Krüger, T.   **Vol. 8:** 681
Krüger, W.   **Vol. 10:** 471n
Kudar, J.   **Vol. 9:** 7, 536, 605–606
Kuenen, J.P.   **Vol. 2:** 537
  Vol. 3: 652
  Vol. 4: 683

Kuhn, T.S.   **Vol. 2:** 105, 114
  Vol. 6: vi, 9, 324–325
  Vol. 8: 4f, 30, 206, 215, 218
Kuhn, W.   **Vol. 5:** 168, 171, 251f, 506
Kultichitsky, L.A.   **Vol. 8:** 633
Kunz, J.   **Vol. 1:** 240, 243, 273, 351, 354,
  379, 490
Küpferer, K.   **Vol. 2:** 237

labour, distribution of   **Vol. 11:** 123
labour, division of   **Vol. 12:** 183, 192, 193,
  438, 440
Ladenburg, R.   **Vol. 3:** 22, 37, 41, 414, 496
  Vol. 4: 4, 5, 28, 33, 203, 238, 239, 244,
  252, 275, 308, 323, 402, 403, 405, 459,
  480, 710–717, 737
  Vol. 5: 21, 26, 27–28, 42–43, 45f, 67,
  87, 106, 113, 256, 260, 263, 267, 279,
  296, 398–403, 440, 443
  Vol. 6: 124, 153, 390
  Vol. 8: 709f
  Vol. 9: 382f, 397, 404
Lagrange, J.L.   **Vol. 5:** 243, 247, 271, 273, 277
Lam, I.   **Vol. 12:** 17p
Lamarckism   **Vol. 10:** 543, 547
Lamb shift   **Vol. 7:** 34n, 49
Lamb, H.   **Vol. 11:** 213, 218
Lamb, W.E.   **Vol. 7:** 48
  Vol. 8: 239, 333, 406, 493, 502, 509,
  540, 556, 567, 645, 739, 748f
  Vol. 12: 38p
Lamettrie, J.   **Vol. 10:** 3n
Land og Folk (Danish newspaper)
  Vol. 11: 80, 189, 328, 331
Landau, E.   **Vol. 1:** xviii, 93
  Vol. 8: 767, 769
Landau, L.D.   **Vol. 7:** 7–12, 19, 20, 28, 59,
  80, 82, 101, 121, 123, 137, 139, 151,
  165, 172, 173, 177, 179, 182, 184, 185,
  188, 189, 191–193, 199, 229–235, 237,
  281n, 441, 442, 464, 466, 468
  Vol. 8: 215, 475
  Vol. 9: 8, 30, 33, 132, 260, 262, 264,
  314, 326, 499, 510, 568ff, 612
  Vol. 10: 407, 531, 532
  Vol. 11: 7, 7p, 10, 11, 223–226, 230,
  298, 302
Landé, A.   **Vol. 1:** xxxi
  Vol. 3: 9, 17, 18, 22, 23, 24, 39, 40,
  180, 181, 183, 278, 446, 527, 528, 530,
  555, 556, 558, 618, 664, 689

Vol. 4: 6, 7, 8, 9, 11, 37, 40, 48, 52, 57, 58, 63, 64, 65, 68, 69, 77, 78, 95, 105, 116, 123, 124, 125, 131, 133, 145, 157, 159, 203, 215, 217, 218, 219, 227, 276, 282, 288, 296, 379, 380, 391, 392, 424, 574, 575, 592, 593, 598, 599, 601, 602, 629, 630, 643, 646, 647, 648, 649, 719–723, 725

Vol. 5: 34, 233, 237, 251*f*, 255, 260*ff*, 266, 278*f*, 288*f*, 332, 422, 434, 437, 493

Vol. 6: 391

Vol. 8: 683

Vol. 10: 339

Lange, C.   **Vol. 10:** xxviii

Lange, H.O.   **Vol. 11:** 328, 331, 406*p*

Langebek, J.   **Vol. 11:** 409*n*

Langevin, P.   **Vol. 1:** 158, 159, 276, 277, 278, 381, 382

Vol. 2: 255, 256, 257, 259, 260, 263, 264, 363

Vol. 3: 32, 54, 147, 361, 389, 390, 392, 393, 394, 433, 525, 553, 613

Vol. 6: 36, 318, 359

Vol. 7: 31, 32*p*, 300, 302, 473, 474, 477

Vol. 9: 134, 583, 585

Vol. 10: 434, 446, 448

Vol. 11: 10, 213, 219–222

Vol. 12: 376*p*

Langmuir, I.   **Vol. 1:** 115

Vol. 2: 130, 221, 222, 481, 538–540, 590, 615

Vol. 4: 9, 48, 52, 53, 57, 80, 81, 105, 203, 204, 238, 244, 252, 276, 277, 308, 315, 323, 423, 457, 459, 479, 480, 699, 725, 726

Vol. 5: 239, 407, 486

Vol. 6: 36, 37, 415

Vol. 8: 780, 782

Vol. 12: 525

Langseth, A.   **Vol. 11:** 353, 406*p*, 478, 480–482, 484, 485

Vol. 12: 101, 127

language (*see also* linguistics)   **Vol. 7:** 273, 286*p*, 388, 404, 417

Vol. 12: 180, 188

common/daily/plain   **Vol. 10:** xxxiii–xl, xlii, xlix, 10, 14, 83, 84, 88, 89, 103, 127–129, 149, 157–160, 165, 177, 212,

243, 276, 277, 279, 307, 308, 404, 405, 428, 444, 545, 549, 574

Danish   **Vol. 10:** 254–256, 263, 264, 266

distinguishing humans from animals   **Vol. 10:** 245, 246

everyday   **Vol. 12:** 62, 69

forms of   **Vol. 12:** 161, 174

mathematics as refinement of   **Vol. 10:** 84, 142, 157, 405

of physics   **Vol. 10:** 151, 276

origin of   **Vol. 10:** xlv, 93, 105, 159, 168, 286, 291, 405

problems of   **Vol. 10:** xxxii–xxxiv, xxxvii, xxxix, xl, xlvii, xlix, 126, 483, 524, 526

rôle of   **Vol. 6:** xx, 41, 52, 158, 195, 205, 214, 262, 271, 283, 297, 360, 418, 448, 452

Lankester, E.R.   **Vol. 12:** 112, 286

Lao-Tse   **Vol. 6:** xx

Vol. 10: 60

Laplace, P.S.   **Vol. 1:** xvii

Vol. 5: 243, 247, 273

Vol. 10: 119

Vol. 12: 233

Lardy, P.   **Vol. 7:** 328

Larmor precession/frequency   **Vol. 2:** 257, 260, 325–326, 363, 367, 446

Vol. 8: 169, 172, 177

Larmor's formula   **Vol. 8:** 169, 173, 177

Larmor's theorem   **Vol. 5:** 289, 412, 414, 423, 427, 430, 432, 447, 451, 459, 461

Larmor, J.   **Vol. 1:** 8–10, 104, 105, 447, 528, 529, 530, 531, 542, 543

Vol. 2: 257, 260, 363, 446

Vol. 3: 54, 57, 146, 148, 150, 152, 157, 184, 361, 420, 421, 431, 434, 435, 445, 446, 525, 553, 558

Vol. 4: 215, 216, 365, 369, 370, 434, 469, 569, 573, 574, 625, 629

Vol. 5: 277

Vol. 6: 359, 375

Vol. 10: 339, 393–395

Laslett, L.J.   **Vol. 11:** 358*p*, 518*n*

Lassen, N.O.   **Vol. 5:** 95

Vol. 8: vi, 232, 235, *244*, 246*f*, 252, 257*ff*, 263*f*, 523, 541, 555*ff*, 567, 576*f*, 581, 583–588, 589, 591*f*, 598*f*, 602*f*, 605*ff*, 615*ff*, 620*f*, 625, 645, 731*f*, 740–748, 789, 824

Vol. 9: 479, 484, 639

Vol. 11: 358*p*, 495, 498, 499

latent heat of evaporation   **Vol. 9:** 210, 252, 277, 313f, 326f, 426, 453

Lattes, C.M.G.   **Vol. 8:** 554, 567

lattice waves/vibrations   **Vol. 12:** 8, 23, 27, 28, 530

Latyshev, G.D.   **Vol. 8:** 633

Laub, J.J.   **Vol. 5:** 27

Laue–Bragg method   **Vol. 10:** 393, 437

Laue, M. von   **Vol. 2:** 428, 578
   Vol. 3: 583, 691
   Vol. 4: 442, 473, 659
   Vol. 6: 387
   Vol. 7: 252
   Vol. 8: 88, 104
   Vol. 10: 387, 435, 436
   Vol. 11: 29
   Vol. 12: 43, 287

Laurits Andersen Foundation
   Vol. 12: 182, 190

Lauritsen, C.C.   **Vol. 9:** 201ff, 635, 637
   Vol. 12: 129p

Lauritsen, S.   **Vol. 12:** 129p

Lauritsen, T.   **Vol. 2:** 10
   Vol. 8: 232, 234f, 237, 239f, 245, 255, 308, 315, 318, 320, 323–326, 328, 331f, 336, 375, 405f, 467, 486, 555, 566, 585, 587, 645, 731f, 748–750, 822f
   Vol. 9: 413, 443, 635, 637

Lauritzen, S.   **Vol. 11:** 726

Lautrup, C.C.   **Vol. 1:** 103, 293, 534, 535

Lavoisier, A.-L.   **Vol. 10:** 54, 208
   Vol. 11: 657, 662

Lawrence, E.O.   **Vol. 7:** 32p
   Vol. 9: 134, 135, 217, 221, 590f
   Vol. 10: 413, 449
   Vol. 11: 436p, 517, 518, 524, 526, 531, 568

Lawrence, T.E.   **Vol. 11:** 692, 697

lead   **Vol. 11:** 574–576, 578, 587, 589–591, 596
   absorption in   **Vol. 2:** 30, 35
   spectrum of   **Vol. 2:** 419

League of Nations   **Vol. 10:** 536n

Leahy, W.D.   **Vol. 11:** 262

Leary, L.   **Vol. 10:** xxxviii$n$, 559n

LeBaron, R.   **Vol. 11:** 563

Lecher, E.   **Vol. 1:** 184, 308

lectures, Bohr's (*see also individual entries*)
   Copenhagen University   **Vol. 8:** 16, 167–179, 245, 250, 391–401, 820, 823

Danish Physical Society   **Vol. 8:** 11, 23f, 73–106, 244, 818, 820

Göttingen   **Vol. 8:** 692, 786

Manchester   **Vol. 8:** 16, 22, 107–114, 718, 819

Princeton University (*see also* Princeton)
   Vol. 8: 232, 252, 301–306, 419–422, 823

lectureship at University of Copenhagen
   Vol. 2: 4

Lederer, E.L.   **Vol. 1:** 290, 344, 395, 399

Lee, T.D.   **Vol. 10:** 454

Leeds, colloquium at   **Vol. 2:** 628

Leeuwen, H. J. van   **Vol. 1:** 93
   Vol. 2: 338
   Vol. 3: 193

Legendre, A.M.   **Vol. 5:** 271

Legros, A.   **Vol. 10:** xxvii$n$

Lehmann, F.E.   **Vol. 7:** 328

Lehmann, I.   **Vol. 1:** 520, 521, 536, 537

Lehmann, J.   **Vol. 12:** 126n, 394, 395

Lehmann, O.   **Vol. 12:** 152, 164

Leibniz, G.W.   **Vol. 7:** 376
   Vol. 10: xliv, 7, 566

Leiden   **Vol. 5:** 28, 68, 227, 228f, 234f, 238, 289, 326, 329ff, 484, 486, 497, 507
   Vol. 6: 21, 44, 415, 418
   Vol. 8: 22, 667, 668, 672f, 675f, 726, 727, 728, 729
   Vol. 9: 36, 567, 602, 603, 604
   lecture in   **Vol. 3:** 15, 16, 201–216

Leiden, University of   **Vol. 12:** 5, 9, 115, 116, 348, 350, 351, 355, 356

Leipunski, A.I.   **Vol. 9:** 652

Leipzig   **Vol. 8:** 650, 651, 655f, 701, 703

Leipzig, University of   **Vol. 11:** 15
   offer of professorship to Heisenberg
   Vol. 5: 376, 378
   physics institute   **Vol. 12:** 5

Leithäuser, G.   **Vol. 2:** 199

Lemmerich, J.   **Vol. 6:** 11

Lenard, P.   **Vol. 1:** 11, 81, 82, 83, 87, 88
   Vol. 2: 308, 309
   Vol. 4: 429, 467
   Vol. 5: 274
   Vol. 6: 374
   Vol. 8: 721f, 725

Lenin, V.I.   **Vol. 11:** 217

Leningrad Lecture, Bohr's   **Vol. 7:** 338, 520

Lennard-Jones, J.E.   **Vol. 10:** 399

Lenz, W. **Vol. 3:** 656, 658, 687, 691
  Vol. 4: 48, 51, 52
  Vol. 5: 256, 262, 278, 346
  Vol. 6: 41, 344, 435
Leon, X. **Vol. 10:** 509, 510
Leprince-Ringuet, L. **Vol. 11:** 711*n*
Lergaard, K. **Vol. 10:** 366*p*
Leslie, M.S. **Vol. 8:** 752
lesson
  from atomic physics **Vol. 10:** xxiii, xxiv,
    xxvi, xxxii, xxxiii, xlix, 6, 39, 60, 70,
    76, 84, 90, 92, 94, 95, 102, 104, 109,
    111, 112, 116, 119, 122, 151, 152, 157,
    159, 174, 179, 182, 184, 189, 193, 197,
    199, 200, 207, 241, 275, 276, 278, 285,
    291, 539, 541, 544, 548, 573, 587, 591
  in other respects **Vol. 10:** 30, 44, 83, 96,
    158, 182, 189, 198, 247, 253, 256, 259,
    262, 265, 270, 314, 318, 320, 322, 573
level
  density **Vol. 9:** 15, 21*f*, 24, 27, 30, 38,
    41*f*, 158, 169, 177, 208, 258–259, 260,
    264, 268*f*, 312, 325, 337, 345, 351,
    353, 357, 361, 375, 384*f*, 389, 392,
    401, 595*f*, 603, 605
  distribution **Vol. 9:** —
  width **Vol. 9:** 15, 21, 32, 43, 47*f*, 154*f*,
    170, 178, 187, 209, 211, 249, 255, 298,
    337, 345, 375*ff*, 392*f*, 398, 421, 449,
    490–491, 492, 501, 507*f*, 512, 519,
    616, 666
Levi, H. **Vol. 7:** 9*p*
  Vol. 9: 28, 46
  Vol. 11: viii, ix, 358*p*, 460*n*
Levin, H. **Vol. 10:** 82
Lewis, E.P. **Vol. 1:** 585
Lewis, G.N. **Vol. 2:** 618, 619
  Vol. 4: 48, 57, 80, 105, 203, 204, 244,
    276, 315, 405, 457, 479, 661
  Vol. 5: 278
  Vol. 6: 46, 387, 393
  Vol. 8: 780, 782
  Vol. 10: 496
  Vol. 12: 45
Liberation Council (Denmark)
  Vol. 11: 702, 703
Licht, E. de F. **Vol. 10:** 74
Lie, T.H. **Vol. 11:** 79, 82, 188, 326, 328,
    330, 331
Liebenow, C. **Vol. 1:** 231, 344
Liebert, G. **Vol. 3:** 135, 182

Liège **Vol. 8:** 704*f*, 714, 716
life
  as elementary fact **Vol. 10:** 5, 10, 11, 23,
    24, 34, 92, 104, 166, 199
  complexity/diversity of **Vol. 10:** 29, 34,
    135, 150–152, 168, 215, 248, 279, 502,
    504, 574
  concept of **Vol. 10:** xxxii, 6, 9, 10, 29, 33,
    45, 47, 53, 96, 118, 151, 152, 167*n*,
    184, 214, 246, 279, 344, 358, 467, 468,
    482, 485, 486, 493, 503, 504, 519, 522,
    544, 545, 548, 549, 553, 559, 560, 582,
    584, 587, 590
  correlation with atomic physics **Vol. 10:**
    10, 19, 60, 61, 180
  daily **Vol. 10:** 85, 93, 105, 141, 148, 152,
    157, 174, 193, 219, 248, 254, 263, 276,
    277, 342, 357, 480, 481, 545, 549, 573
  definition/description/explanation of
    Vol. 10: xlii, xlviii, 3–5, 8–11, 13, 14,
    17, 18, 23, 24, 33, 35, 45, 55, 68, 70,
    71, 77, 104, 116, 122, 150, 151, 167,
    168, 174, 179–181, 199, 231, 235, 253,
    262, 501, 502, 504, 539, 542, 555, 559,
    560, 574
  force (*see also* entelechy) **Vol. 10:** 117
lifetime **Vol. 5:** 17, 32*f*, 37*f*, 53, 56, 58–59,
    64, 73, 106, 110–112, 124, 147,
    153–154, 189, 203, 301, 317, 344, 358*f*,
    396, 416*f*, 419, 421, 491, 505–506
  Vol. 6: 13, 133, 138, 143, 156, 405,
    458, 460
  Vol. 8: 25, 204, 261, 571*f*, 575, 577,
    585, 587, 592, 598, 605, 619, 621*f*,
    661, 734, 736
Lifshitz, E.M. **Vol. 7:** 281*n*
light
  as tool of observation **Vol. 10:** 29
  cone **Vol. 7:** 494
  monochromatic **Vol. 7:** 344, 476, 479
  nature of **Vol. 10:** 29, 33, 120, 175, 244,
    350, 363, 432, 496, 503, 504
    Vol. 11: 726
  propagation of **Vol. 7:** 111, 158, 348, 400,
    401, 406, 414, 415, 419
    Vol. 10: 85, 211
  quanta, hypothesis of **Vol. 3:** 209, 234*ff*,
    244, 374, 412, 491*ff*, 570*ff*
    Vol. 4: 436, 470
  quantum, *see* photon
  rays **Vol. 11:** 646, 648

[103]

scattering, *see* scattering of light
signal   **Vol. 7:** 109, 117, 157, 163
simplicity of   **Vol. 10:** 29, 33
treatment   **Vol. 10:** 21, 29, 35
velocity of   **Vol. 7:** 27, 43, 60, 63, 66, 86,
   87, 123, 124, 126, 127, 141, 142, 198,
   219, 225, 267, 317, 364
   Vol. 10: 42, 86, 104, 158, 277, 350,
   363, 407
   Vol. 11: 670
   Vol. 12: 14, 378, 428, 431
wave, *see* wave
Light Therapy, International Congress on, 1932
   Vol. 10: 21, 35, 164, 530, 532
   Vol. 11: 337
Lilienthal, D.E.   **Vol. 11:** 66
Lillie, R.S.   **Vol. 10:** 17, 19
Lindemann, F.A. *see* Cherwell, Lord
Linder, A.   **Vol. 7:** 322
Linderstrøm-Lang, K.U.   **Vol. 11:** 297n, 353,
   470–475, 480, 484, 729
   Vol. 12: 101, 127
Lindh, A.E.   **Vol. 4:** 499, 532
Lindhard, J.   **Vol. 7:** 229, 234–238, 489,
   492, 522
   Vol. 8: vf, 24, 258–265, 574, 581,
   584f, 587, 589, 592, 593–625, 823f
   Vol. 9: 51, 79, 616f, 621, 670, 679
   Vol. 10: 589
   Vol. 12: 530
Lindsay, R.B. **Vol. 3:** 601
   Vol. 4: 493, 530
line
   shape   **Vol. 9:** 45, 611–612
   spectra (*see also* spectral lines)   **Vol. 6:** 52,
      94, 124, 138, 143, 152f, 173, 226, 228,
      240, 242, 311f, 333, 349, 379f, 382ff,
      392, 395f, 457
      Vol. 7: 366, 402, 415
      Vol. 12: 43–45, 344, 345, 357, 381, 383
   spectra and atomic structure   **Vol. 4:** 34ff,
      549–610, 611–656
   width   **Vol. 5:** 32f, 58, 64, 111f, 114,
      323f, 358f, 416f, 419, 421, 491, 496,
      497, 504, 506
      Vol. 6: 13, 133, 138, 143, 156, 173,
      396, 460
      Vol. 7: 36, 177, 187, 188, 474, 478
      Vol. 9: 45, 108, 298, 611
Line-Spectra, On the Quantum Theory of
   Vol. 3: 3ff, 65–200

correspondence related to   **Vol. 3:** 10ff
German translation of   **Vol. 3:** 9
linear harmonic vibrator   **Vol. 3:** 49, 78
linguistics (*see also* language)   **Vol. 10:** 141,
   259, 269, 300
Linstead, H.   **Vol. 10:** 146p
liquid helium   **Vol. 9:** 34, 423, 450,
   582–583, 584
liquid-drop model   **Vol. 9:** 16f, 27, 29f, 32, 34,
   36–37, 40, 44f, 47, 56, 67, 69–70, 74, 76,
   79, 81f, 141, 193–194, 216, 220, 241,
   261, 312, 325, 326, 336, 342, 344, 365f,
   369–374, 386, 389, 401, 407f, 416, 419,
   421, 423–427, 432f, 445, 448ff,
   451–454, 458f, 523f, 544, 559, 567,
   582–583, 584, 607f, 616, 620, 666ff, 671
   Vol. 12: 12, 54
literature   **Vol. 10:** 255, 259, 264, 265,
   270, 300
lithium   **Vol. 10:** 20, 409–411, 413, 449
   Vol. 11: 568, 614, 623
   atom, structure of   **Vol. 2:** 134, 202–204,
   229, 315, 405, 625
   spectrum of (*see also* alkali metals, spectra
      of; alkali spectra)   **Vol. 2:** 202–203,
      361, 370, 405, 407, 559, 562, 596
      Vol. 3: 276ff
   stopping in   **Vol. 5:** 181, 197
Littlewood, J.E.   **Vol. 1:** xviii, 520, 521, 524,
   525, 532, 533, 536, 537, 544, 545, 562,
   563, 564, 565
Litvinov, M.   **Vol. 11:** 10, 213, 217, 219–222
Liveing, G.D.   **Vol. 10:** 373
Livens, G.H.   **Vol. 1:** 116, 121, 122, 474,
   476–480
Liverpool Meeting, *see* British Association for
   the Advancement of Science
Liverpool, University of   **Vol. 11:** 227
living organisms   **Vol. 7:** 394
Livingston, M.S.   **Vol. 8:** 257, 428, 519, 524,
   549, 567
   Vol. 9: 248
Livingstone, R.   **Vol. 10:** 82
Lloyd George, D.   **Vol. 8:** 673
Lo Surdo, A.   **Vol. 2:** 321, 511
   Vol. 3: 175, 599
Lobachevsky, N.I.   **Vol. 12:** 372
Lockyer, N.   **Vol. 2:** 276
Loewenfeld   **Vol. 1:** 65

logic (*see also* multivalued logics) **Vol. 10:**
    xxxiv, xxxv, xxxvii, 84, 129, 130, 141,
    142, 157, 298, 316, 321, 483, 485, 486,
    502–505, 510, 512, 513, 544, 546, 548,
    550, 555, 562, 563, 566–569, 571, 572
logical empiricism/positivism
    Vol. 10: xli, 14–16, 18–20
Lögstrup, M.   **Vol. 8:** 787
Lohuizen, T. van   **Vol. 5:** 288
Lomholt, A.   **Vol. 8:** 250
    Vol. 11: 399, 409
    Vol. 12: 319, 323
London lecture, Bohr's
    1936   **Vol. 9:** 19, 21f, 157–158, 641, 643
    1939   **Vol. 9:** 360
London, F.   **Vol. 6:** 397
    Vol. 10: 403, 427, 443, 447
    Vol. 12: 38p
Longinus, D.   **Vol. 12:** 127, 514
longitudinal mass, *see* relativistic mass
Lord Mayor of London   **Vol. 11:** 726
Lorentz
    dispersion formula   **Vol. 5:** 318
    force   **Vol. 6:** 309, 334ff, 339, 341ff
    jubilee   **Vol. 5:** 68, 226, 227, 229, 234,
        238, 327, 329, 406, 484, 486
    Jubilee Meeting, 1925   **Vol. 6:** 21, 418
    triplet   **Vol. 3:** 150, 158, 161, 269, 280,
        526, 554
        Vol. 4: 370, 569, 574, 626, 629
Lorentz Kamerlingh Onnes Memorial
    Conference, 1953   **Vol. 12:** 5,
    38p, 530
Lorentz–Lorenz formula   **Vol. 5:** 137
Lorentz theory of electrons, *see* electron theory
Lorentz, H.A.   **Vol. 1:** 94, 95, 97, 102, 124,
    132, 134, 136, 138, 139, 141, 142, 143,
    144, 146, 150, 159, 173, 174, 175, 176,
    179, 197, 198, 199, 203, 214, 215, 218,
    221, 222, 223, 224, 225, 226, 227, 228,
    229, 230, 231, 233, 236, 237, 239, 242,
    244, 246, 247, 259, 260, 272, 273, 284,
    285, 286, 287, 298, 299, 300, 301, 303,
    309, 318, 319, 320, 321, 329, 333, 335,
    336, 337, 339, 340, 341, 342, 343, 344,
    345, 348, 349, 351, 354, 355, 357, 358,
    365, 367, 368, 378, 389, 390, 391, 393,
    401, 402, 404, 407, 413, 414, 415, 416,
    417, 418, 446, 447, 448, 449, 451, 453,
    458, 459, 460, 462, 474, 476,
    502, 503

Vol. 2: 124, 237, 287, 288, 295, 325,
    362, 367, 446, 537, 569, 570, 571, 572
Vol. 3: 28, 29, 30, 32, 33, 34, 35, 50,
    146, 150, 158, 161, 202, 231, 265, 281,
    361, 363, 367, 388, 390, 391, 419, 420,
    421, 492, 609, 610, 612, 614, 615, 622,
    623, 624, 625, 626, 628, 629, 652,
    666–668
Vol. 4: 15, 29, 44, 214, 366, 434, 435,
    451, 469, 477, 562, 569, 620, 658
Vol. 5: 68, 93, 130, 138, 227, 274,
    277, 326
Vol. 6: 35, 36, 37f, 50, 359, 375,
    393, 415
Vol. 8: 22, 668, 670, 673, 675
Vol. 10: 337, 338, 375, 393, 394, 433,
    437, 440, 441, 445, 446
Vol. 12: 5, 8, 39, 42, 355, 356, 358,
    359, 371
Lorenz law   **Vol. 1:** 144, 339
    Vol. 5: 137
    deviations from   **Vol. 1:** 342
Lorenz, K.   **Vol. 7:** 411
Lorenz, L.V.   **Vol. 1:** 145
    Vol. 5: 130, *137*
    Vol. 10: 350, 363
Lorenzen   **Vol. 11:** 358
Lorrain Smith, J.   **Vol. 1:** 106, 534, 535
    Vol. 12: 108
Los Alamos   **Vol. 9:** 50
Los Alamos National Laboratory   **Vol. 11:** 16,
    17, 22, 33, 34, 39, 41–43, 86, 87, 89,
    250, 258, 259, 267, 269, 270, 273, 274,
    286, 287, 310n, 361, 564
loss of electrons, *see* capture/loss of electrons
Louw, A.   **Vol. 10:** 250p
love (*see also* charity) **Vol. 10:** 188, 202, 279,
    286, 292
Lucky Lottery (Lykke-Lotteriet)   **Vol. 11:** 735
Lucretius   **Vol. 10:** 165, 207
Ludwig, C.   **Vol. 1:** xviii
    Vol. 10: 5
Ludwig, K.F.W.   **Vol. 6:** xxi
Lugeon, J.   **Vol. 7:** 328
Lummer, O.   **Vol. 1:** 273, 378
Lund, A.   **Vol. 10:** 299n
Lund, conference in   **Vol. 3:** 16ff
Lund, G.   **Vol. 12:** 17p
Lund, H.   **Vol. 5:** 387, 389
Lund, T.   **Vol. 3:** 286, 295

Lund, University of   **Vol. 5:** 383, 385
    Vol. 11: 733
Lunding, E.   **Vol. 10:** 299*n*
Lundsgaard, E.   **Vol. 11:** 406*p*, 516, 524
Luria, S.   **Vol. 10:** 23
Luttinger, J.M.   **Vol. 7:** 216, 485
Lyell, C.   **Vol. 11:** 445
Lyman series   **Vol. 2:** 116, 169, 386
    Vol. 3: 207
Lyman, T.   **Vol. 2:** 386, 399
    Vol. 3: 207, 424
    Vol. 4: 16, 37, 133, 440, 472, 594, 644, 659
    Vol. 5: 6*f*, 278, 297, 404
    Vol. 12: 43
Lyshede, J.M.   **Vol. 9:** 621

Maaløe, O.U.   **Vol. 10:** 487, 488
Maar, Mrs   **Vol. 5:** 383, 385, 502
    Vol. 6: 24, 194, 425, 446*f*
MacDonald, D.K.C.   **Vol. 12:** 38*p*
Mach, E. **Vol. 2:** 531
    Vol. 10: 15, 17
    Vol. 12: 101*n*, 526
Machin, K.E.   **Vol. 10:** 126
machine, *see* automata/machines
MacIver, R.M.   **Vol. 7:** 411
Mackeprang, M.   **Vol. 10:** 250*p*
MacKinnon, E.   **Vol. 5:** 61*f*
MacLeish, A.   **Vol. 7:** 411
    Vol. 10: 82
Madelung, E.   **Vol. 1:** 566, 567
    Vol. 2: 127, 237
    Vol. 4: 720
Madsen, B.S.   **Vol. 8:** 428, 565, 567
Madsen, C.B.   **Vol. 8:** 519
Madsen, T.   **Vol. 10:** 221
Madsen-Mygdal, T.   **Vol. 10:** 250*p*
Madvig, J.N.   **Vol. 11:** 339
    Vol. 12: 107, 248, 249
magic numbers   **Vol. 9:** 77*f*, 80, 619
magnesium   **Vol. 12:** 45
magnesium, spectrum of   **Vol. 2:** 503, 507
magnetic
  charge   **Vol. 7:** 27, 198
  field (*see also* field, electromagnetic)
    Vol. 7: 9, 83, 139, 213, 214, 233, 330, 346
    Vol. 11: 517, 524, 566, 568, 576, 591
    Vol. 12: 24, 28, 32, 34, 35, 42, 45, 60, 66, 340, 341, 356, 368
  field, effect of   **Vol. 1:** 158, 382*ff*, 395
  force/intensity   **Vol. 7:** 30, 103, 153, 201, 203, 224
  pole strength distribution   **Vol. 7:** 75, 91, 117, 134, 145, 162
  properties of ions   **Vol. 4:** 170*ff*, 239*ff*, 308*ff*
magnetic moment   **Vol. 2:** 96, 99, 128–130, 233, 377, 424
    Vol. 5: 233, 236, 238*f*, 283, 285, 288–289, 329*f*, 339, 406, 412, 414, 423*f*, 427*f*, 484, 486, 508
    Vol. 7: 175, 176, 181, 187, 190
    Vol. 9: 13, 94, 105–106, 112, 125*f*, 131, 560, 635, 637
    Vol. 12: 45, 48, 386, 387
  of atom   **Vol. 6:** 166, 267, 291, 311, 333, 341, 348*f*, 392
  of electron (*see also* electron spin)
    Vol. 6: 24, 52, 136, 144, 158, 267, 291, 307, 309, 310*f*, 331–336, 337–349, 392, 394–395, 425, 442*ff*, 446, 479
  of nucleus   **Vol. 6:** 392, 404
  of proton (*see also* spin of proton)
    Vol. 6: 136, 158
magnetism   **Vol. 10:** 342, 344, 346, 347, 351, 357, 358, 360, 364, 446, 447
    Vol. 11: 311, 313, 568, 657, 662, 670
    Vol. 12: 20, 220, 348, 351
  molecular   **Vol. 2:** 237, 259–262, 264–265
  no classical explanation
    Vol. 1: 382, 395, 519, 565, 581
  theory of   **Vol. 2:** 111, 115, 128–130, 233, 237, 253–265, 363, 377–379, 395, 530, 581, 588, 626, 631
magneton   **Vol. 2:** 128–130, 233, 237, 254–265, 328, 377, 588, 626
    Vol. 7: 181, 191
    Vol. 9: 13, 106, 126, 131
Maier-Leibnitz, H.   **Vol. 5:** 95, 215
    Vol. 9: 13
Main Smith, J.D.   **Vol. 4:** 40, 41, 42
    Vol. 5: 260, 266, 279
    Vol. 6: 391
    Vol. 12: 48
Maiski, I.M.   **Vol. 11:** 212*n*
Makower, W.   **Vol. 1:** 568, 569, 570, 571
    Vol. 2: 374, 428, 501, 619
    Vol. 3: 239
    Vol. 8: 4, 13, 21, 667, 672, 696*f*

Vol. 10: 384, 396, 397
Vol. 12: 286, 287
Malik, C.H.   **Vol. 10:** 82
Malmer, I.   **Vol. 2:** 336, 412, 630
Malpighi, M.   **Vol. 10:** 55
Manchester   **Vol. 6:** 320, 377f, 385
    Vol. 8: 4f, 7, 13, 21, 24, 71, 160,
    212ff, 216, 223, 226, 649, 664, 666f,
    670, 675, 698, 752, 754, 755, 791, 794,
    804, 806ff, 818
    Vol. 10: 383, 384, 387, 393, 396–400,
    408, 409, 413, 435, 592, 593
  Bohr's colloquium at   **Vol. 2:** 628, 629, 631
    Vol. 8: 16, 22, 107–114, 819
  Bohr's experiments in   **Vol. 2:** 3–4, 11–14,
    576, 624
    Vol. 8: 4f, 13, 21, 37–41, 818
Manchester Guardian   **Vol. 11:** 328, 331
Manchester Hospital   **Vol. 11:** 726
Manchester, University of   **Vol. 2:** 8, 39, 90,
    330–333, 425, 428f, 433, 508, 542,
    544, 558, 561, 564, 566, 576, 589, 593,
    603, 624
    Vol. 11: 692, 697
    Vol. 12: 108, 111, 112, 114, 122, 144,
    149, 150, 160, 173, 209, 210, 213, 214,
    265, 266, 281, 285, 287
  award of honorary degree   **Vol. 5:** 88
  physical laboratory   **Vol. 12:** 108, 119, 138,
    140, 149, 150, 215n, 274, 282, 286
Mandelshtam, L.I.   **Vol. 11:** 230
Mandelstam, S.   **Vol. 10:** 455p
Manhattan Project, *see* atomic bomb project
  (American–British)
Manneback, C.   **Vol. 7:** 520
    Vol. 12: 17p
Mantere, O.   **Vol. 11:** 505n
March, A.   **Vol. 8:** 811f
Margaretha, Princess   **Vol. 11:** 289n
marine zoology   **Vol. 12:** 286
Marke, A.W.   **Vol. 2:** 606
Marks, H.S.   **Vol. 11:** 325–327, 329, 330
Marsden, E.   **Vol. 2:** 75, 76, 96, 97, 98, 99,
    161, 501, 580, 614, 615, 617
    Vol. 6: 379
    Vol. 8: 4, 8, 110, 114, 145f, 163, 170,
    174, 178, 645, 667f, 672, 676, 751–752
    Vol. 10: 330, 384, 398
    Vol. 12: 111, 280p
Marshall Plan   **Vol. 11:** 69

Marshall, G.C.   **Vol. 11:** 4, 5, 65, 69, 71, 72,
    74–76, 159, 160, 163, 169, 320, 321,
    326, 327, 330
Märtha, Crown Princess   **Vol. 11:** 282
Marvel, J.   **Vol. 11:** 67, 68, 77, 156, 322, 323,
    326, 327, 329, 330
Marx, K.   **Vol. 11:** 196, 200
Mason, M.   **Vol. 8:** 753
mass
  correction, *see* reduced mass
  defect   **Vol. 9:** 12, 34, 52, 67, 93f,
    112–113, 153, 208, 231, 234ff, 307,
    321, 367ff, 372, 470, 576, 582, 584
  equivalence with energy   **Vol. 7:** 236,
    365, 371
    Vol. 11: 568, 576, 591, 613, 622, 659,
    664, 670
    Vol. 12: 374, 375, 378
  inertial (*see also* inertial mass)
    Vol. 11: 672
  number (*see also* atomic weight)   **Vol. 9:**
    42, 66, 193, 197, 274, 280, 306f, 311,
    313, 317, 321, 324ff, 329, 344f, 350ff,
    357, 367f, 371f, 380, 384, 414f, 417,
    428, 431, 438, 440–441, 444ff, 455,
    457, 463ff, 478, 523, 553
    Vol. 11: 576, 591
  point   **Vol. 7:** 60, 124, 178, 188, 219
  ratio in fission, *see* fission, mass ratio in;
    symmetrical/asymmetrical
  ratios   **Vol. 12:** 59, 60, 64, 65
  relativistic, *see* relativistic mass
  transport   **Vol. 7:** 182, 191
Massachusetts Institute of Technology
    Vol. 11: 367, 518n, 716n, 720–722, 735
    Vol. 12: 15
Massey, H.S.W.   **Vol. 6:** 347
    Vol. 7: 242n
    Vol. 8: 440, 452, 567, 634, 709f
    Vol. 9: 51, 494, 513
master's degree (Bohr's 1909)   **Vol. 12:** 4n,
    135, 136, 138, 140, 144, 149, 150,
    209, 213
materialism   **Vol. 10:** 64
mathematical physics   **Vol. 10:** 406, 440
    Vol. 11: 670
    Vol. 12: 363
mathematics   **Vol. 10:** xl, 84, 141, 142, 157,
    174, 175, 178, 189, 194, 316, 321, 405,
    415, 426, 479, 481, 502, 503, 509, 510,
    547, 551, 556, 564–566, 570, 571, 574

Vol. 11: 369, 372, 541, 549, 669, 671, 672, 734
Vol. 12: 90, 233, 382, 384
Mathieu, E.   **Vol. 1:** 17
matrix
   mechanics (*see also* quantum mechanics)
      Vol. 6: 3, 8, 14*f*, 29, 31, 33, 39*f*, 52, 71, 95*ff*, 125*ff*, 137, 152*ff*, 157, 180, 184, 393, 417*f*, 424, 433, 462*f*
      Vol. 7: 4
      Vol. 12: 11
   theory   **Vol. 12:** 10
Mattauch, J.   **Vol. 11:** 243, 244
matter
   constitution/constituents/structure of
      Vol. 7: 307
      Vol. 11: 123, 133, 151, 153, 515, 523, 530, 531, 565, 566, 643, 646, 648, 670, 672, 726
      Vol. 12: 47, 51, 52, 54, 221, 234, 277, 526
   divisibility of   **Vol. 7:** 303, 341, 389, 398, 400, 412, 414
   instability of   **Vol. 7:** 321
   properties of   **Vol. 7:** 394, 403, 416
      Vol. 12: 4, 20, 42, 51, 105, 112, 233, 281, 282, 378
   waves (*see also* wave function)
      Vol. 12: 358
Matthée, H.   **Vol. 9:** 545
Maupertuis, P.L.M. de   **Vol. 5:** 244, 247
Max Planck Institute for Chemistry
   Vol. 12: 84
maximum energy transfer   **Vol. 8:** 122, 432, 466, 473*f*, 609, 669, 773, 776
maximum velocity of propagation, *see* light velocity
Maxwell
   Centenary Celebrations, 1931
      Vol. 6: 317, 357, 359
   distribution (*see also* statistical mechanics)
      Vol. 5: 5, 382
      Vol. 6: 143, 322*ff*, 400
      Vol. 9: 210, 243, 277, 313, 326
      Vol. 12: 24, 28
   distribution law   **Vol. 1:** 134, 298*ff*
   equations   **Vol. 7:** 81, 138
   Lecture, Bohr's, 1931   **Vol. 6:** 316*f*, 357–360

theory (*see also* electromagnetic theory)
   Vol. 7: 449
Maxwell, J.C.   **Vol. 1:** xvii, 134, 144, 147, 173, 174, 198, 200, 201, 206, 224, 242, 298, 299, 300, 318, 320, 323, 337, 354, 419, 425, 430, 448, 451, 459, 461
   Vol. 3: 20, 45, 230, 231
   Vol. 4: 434, 436, 469, 470
   Vol. 5: 129–130, 136–137, 273
   Vol. 6: 317, 357, 359*f*, 374*f*
   Vol. 10: 54, 209, 337, 350, 363, 386, 404, 432
   Vol. 11: 670
   Vol. 12: 100, 105, 106, 220, 221, 223, 233, 371, 372
Maxwell–Lorentz theory, *see* electrodynamics, classical
Mayer, J.   **Vol. 8:** 737*f*
Mayer, M.G.   **Vol. 9:** 77*ff*, 619
Mayer, R.   **Vol. 10:** 68
Mazur, P.   **Vol. 10:** 493
McCloy, J.J.   **Vol. 11:** 68, 69, 71, 159, 320–322, 326, 329, 330
McCormmach, R.   **Vol. 5:** 24, 62
McElroy, W.D.   **Vol. 10:** 134, 483*n*
McGill University (Montreal)
   Vol. 11: 326, 329
   Vol. 12: 265
McKay, H.A.C.   **Vol. 9:** 28, 156
McLaren, S.B.   **Vol. 1:** 105, 106, 422, 432, 434, 474, 476, 528, 529
   Vol. 2: 569, 571, 588, 614, 615
   Vol. 5: *91*, 297
   Vol. 6: 52
McLennan, J.C.   **Vol. 2:** 409
   Vol. 3: 45, 576, 592
   Vol. 4: 34, 399
   Vol. 12: 31, 34
McMahon, B.   **Vol. 11:** 327, 330
McNaughton, A.G.L.   **Vol. 11:** 325, 329
mean
   energy/velocity loss   **Vol. 8:** 7, 14*f*, 22, 111, 114, 121, 124, 129, 135, 137*ff*, 208, 213, 219, 225, 236, 255*f*, 274, 283, 371, 379*f*, 393, 396*f*, 400, 411*f*, 470, 473*f*, 476, 483, 489, 494, 504*f*, 511, 516*f*, 523, 546, 636
   square deviation (*see also* Gaussian distribution; straggling)   **Vol. 8:** 15, 321, 440, 470, 473, 482*f*, 518, 520, 527, 551, 601, 633

mean free path **Vol. 1:** 146, 308, 318, 331
　　Vol. 8: 29, 169, 173, 177, 269, 279,
　　533, 535, 539, 614*f*, 619, 689, 693
　　Vol. 9: 17, 81, 423, 451, 649, 651
　effective **Vol. 1:** 146, 321
measurement problem **Vol. 11:** 657, 660,
　　662, 664, 665
　　Vol. 12: 8, 62, 69, 224, 534
mechanical
　conception of nature **Vol. 10:** xxxiii, 85,
　　104, 111, 119, 121, 276, 277, 279, 284,
　　290, 343, 345, 357, 358, 539, 542,
　　544, 548
　　models/pictures **Vol. 10:** 54, 401, 402
　　Vol. 12: 44, 156, 168, 233, 358, 371,
　　439, 441
　theory of heat (*see also* statistical mechanics;
　　thermodynamics) **Vol. 10:** 149
　　Vol. 11: 612, 620
　　Vol. 12: 31, 34, 221, 255, 259, 290, 292
　transformability, principle of (*see also*
　　adiabatic invariance; adiabatic
　　principle/process/transformation)
　　Vol. 3: 7, 74, 91, 369, 405
"mechanical" state **Vol. 2:** 177
mechanicism/mechanism **Vol. 10:** ix, xxviii,
　　3–6, 8, 9, 15, 45, 46, 61, 68, 70, 71, 77,
　　91, 117, 122, 175, 184, 198, 199, 279,
　　538, 541, 544, 548, 559, 560, 574
　　Vol. 11: 134, 669–671
　　Vol. 12: 157, 170, 233
mechanics
　classical/ordinary (*see also* electrodynamics,
　　classical/ordinary; classical mechanics)
　　Vol. 2: 8, 104, 112, 128, 136, 145, 167,
　　172*ff*, 192*ff*, 217*ff*, 243, 254–255,
　　257–258, 265, 270, 294, 312, 315, 322,
　　352*ff*, 394*ff*, 435*ff*, 464*ff*, 529, 560, 563
　statistical **Vol. 2:** 257, 264, 420–422,
　　450, 461
Medical Association (Denmark)
　　Vol. 11: 682, 683, 685, 686
medical science **Vol. 11:** 133, 151, 153, 184,
　　565, 726, 731
　　Vol. 12: 60, 65, 84, 126, 181, 189, 212,
　　217, 326, 327, 329, 330, 454, 459
medicine **Vol. 7:** 408, 421
　　Vol. 10: 35, 67, 70, 72, 76–78, 104,
　　157, 180, 259, 269, 287, 292, 566
Meggers, W.F. **Vol. 3:** 38, 448
Meisen, V. **Vol. 5:** 137

Meissner, W. **Vol. 8:** 679
　　Vol. 12: 24, 28
Meitner, L. **Vol. 1:** xliv
　　Vol. 6: 445, 447
　　Vol. 7: 4, 32*p*, 278
　　Vol. 8: 221*f*, 224, 791*f*
　　Vol. 9: 6, 9, 10, 28, 42, 52*ff*, 55, 56*ff*,
　　62*f*, 66*f*, 69, 100, 134, 285, 317, 329,
　　342, 344*f*, 361, 365*f*, 369, 374, 378*f*,
　　382, 407*f*, 432, 438, 458, 462, 469,
　　473, 550*ff*, 574*f*, 589, 622*ff*, 628*f*,
　　631*ff*, 636, 638*ff*, 659, 662
　　Vol. 10: 22, 412
　　Vol. 11: 12, 28, 29, 569, 579, 598,
　　646, 649
　　Vol. 12: 12, 17*p*
melting point **Vol. 2:** 416, 421–422
Memorandum (to Roosevelt) **Vol. 11:** 4, 5,
　　30–34, 41, 43, 45, 47, 48, 69, 101, 111,
　　113, 118, 119, 159, 176, 179, 247–249,
　　251, 252, 255, 271, 273–275, 278–281,
　　289, 292, 320
memory **Vol. 10:** 93, 105, 187
Mendeléev, D.I. **Vol. 1:** xxii, xxxiii
　　Vol. 3: 204, 582
　　Vol. 4: 422, 431, 443, 468, 473, 658
　　Vol. 6: 375, 387
　　Vol. 10: 58, 103, 373, 376, 385, 387,
　　393, 394, 396, 403, 437, 438
　　Vol. 11: 212, 214, 216
　　Vol. 12: 42, 288
Mendelssohn, K. **Vol. 12:** 38*p*
Menzies, S. (*see also* "C" (head of British
　　intelligence)) **Vol. 11:** 25*n*
Mercier, A. **Vol. 7:** 328
mercury spectrum **Vol. 2:** 408–409
　　Vol. 5: 39, 48, 53*f*, 56, 122, 145*ff*,
　　148, 152*f*, 344, 396, 505–506
　　Vol. 12: 368
Meredith, J.C. **Vol. 10:** 4*n*
Merton, T.R. **Vol. 2:** 399, 419
meson **Vol. 8:** 257, 554, 571–572, 633, 730,
　　732*f*, 735, 765, 810, 813
　　Vol. 9: 77, 618, 634*f*, 637
　　Vol. 10: 452
　　Vol. 11: 569
　　Vol. 12: 12, 54
　field **Vol. 7:** 221
　particle **Vol. 7:** 221, 222, 228
　theory **Vol. 7:** 222
mesothorium **Vol. 9:** 350, 353, 589, 634, 636

Meštrović, I.   **Vol. 12:** 234
metabolism   **Vol. 10:** 6, 8, 11–13, 19, 22, 46, 69, 76, 104, 131, 136, 137, 150, 207, 544, 548
metal atoms   **Vol. 1:** 297
    action upon electrons from stationary field **Vol. 1:** 325
    close together   **Vol. 1:** 334
    electric doublets   **Vol. 1:** 321, 365, 390
    elementary magnets   **Vol. 1:** 322, 365, 390
    fixed centres of force   **Vol. 1:** 320
    hard elastic bodies   **Vol. 1:** 324, 342, 364
    hard elastic spheres   **Vol. 1:** 318
    not small compared to mutual distances **Vol. 1:** 324ff, 342, 374, 391
metal lattice   **Vol. 12:** 499, 501
metallic conduction   **Vol. 11:** 566
    Vol. 12: 530
metallurgy   **Vol. 10:** 157
metals   **Vol. 12:** 4, 8, 22–25, 27, 28, 30, 31, 33, 34, 39, 51, 54, 144, 499–501
    conductivity of   **Vol. 12:** 4, 5, 8, 20, 22, 23, 25–27, 30, 31, 33, 34, 39, 344, 346, 498, 500
    interior   **Vol. 1:** 302ff
metaphysics   **Vol. 10:** 35, 46
methane molecule, structure of   **Vol. 2:** 138, 150, 232
Meulengracht, J.E.   **Vol. 11:** 673p
Meyenn, K. von   **Vol. 5:** x, 23, 220
    Vol. 6: 10
    Vol. 7: 10n
    Vol. 8: 699
Meyer, E.   **Vol. 5:** 341
Meyer, G.   **Vol. 1:** 30
Meyer, K.   **Vol. 2:** 613, 615, 616
    Vol. 5: 133, *140–141*
    Vol. 10: 330, 348, 361
    Vol. 11: 354, 494–499, 729
    Vol. 12: 122, 123, 125, 310n, 427p, 428–431, 511–513, 527
Meyer, L.   **Vol. 3:** 582
    Vol. 4: 422, 431, 444, 468, 474
Meyer, P.H.E.   **Vol. 12:** 38p
Meyer, R.C.   **Vol. 9:** 31, 56, 357, 365, 383f, 389
Meyer, S.   **Vol. 2:** 597
Meyer, T.   **Vol. 3:** 290, 299
Meyer, V.   **Vol. 11:** 517, 524, 557, 560
Meyer-Abich, K.M.   **Vol. 6:** 27
Meyerhof, O.   **Vol. 10:** 19, 20, 461

Meyerson, E.   **Vol. 10:** 301, 302
MGM studios   **Vol. 12:** 129p
Michaelsen Foundation   **Vol. 12:** 182, 190
Michelson, A.A.   **Vol. 3:** 362
    Vol. 5: 6, 10, 297, 404–405, 487
    Vol. 12: 371
Michigan, University of   **Vol. 5:** 28f, 334, 382, 384, 386, 394
    Vol. 8: 702f
    Vol. 9: 39, 577
microbiology   **Vol. 10:** 168, 487
    Vol. 12: 534
microcanonical ensemble   **Vol. 9:** 375f, 392
microscope (*see also* optical instruments)
    Vol. 6: 16–20, 29, 31, 37, 40, 121, 142, 151f, 162–163, 186, 224, 236, 316, 417
    Vol. 12: 60, 64
Middelboe, V.   **Vol. 9:** 621
Middle Ages   **Vol. 10:** 157
Mie, G.   **Vol. 3:** 486, 691
    Vol. 5: 39, 87, 121, 137, 321, 325, 440, 443
Miers, H.A.   **Vol. 2:** 333, 541–542, 619
Miethe, A.   **Vol. 5:** 352
Miguez, A.P.   **Vol. 3:** 497
Milch, L.   **Vol. 5:** *401*
Mill, J.S.   **Vol. 10:** xxxiv, 502, 503
Millard Scholarship   **Vol. 12:** 285
Miller, H.   **Vol. 9:** 185, 188
Miller, H.M.   **Vol. 8:** 223, 645, 664, 753
Millikan, R.A.   **Vol. 2:** 112, 199, 582
    Vol. 3: 32, 61, 212, 361, 363, 426, 584, 587, 588
    Vol. 4: 55, 411, 441, 472
    Vol. 5: 233, 251f, 278, 289
    Vol. 6: 311, 384, 386
    Vol. 9: 100, 126f, 652
    Vol. 10: 438
    Vol. 11: 530
Millington, G.   **Vol. 8:** 792
mind–body problem   **Vol. 10:** xliii, xliv, xlvii, xlviii
Minkowski, H.   **Vol. 12:** 372
Minkowski, R.   **Vol. 4:** 23
    Vol. 5: 497
Miskel, J.A.   **Vol. 8:** 562
Mitchell, Miss   **Vol. 9:** 553
Mittweida Ball Club   **Vol. 11:** 676p
mobility
    of electrons, *see* electrons, mobility of
    of ions, *see* ions, mobility of

[110]

model of atom, *see* atomic model
moderator   **Vol. 11:** 578, 581, 582, 596, 602, 603, 615, 626
Moeller, C., *see* Møller, C.
Mohler, F.L.   **Vol. 3:** 38, 448, 468, 592
   Vol. 5: 145, 148
Mohr, C.B.O.   **Vol. 8:** 452, 567, 709*ff*
molecular
   biology   **Vol. 10:** 131, 164, 167, 168, 500
   chaos/disorder   **Vol. 6:** 323*f*, 439*f*
   formation   **Vol. 2:** 105, 138*ff*, 151*ff*, 200, 225–232, 409, 481, 485*f*, 530, 580
      Vol. 5: 72, 83, 189, 203, 346
      Vol. 8: 33, 661
   formation/structure   **Vol. 6:** 9, 154, 376, 392–393, 394, 395–398, 423, 463
   rotation   **Vol. 7:** 524
   spectra (*see also* band spectra; spectra, molecular)   **Vol. 6:** 395–398
   structure   **Vol. 1:** xxv
      Vol. 2: 6, 104*ff*, 138*ff*, 150*ff*, 163, 215–233, 396, 420, 447, 480–488, 565, 567, 577, 580, 624*ff*, 634
      Vol. 8: 63, 64, 66, 85, 87*f*, 100, 103, 144, 675, 757*f*
molecule of HCl   **Vol. 3:** 192
Møller scattering   **Vol. 8:** 214*ff*
Møller's stopping formula, *see* stopping formula, Møller's
Møller, C.   **Vol. 1:** xliii
      Vol. 6: 47*f*, 195, 304, 463*ff*, 479
      Vol. 7: 4, 222, 302
      Vol. 8: 213*ff*, 220, 223*f*, 262, 268*f*, 273, 275, 278, 282, 284, 502, 517*f*, 522, 567, 590, 645*f*, 651, 655*f*, 701, 703, 710*ff*, 728*ff*, 732*ff*, 749*f*, 764*f*, 792, 794, 821
      Vol. 9: 17, 28, 528, 627, 629, 634, 637, 668
      Vol. 10: 283, 289, 455*p*
      Vol. 11: 29, 264, 358*p*
      Vol. 12: 13, 71*p*, 111, 277, 277*p*, 526, 529
Møller, N.   **Vol. 12:** 127, 463–465, 513–515
Møller, P.F.   **Vol. 10:** 74
Møller, P.M.   **Vol. 6:** xx, xxii
      Vol. 10: xxxi, xliv, xlv, 159, 230, 234, 256, 266, 480*n*, 483*n*, 485, 487
      Vol. 12: 152, 155, 163, 166
Møllgaard, E.H.   **Vol. 1:** 500, 501
moment, magnetic, *see* magnetic moment

momentum
   and energy   **Vol. 7:** 405, 418
   angular, *see* angular momentum, quantization of
   conservation/balance of, *see* conservation of momentum
   determination/control   **Vol. 7:** of 15, 360, 513
   exchange/transfer   **Vol. 7:** 9, 75, 81, 84, 90, 99, 101, 109, 117, 134, 138, 140, 144, 150–152, 157, 163, 198, 207, 208, 213–215, 237, 238, 256, 293, 295, 296, 298, 304, 314, 333, 346, 349, 354–357, 359, 360, 362, 364, 365, 461, 462
   fixation of   **Vol. 7:** 310
   of atomic object   **Vol. 7:** 351
   of charged body   **Vol. 7:** 19, 20, 86, 109, 141, 157, 232
   of particle   **Vol. 7:** 293, 428
   of photon   **Vol. 7:** 73, 132, 342, 343, 345
   of radiation, *see* radiation momentum
Mond Laboratory   **Vol. 11:** 212, 213, 219
Monnier, M.   **Vol. 7:** 328
monomolecular reaction   **Vol. 9:** 375–378, 434, 459
More, H.   **Vol. 10:** 555
Morgan, C.L.   **Vol. 10:** 6
Morgan, T.H.   **Vol. 10:** 23, 553
Morgenstern, C.   **Vol. 7:** 251, 480
Morgenthau, H.   **Vol. 11:** 255*n*
Morse, D.A. and Mrs   **Vol. 11:** 327, 330
Morse, M.   **Vol. 9:** 70
Moscow lecture, Bohr's, 1937
      Vol. 9: 41, 677
Moscow University   **Vol. 11:** 57, 203, 204, 299
Moscow, Treaty of   **Vol. 11:** 341
Moseley's law   **Vol. 2:** 131–134, 335, 410, 429
      Vol. 12: 288
Moseley, H.   **Vol. 12:** 286
Moseley, H.G.J.   **Vol. 1:** xxxi
      Vol. 2: 131, 132, 133, 134, 237, 304, 314, 315, 410, 411, 412, 419, 427–430, 501, 543–547, 578, 580, 589, 590, 615, 629, 630
      Vol. 3: 204, 282, 336, 567, 568, 583, 589, 590, 591, 601
      Vol. 4: 33, 48, 52, 54, 55, 93, 103, 194, 195, 200, 249, 268, 269, 273, 320, 337, 350, 407, 422, 433, 442, 457, 460, 463, 468, 473, 479, 481, 482, 515, 546, 560, 619, 659

[111]

Vol. 5: 251*f*, 278
Vol. 6: 379, 387*f*, 390*f*
Vol. 8: 672, 819
Vol. 10: 376, 379, 387, 393, 394, 397, 400, 418, 437, 463
Vol. 11: 567
Vol. 12: 43, 112, 265, 285–288, 524
Moseley, H.N.   **Vol. 12:** 285
Moseley, M.   **Vol. 12:** 286
Moseley diagrams   **Vol. 4:** 336, 442, 461, 473, 481, 506, 513, 538, 544
most probable energy loss   **Vol. 8:** 15, 111, 114, 140, 274, 276*f*, 283, 285*f*, 474, 479, 494, 503, 520, 547, 792
Mott scattering   **Vol. 8:** 210*ff*, 306, 437, 442, 445, 678, 755
Mott, N.F. **Vol. 6:** 305, 309, 341, 346*f*, 398
   Vol. 7: 32*p*, 272*n*
   Vol. 8: 210*ff*, 216, 219, 269, 272, 278, 281, 306, 379, 437, 440, 442, 445, 503, 567, 634, 636*f*, 645*f*, 678, 698, 709*f*, 754–755
   Vol. 9: 51, 100, 134, 137*f*, 186, 254, 494, 513
   Vol. 10: 400, 403, 443
Mottelson, B.R.   **Vol. 8:** 590
   Vol. 9: 529
   Vol. 12: 534
Mowinckel, S.   **Vol. 10:** 536, 537
M.Sc. examination paper   **Vol. 1:** 93
   translation of   **Vol. 1:** 131–161
Müller, C.   **Vol. 2:** 199
Müller, F.   **Vol. 1:** 520, 521
Muller, H.J.   **Vol. 10:** 20, 22, 565
Müller, J.J.C.   **Vol. 1:** 17
Müller, O.   **Vol. 11:** 358*p*
multi-dimensional geometry (*see also* geometry)   **Vol. 7:** 318, 378
multiple scattering, *see* scattering, multiple
multiple-mode excitation   **Vol. 9:** 293*f*
multiplet structure (*see also* fine structure; complex structure)   **Vol. 4:** 661
   Vol. 5: 166, 168*f*, 171, 255, 258, 261, 264, 272, 278, 288*f*, 332, 333, 447–448, 452, 470, 472, 478, 481, 499*ff*
multiply periodic systems
   Vol. 1: xxxii–xxxiv
   Vol. 3: 371, 595*ff*
   Vol. 5: 45*f*, 52, 54, 62*f*, 92, 145*ff*, 158–165, 223*f*, 243, 247, 251*f*, 255*ff*, 261*ff*, 271, 277*ff*, 299, 303, 357, 395,

397, 411, 413, 425, 429, 446, 450, 451, 454*f*, 465
   quantum theory of   **Vol. 4:** 447, 475
multipole radiation (*see also* quadrupole)
   Vol. 9: 194, 278, 315, 327
multivalued logics   **Vol. 7:** 285, 393
Munich   **Vol. 2:** 331
   Vol. 5: 57, 61, 88, 304, 306, 354, 355*f*, 357, 359, 375, 377, 431, 433, 435, 437, 441, 444, 505
   Vol. 6: 10, 17, 325
Murcheson, C.   **Vol. 10:** 320*n*
Murchison, I.   **Vol. 11:** 445
Murphy, G.   **Vol. 10:** 82
music   **Vol. 10:** xlvi, 160, 255, 264, 346, 360, 584
   Vol. 12: 359
musical harmonies   **Vol. 11:** 669
mutation   **Vol. 10:** 22–24, 60, 70, 91, 104, 122, 168, 180, 198, 471, 472, 552, 554
mysticism   **Vol. 10:** 39, 60, 64, 71, 77, 160, 244, 502, 504, 567, 572
   Vol. 11: 197, 200
   Vol. 12: 224

Nagaoka's atomic model, *see* atomic model, Nagaoka's
Nagaoka, H.   **Vol. 2:** 127, 548–549, 615
Nagel, W.   **Vol. 10:** 5*n*
Nambu, Y.   **Vol. 10:** 455*p*
Nansen, F.   **Vol. 1:** 62
Napoleon, *see* Bonaparte, N.
Nardroff, R. von   **Vol. 5:** 318
Nærumgaard   **Vol. 12:** 120, 121*p*, 126*n*, 394, 395, 400, 403
National Academy of Sciences (U.S.)
   Vol. 11: 52, 133, 291, 325, 329, 638
National Association for Combating Cancer (Denmark)   **Vol. 11:** 373, 375, 673*p*, 683, 684, 686–688, 727, 729, 731, 734
   Vol. 12: 327, 330
National Association for Combating Tuberculosis (Denmark)
   Vol. 12: 330*n*
National Industrial Conference Board (U.S.)
   Vol. 11: vii, 361, 562, 563, 565, 717–720, 734
National Meeting of Industry (Denmark)
   Vol. 11: 573, 584–586, 607
National Museum (Denmark)   **Vol. 11:** 338

National Science Foundation (Denmark)
Vol. 11: 348, 352, 464, 465, 467, 733
Vol. 12: 189*n*
National Science Foundation (Sweden)
Vol. 11: 466*n*
National Science Foundation (U.S.)
Vol. 11: 466*n*
nations, relationship between **Vol. 10:** 102, 143, 160, 206, 247
NATO **Vol. 11:** 77, 323, 324
natural
laws **Vol. 7:** 317
phenomena **Vol. 11:** 62, 123, 133, 138, 141, 151, 153, 311, 313, 536, 541, 547, 549, 612, 619, 657, 658, 662, 663, 670
Vol. 12: 43, 44, 220, 254, 258, 438, 440
philosophy **Vol. 10:** 67, 76, 195, 240, 241, 243, 277
Vol. 11: 530, 567, 669, 672
Vol. 12: 42, 90, 105, 221, 233, 281, 371, 516
state, *see* permanent state
system of elements **Vol. 4:** 453, 478
Nature (journal) **Vol. 11:** 50, 328, 331
Vol. 12: 4, 6, 108, 109, 111, 112, 252, 265, 274
letters to **Vol. 3:** 9, 30, 32, 34, 35, 38, 39
Vol. 4: 10*ff*, 19*ff*, 72–82, 177–180
nature, laws of **Vol. 11:** 434, 435, 448, 450, 512, 520, 530, 536, 539, 547, 548
Vol. 12: 222, 228, 229, 254, 258, 374, 375, 378, 438, 440
Naturphilosophie **Vol. 10:** 345, 346, 358–360
Naturwissenschaften (journal) **Vol. 12:** 7, 498, 525
Nazism **Vol. 10:** 220, 471*n*
nebulium **Vol. 2:** 107
Neddermeyer, S.H. **Vol. 8:** 235, 810
Needham, J. **Vol. 10:** 11, 19
Neergaard, N. **Vol. 11:** 525*p*
Vol. 12: 503, 505
negative-energy states **Vol. 9:** 6*f*, 107, 109, 549
negaton (*see also* electron) **Vol. 7:** 485
Vol. 12: 11, 51
Nehru, J. **Vol. 11:** 82
Neisig, K. **Vol. 11:** 358*p*
neon
dispersion in **Vol. 2:** 53
spectrum of isotopes of **Vol. 2:** 419

structure of atom of **Vol. 2:** 207
Vol. 11: 577, 594
neptunium **Vol. 11:** 579, 597
Nernst, W. **Vol. 1:** 101, 125, 448
Vol. 2: 150
Vol. 3: 586, 677
Vol. 5: 15
Vol. 10: 386, 433, 434
nervous
action **Vol. 10:** 54, 167
message **Vol. 10:** 33, 55, 91, 104
system **Vol. 10:** 68, 70, 93, 94, 105, 169, 545, 549
Neumann, J. von **Vol. 6:** 138, 327, 453
Vol. 7: 260, 262, 263*n*, 302, 308, 458, 459
Vol. 10: 82, 499
Vol. 11: 713
Neurath, O. von **Vol. 10:** xli, 36*p*
neurophysiology **Vol. 10:** 68
neurosis **Vol. 10:** 93, 566
neutralization
body **Vol. 7:** 91, 92, 145
distribution **Vol. 7:** 93, 146
neutrino **Vol. 7:** 182, 191, 448, 450, 476, 479
Vol. 9: 8–9, 13–14, 132, 275, 280, 307, 321, 540*f*, 593*f*, 597*f*
Vol. 10: 412, 451, 452, 454
Vol. 12: 6*n*, 365
hypothesis **Vol. 5:** 216
neutron **Vol. 7:** 180, 181, 190
Vol. 9: —
Vol. 10: 409–412, 449–451
Vol. 11: 43, 91, 102, 229, 515, 523, 567, 569, 577–583, 593–597, 599, 600, 602–606, 613–616, 622–627, 646, 649, 650
Vol. 12: 51, 211, 216
collisions **Vol. 8:** 219, 304, 631, 702*f*, 761
diameter **Vol. 9:** 11*f*, 131
delayed **Vol. 11:** 580, 600, 615, 626
emission **Vol. 11:** 86
properties of **Vol. 9:** 12, 115–121, 415, 445, 676
scattering, *see* scattering of neutrons
New York Herald Tribune **Vol. 11:** 328, 331
New York meeting, 1939, *see* American Physical Society
New York Times **Vol. 10:** 238*p*
Vol. 11: 11
New York University **Vol. 11:** 371

Institute of Mathematical Sciences
 Vol. 11: 370, 669, 672, 710–714, 734
Newton's postulate of action/reaction
 Vol. 12: 223
Newton, I. **Vol. 1:** xvii
 Vol. 3: 20, 45, 225, 228, 229, 231, 234
 Vol. 5: 24, 129, 133, 136, 140, 243,
 247, 273, 310
 Vol. 6: 359f, 374
 Vol. 7: 399, 413
 Vol. 8: 186, 192, 270, 279
 Vol. 10: xxxvii, 34, 85, 103, 117, 174,
 194, 208, 404, 415, 432, 555
 Vol. 11: 348, 372, 444, 543, 550, 657,
 662, 669, 670
 Vol. 12: 61, 68, 103–106, 155, 160,
 167, 173, 220–225, 371, 378, 528
Newtonian mechanics (*see also* classical
 mechanics) **Vol. 7:** 257n, 285, 388
 Vol. 10: 30, 54, 57, 85, 127, 142, 148,
 157, 196, 209, 386, 395, 555
 Vol. 11: 515, 523, 567, 657, 659,
 662–664
 Vol. 12: 200, 204, 220–224, 233
Nexø, M.A. **Vol. 12:** 103
Nicholson's theory **Vol. 1:** 563
Nicholson, J.W. **Vol. 1:** 423, 427, 562, 563,
 568, 569, 572, 573, 578, 579
 Vol. 2: 107, 108, 109, 111, 127, 133,
 134, 166, 167, 175, 182, 183, 184, 193,
 194, 202, 203, 269, 270, 312, 313, 314,
 315, 331, 333, 334, 354, 385, 386, 387,
 395, 397, 404, 409, 411, 437, 508, 509,
 516, 517, 545, 559, 562, 579, 580, 584,
 593, 626, 627
 Vol. 3: 274
 Vol. 4: 196, 270, 436, 470
 Vol. 10: 386, 392, 417
Niel, C.B. van **Vol. 10:** 134
Niels Bohr Archive **Vol. 7:** 4, 18, 35, 39,
 169, 196, 519
 Vol. 10: 299n, 330n, 460, 501n,
 516n, 581
Niels Bohr Institute, *see* Institute for
 Theoretical Physics (Copenhagen)
Nielsen, C. **Vol. 12:** 463, 465
Nielsen, H.H. **Vol. 12:** 17p
Nielsen, J. Rud, *see* Rud Nielsen, J.
Nielsen, J.N. **Vol. 6:** 308
 Vol. 8: 704

 Vol. 10: 74
 Vol. 11: 350, 351, 406p, 453
Nielsen, K.O. **Vol. 8:** 235
Nielsen, R. **Vol. 10:** 4
 Vol. 12: 152, 164
Nier, A.O. **Vol. 9:** 71, 378, 441, 465
Nietzsche, F.W. **Vol. 10:** 266n
Niggli, P. **Vol. 7:** 328
Nijboer, B.R.A. **Vol. 12:** 38p
Nishina, Y. **Vol. 3:** 660, 662
 Vol. 5: 483, 485
 Vol. 7: 4, 175, 186
 Vol. 8: 26, 233, 646
 Vol. 9: 76, 540f
 Vol. 11: 514, 522
 Vol. 12: 117p
Nissen, K.A. **Vol. 2:** 619
 Vol. 4: 591, 641
nitrogen **Vol. 11:** 576, 577, 592, 593
 Vol. 12: 54
 active **Vol. 2:** 238–239
 dispersion in **Vol. 2:** 53–54, 336
Nobel
 Committee **Vol. 5:** 239f, 407
 Foundation **Vol. 12:** 102
 Prize **Vol. 5:** 239–240, 407, 502–503
 Vol. 10: 23, 157, 192
 Vol. 11: 281, 348, 357, 516n, 563,
 601n, 733
Nobel Prize, Bohr's, 1922 **Vol. 1:** xxxv
 Vol. 4: 25ff
 Vol. 12: 10, 102, 118, 152, 164, 212,
 217, 487
 Lecture **Vol. 4:** 25, 425–465, 467–482
 Vol. 6: 262, 284
 speech at banquet **Vol. 4:** 26, 27
Nobel, A. **Vol. 4:** 26
Nobel, J. de **Vol. 12:** 38p
Nolfi, P. **Vol. 7:** 328
non-commutability **Vol. 10:** 42, 87, 103, 158,
 401, 403, 406, 425, 441, 531, 533
non-commutativity (*see also* commutation
 relations) **Vol. 7:** 64, 126, 177, 188,
 252, 257, 318, 332, 347, 373, 378,
 429, 480
non-conservation, hypothesis of, *see*
 conservation of energy and momentum
non-Euclidean geometry **Vol. 10:** 158,
 176, 209
non-Euclidean metric **Vol. 7:** 389

non-isolated system, radiation from
Vol. 3: 587*ff*
non-stationary problems **Vol. 1:** 357*ff*
Nordheim, L.W. **Vol. 5:** 303
Vol. 6: 304, 306, 308
Vol. 7: 489, 492
Vol. 9: 79*ff*, 384, 646*ff*, 650
Vol. 12: 24, 28
Nordic Association **Vol. 11:** 726, 735
Nordic Electrical Engineers' Meeting, 1937
Vol. 9: 41, 213–221
Nordic Insulin Foundation **Vol. 11:** 516, 524
Nordic Interparliamentary Council
Vol. 11: 707–709
meeting, 1955 **Vol. 11:** 364, 365, 611,
618, 707, 708, 735
Nordic Physicists, First Meeting of, 1920
Vol. 8: 25, 181–200, 820
Vol. 12: 307, 310
Nordic Physics Conference, second
Vol. 4: 25
abstract of lecture before **Vol. 4:** 421–424
Nordic Scientists' Meeting, 1936, *see* Helsinki
conference, 1936
NORDITA **Vol. 11:** 338, 356
Nørgaard, F. **Vol. 10:** 74
Nørlund, A.C. **Vol. 12:** 138, 140, 209, 213,
412, 420
Nørlund, Margrethe *see* Bohr, Margrethe
Nørlund, N.E. **Vol. 1:** xviii, 96, 504, 505,
512, 513, 514, 515, 544, 545, 546, 547
Vol. 2: 615, 617, 618, 619, 620
Vol. 4: 670
Vol. 10: xxx*p*, 299*n*, 300
Vol. 11: 347, 369, 370, 400, 406*p*,
410, 656*p*
Vol. 12: 100
Nørlund, P. **Vol. 10:** 299*n*, 300
Nørlund, S. **Vol. 12:** 138, 140, 209, 213,
412, 420
normal distribution **Vol. 7:** 246
normal state (*see also* permanent state)
Vol. 3: 254
Vol. 5: 38*f*, 43, 45, 52, 55, 56, 65*f*,
114, 122*f*, 145*f*, 149*ff*, 181, 196,
276*ff*, 309, 344, 395, 445*ff*, 450*f*, 491,
505–506
Vol. 6: 177*ff*, 311*f*, 383*f*, 386, 395
Vol. 8: 20*f*, 23, 63, 117, 208, 238, 255,
259, 263*f*, 344, 349, 366, 368, 381,

411*f*, 598, 600, 603*f*, 606, 610, 613,
620*f*, 657, 672, 683
Norsk Hydro **Vol. 11:** 616, 628
northern lights **Vol. 12:** 139*n*, 524
Norwegian Academy of Science and Letters
Vol. 11: 399, 408
Norwegian Society of Engineers, *see* Oslo
lecture, Bohr's, 1940
*n*th power law (*see also* inverse cube law of
force) **Vol. 8:** 230, 305, 452–455,
631, 634–635, 806, 811, 813
nuclear
arms race **Vol. 7:** vii
atom, *see* atomic model, Rutherford's
charge (*see also* charge) **Vol. 11:** 575, 577,
579, 590, 593, 597
Vol. 12: 42–45, 364
constitution/structure (*see also* compound
nucleus) **Vol. 2:** 96–99, 118–119,
212–214, 324–325, 328, 418, 423*f*,
530, 591, 593
Vol. 6: 158, 319, 330, 356, 360, 377*f*,
381, 403–407, 458
Vol. 7: 180, 181, 190, 216, 221, 227,
304, 343, 370, 377, 401, 415, 449, 489,
492, 495
Vol. 8: 221, 269, 278, 651, 678,
679–680, 701*ff*, 709*f*, 712*f*
Vol. 10: 87, 103, 340, 384, 399, 408,
410, 450, 451, 473, 498, 500
Vol. 11: 91, 101, 102, 270, 529, 568,
646, 649
Vol. 12: 42, 51, 54, 61, 67, 211, 215,
265, 268, 340, 342, 381, 383
density (*see also* packing of particles in
nuclei) **Vol. 9:** 167, 175, 216, 220,
232*f*, 416, 420, 424, 445, 448, 452, 523
dimensions, *see* dimensions, nuclear
disintegration **Vol. 10:** 179, 398, 399, 409,
411–413, 449, 450
energy (*see also* atomic energy, as power
source) **Vol. 9:** 60, 75*f*, 437,
441–442, 462, 465–466
Vol. 10: 76, 287, 292, 413, 452, 589
fission **Vol. 1:** xliii–xliv
forces **Vol. 11:** 569
Vol. 12: 12, 42, 44, 51, 54
mass (*see also* isotopes; reduced mass)
Vol. 2: 119, 213, 416, 418–425
model (of the atom) (*see also* atomic model,
Rutherford's; independent-particle

[115]

model, liquid-drop model; shell model)
  Vol. 1: xlii–xliii
  Vol. 9: 19, 27, 29, 44, 51, 67, 76–83,
    95, 141, 145, 147, 153*ff*, 181, 186, 194,
    236, 239, 246, 259, 274*f*, 280*f*, 307,
    321, 336, 507*f*, 539, 607*f*, 614*f*, 619,
    642, 644, 646*ff*
  Vol. 12: 42–44, 51, 108, 281
motion   Vol. 5: 187, 189, 201–202, 203,
    346, 349, 491, 503
particles   Vol. 7: 221, 222, 227, 228
physics (*see also* atomic physics)
  Vol. 1: xlii–xlv
  Vol. 4: 86
  Vol. 11: 11, 55, 196, 197, 199, 201,
    228, 296, 298, 303, 356, 361, 363,
    514*n*, 531, 568, 646, 649, 731
  Vol. 12: 5, 6, 12, 54, 84, 109, 119, 211,
    216, 526
reactions/processes   Vol. 1: xliii
  Vol. 7: 182, 191, 371, 403, 416
  Vol. 11: 11, 220, 241, 355, 568, 569,
    577–579, 593–597
  Vol. 12: 6, 54, 211, 216, 386, 387
spin (*see also* spin of nucleus)
  Vol. 12: 365
stability   Vol. 6: 145, 225, 238, 403
states   Vol. 12: 51
statistics   Vol. 6: 330, 404, 406
temperature, *see* temperature of nucleus
transformations, *see* transformation/
  transmutation
transmutations (*see also* β-decay)
  Vol. 6: 225, 228, 230–231, 238–239,
    243, 246, 375–376, 378, 403–407,
    445, 447
  Vol. 10: 56, 86, 102, 338, 409–412, 450
weapons, *see* atomic bomb
nucleon spin, *see* spin of nucleon
Nucleonics (journal)   Vol. 12: 14, 84, 532
nucleus
  compound   Vol. 9: —
    Vol. 12: 51
  discovery of   Vol. 11: 138, 142, 363, 511,
    520, 529, 530, 566, 575, 589, 612, 621,
    646, 649
    Vol. 12: 43, 51, 61, 67, 84, 111, 112,
    145, 199, 203, 209, 211, 213, 216, 221,
    256, 260, 261, 268, 277, 281,
    282, 356
  numbers, rational/irrational   Vol. 11: 669

nutrition   Vol. 10: 10, 91, 104, 119, 137,
    151, 166
Nuttall, J.M.   Vol. 6: 405
  Vol. 10: 409
Nyquist, H.   Vol. 3: 144, 175

Oak Ridge   Vol. 11: 89, 630*n*
objective description   Vol. 10: xxxvi, 83–86,
    88–90, 92–95, 112, 131, 159, 160, 168,
    185, 213, 276–279, 284, 286, 290, 292,
    405, 546, 550, 565, 567, 568, 570, 571,
    573, 574
objectivity (*see also* subjectivity; subject/object
    distinction/relation)   Vol. 6: 190, 192,
    196, 204*f*, 209, 213, 233, 249*f*, 259,
    279, 431, 479
object/subject, *see* subject/object
Obreimov, I.V.   Vol. 11: 59*p*
observability   Vol. 6: 21*ff*, 24, 32, 37, 40, 47, 49,
    52, 70, 80, 88, 92, 95, 104, 116, 125, 127,
    132, 148, 149, 156, 157, 204, 209, 224,
    236, 260, 281, 305, 310, 315, 326, 333,
    339, 343, 346*f*, 421, 425, 450, 464, 466
observational problem, *see* measurement
    problem
Occhialini, G.P.S.   Vol. 8: 554, 567
occupation, German   Vol. 11: 12, 13, 29, 32,
    60, 103, 227, 250*n*, 297, 341, 342, 346,
    353, 357, 376, 427, 430, 564, 616, 628,
    693, 699
  Vol. 12: 106, 113, 125, 128, 211, 216,
    245*n*, 294*n*, 402, 405, 453*p*, 456*n*, 470,
    472, 483*n*
oceanic research   Vol. 12: 286, 290–292, 294,
    300, 301, 316, 321
Oehlenschläger, A.   Vol. 10: 256, 266
Oersted, H.C., *see* Ørsted, H.C.
Oersted Meeting, *see* Nordic Physicists, First
    Meeting
Öhnell   Vol. 9: 638, 640
Oken, L.   Vol. 10: 4*n*
Oldenberg, O.   Vol. 3: 183
  Vol. 5: *495*
Ølgaard, P.L.   Vol. 11: vii, 630*n*
Oliphant, M.   Vol. 9: 28, 201
  Vol. 10: 413, 449
  Vol. 11: 87
Olrik, D.   Vol. 10: 250*p*
Olsen, H.W.   Vol. 8: 247
  Vol. 9: *630*, 631
  Vol. 11: 358*p*, 518, 525

Olsen, J.   **Vol. 8:** *24*, 25, 707*f*
Olympic Games, London, 1908   **Vol. 11:** 372
Onnes, H. Kamerlingh, *see* Kamerlingh
      Onnes, H.
Onsager, L.   **Vol. 8:** 737*f*
      Vol. 10: 492
Open Letter to the United Nations
      Vol. 11: 3–5, 49, 77–81, 101, 113, 160,
      171, 172, 175, 186*p*, 188, 189, 327,
      328*n*, 330, 331
      Vol. 12: 211, 216
Open Skies programme   **Vol. 11:** 366
open world (*see also* openness)
      Vol. 10: 287, 293
      Vol. 11: v, 3–5, 44, 60, 70, 72, 79, 81,
      161, 164, 165, 168, 181, 182, 184, 185,
      189, 192, 320, 327*n*, 365, 367, 368,
      371, 377, 735
      Vol. 12: 106
openness (*see also* open world)   **Vol. 11:** 8,
      17, 18, 32, 51, 53, 56, 68–79, 82, 83,
      106, 116, 161, 162, 164, 165, 167–170,
      178, 181–185, 188, 189, 192, 301, 302,
      320, 357
operator   **Vol. 6:** 128, 138, 154, 169, 175, 461
      Vol. 7: 252, 332, 389, 428–430
ophthalmology   **Vol. 10:** 33
Oppenheimer, F.   **Vol. 8:** 765
Oppenheimer, J.R.   **Vol. 1:** xliii
      Vol. 6: 138, 446*f*
      Vol. 7: 43, 46, 204, 270, 278, 279,
      451, 453
      Vol. 8: 205, 210, 227, 372, 530, 567,
      570, 663, 798*ff*
      Vol. 9: 30, 48, 211, 257, 261, 263,
      387*f*, 484, 493, 499, 510
      Vol. 10: 413, 414, 455*p*, 547, 550, 554,
      556, 561, 562
      Vol. 11: 17, 51, 66, 67, 71, 77, 306,
      307, 309, 310, 320, 326, 327, 330, 361,
      698*n*, 710, 712
      Vol. 12: 14, 86, 87, 386, 388
Oppenheimer, K.   **Vol. 10:** 547, 550, 561, 562
      Vol. 11: 326, 330
optical
   instruments (*see also* microscope)
      Vol. 6: 22, 71, 77, 84, 93, 98, 118,
      120*f*, 132, 150*f*, 156, 224, 236, 419
   phenomena (*see also* absorption of radiation;
      diffraction; dispersion of radiation;
      emission of radiation; refraction;

   spectra)   **Vol. 8:** 12, 34, 84, 99, 117,
      160, 271, 280, 292, 343, 348, 381, 452,
      496, 567, 662, 681, 685, 691
   spectrum, *see* spectra, optical
   theorem   **Vol. 9:** 51, 492–493, 496, 499,
      508, 512–513, 614
optics   **Vol. 10:** 29, 85, 175, 209, 402, 426
      Vol. 12: 47, 220, 348, 351, 438, 441
orbit   **Vol. 9:** 80, 108, 112*f*, 145*f*, 197, 217,
      221, 311, 325, 429, 456, 648*ff*
   elliptical, *see* elliptical orbit; Keplerian
      laws/orbit
   of electron, *see* electron orbit
orbital
   method   **Vol. 8:** 226, 228*ff*, 254, 271*f*, 274,
      276*f*, 280*f*, 283, 285*f*, 289*ff*, 294,
      297*ff*, 306, 343, 348, 387, 389, 629*ff*,
      805, 811
   moment, *see* angular momentum
   pictures   **Vol. 8:** 221, 238*f*, 253, 270, 277,
      279, 286, 289*f*, 330, 338, 343, 348,
      355, 372, 379*f*, 417, 426, 429, 436*f*,
      438–443, 444, 447*f*, 451*f*, 456, 502,
      506, 509, 514, 535, 538, 603, 634
   velocity   **Vol. 8:** 28, 34, 54, 60, 133, 209,
      236, 238, 240, 309, 320, 329, 332, 376,
      381*f*, 406, 462, 491, 494, 506, 517,
      523*ff*, 530, 536, 538*ff*, 543, 550, 603*f*,
      606, 609*ff*, 614, 618, 622*f*, 633, 635,
      772, 775, 779*ff*
orbits of first and second kinds
      Vol. 4: 579, 633
Ord och Bild (Scandinavian journal)
      Vol. 12: 126
Order of the Elephant (Denmark)
      Vol. 11: 347, 427, 431, 564
Order of the Red Banner (U.S.S.R.)
      Vol. 11: 298
ordinary electrodynamics, *see* electrodynamics,
      classical/ordinary
ordinary mechanics, *see* mechanics,
      classical/ordinary
organic
   chemistry   **Vol. 10:** 54, 150
   development, *see* evolution
   functions   **Vol. 10:** 68, 77
   life   **Vol. 11:** 91, 102, 344, 539, 548, 568,
      577, 594, 611, 614, 619, 623
   life, description of   **Vol. 7:** 409, 421
   metabolism   **Vol. 11:** 516, 523, 567, 614, 624
organicism   **Vol. 10:** 6

Oriental philosophy **Vol. 10:** 97

Orla-Jensen, S. **Vol. 11:** 406*p*

Ornstein, L.S. **Vol. 5:** 230, 251*f*, 279, 329

Ørskov, J. **Vol. 11:** 406*p*

Ørsted, A.S. **Vol. 10:** 259, 269, 352, 365

Ørsted, G. **Vol. 11:** 478, 482

Ørsted, H.C. **Vol. 1:** 99
    Vol. 4: 434, 469
    Vol. 5: 47*ff*, 129*f*, 133, 135, 136*f*,
    140*ff*, 273
    Vol. 6: 374
    Vol. 10: 54, 148, 152, 209, 258, 260,
    269, 272, 329, 330, 342–365, 366*p*,
    367–369, 587
    Vol. 11: 311, 313, 339, 352–354, 402,
    412, 470–475, 478–484, 488, 490, 494,
    495, 497, 498, 530, 544, 545, 551
    Vol. 12: 100, 107, 122, 123, 152, 164,
    182, 184, 190, 193, 220, 291, 293, 300,
    302, 306, 307, 309, 310, 317, 318, 321,
    322, 371, 381, 383, 428, 429, 431

Ørsted, H.C. (H.C. Ørsted's grandson)
    Vol. 11: 478, 482

Ørsted Medal **Vol. 5:** 47*ff*, 125, 127*f*, 135,
    136, 142, 379

Ørsted Meeting, *see* Nordic Physicists, First
    Meeting

Ortega y Gasset, J. **Vol. 12:** 14

orthogonal functions **Vol. 7:** 429

Osborn, F.H. **Vol. 11:** 326, 330

oscillator (*see also* harmonic oscillator/
    oscillation; virtual oscillator/radiation)
    Vol. 7: 3, 4, 306, 449
  harmonic, *see* harmonic oscillator model of
    atom; Planck's theory
  strength **Vol. 5:** 42, 44–45
    Vol. 8: 10, 209, 396, 400, 498,
    500*f*, 636
  virtual **Vol. 12:** 357

Oseen, C.W. **Vol. 1:** 101, 102, 103, 106, 109,
    114, 125–128, 422, 426, 430, 432, 434,
    534, 535, 556, 557
    Vol. 2: 109, 118, 119, 126, 128, 324,
    329, 331, 332, 337, 339, 343, 448,
    550–574, 613, 614, 615, 616, 617, 618,
    619, 620
    Vol. 3: 10, 12, 193, 495, 653, 655,
    670, 671
    Vol. 4: 25, 51, 731
    Vol. 5: 91*l*, 238–240, 297, 405–408
    Vol. 6: 189, 192, 322, 324, 413, 430–432

Vol. 8: 12, 19, 646, 755–758
    Vol. 10: xxxi, 462

Oslo lecture, Bohr's, 1940 **Vol. 9:** 411–466

Oslo Meeting, 1925 **Vol. 5:** 219, 238,
    249–252, 281–285, 405, 483, 485

Oslo, University of **Vol. 12:** 139, 141

Ostenfeld, J. **Vol. 12:** 400, 404

Ostermann, H. **Vol. 11:** 358*p*

Ostwald, W. **Vol. 1:** 231, 344
    Vol. 12: 139*n*

Otto Mønsted Foundation **Vol. 12:** 182, 190

Owen, E.A. **Vol. 1:** 106, 108, 110, 111, 530,
    531, 542, 543, 550, 551
    Vol. 2: 613, 614, 615, 617, 618

Owen, R. **Vol. 11:** 445

oxygen **Vol. 8:** 10*f*, 65–68, 87–89, 102–104,
    145, 157, 315, 564, 665, 695, 697, 740,
    743, 746, 748, 767, 780, 782, 792
    Vol. 11: 577, 593, 594, 612, 616,
    620, 628
    Vol. 12: 59, 60, 64
  absorption in **Vol. 2:** 33–35
  dispersion in **Vol. 2:** 34, 53–54, 336
  structure of atom of **Vol. 2:** 189, 231
  structure of molecule of **Vol. 2:** 34,
    138*ff*, 150*ff*

ozone molecule **Vol. 2:** 138, 150, 238

packing fraction, *see* mass defect

packing of particles in nuclei (*see also* nuclear
    density) **Vol. 9:** 152, 164, 173, 186,
    193, 197, 211, 215, 219, 229, 239, 274,
    282, 289, 293, 335*f*, 414, 444, 489,
    505, 523

page proofs, unused **Vol. 3:** 7, 8, 53–64

pair
  creation **Vol. 7:** 48, 174, 175, 185, 186,
    203, 206–208, 377, 457, 459, 497,
    501, 502
    Vol. 9: 5, 11, 305, 319, 668
  production (*see also* positron) **Vol. 8:** 241,
    339–340, 637, 795, 797, 798, 801–803
  theory **Vol. 7:** 41, 43, 197, 204, 209,
    215, 485

Pais, A. **Vol. 7:** 14*n*, 40, 45, 50, 220, 270,
    483–485, 496
    Vol. 8: 250*f*, 646
    Vol. 10: 455*p*
    Vol. 12: 38*p*

Paneth, F.A. **Vol. 1:** 580, 581
  Vol. 2: 419, 532, 616, 618
  Vol. 10: 393
Pantin, C.F.A. **Vol. 10:** 126
Panum, P.L. **Vol. 10:** 5
paradox/dilemma **Vol. 6:** xxi, 15, 21f, 45f,
  50, 52, 69ff, 76, 80, 82, 88, 92, 96, 98,
  116, 119, 125, 126, 131, 139, 148,
  149f, 153, 155f, 182f, 190, 204, 211,
  227, 232, 242, 248, 265, 269, 288, 294,
  322, 324f, 327ff, 365f, 368f, 419f,
  430, 453ff, 466, 473
  Vol. 8: 214, 219, 231, 269, 273, 278,
  282, 304, 306, 379, 385, 411f, 438,
  442, 498, 649, 698, 700ff, 724, 731f,
  803f
paramagnetism **Vol. 1:** 393
  Vol. 3: 147
  Vol. 6: 34, 144, 349, 434
parapsychology **Vol. 10:** 540, 542
Paris lecture, Bohr's
  January 1937 **Vol. 9:** 39
  October 1937 **Vol. 9:** 41, 229,
  265–282, 678
particle **Vol. 11:** 101, 515, 517, 523, 524,
  568, 574–579, 588, 589, 591, 593–596,
  649, 660, 665, 671, 672
  and field **Vol. 7:** 228, 488, 491, 497, 499
  and wave dilemma **Vol. 7:** 331, 350
  charged (*see also* charge) **Vol. 7:** 38, 199,
  212, 235, 236
  concept **Vol. 7:** 222, 463, 466
  elementary **Vol. 12:** 51, 54, 344, 346, 365
  high-speed **Vol. 11:** 90, 102, 515, 523, 531
  Vol. 12: 42, 60, 66
  interacting with radiation, *see* radiation,
  interacting with particles
  movement **Vol. 7:** 221, 347, 353, 354,
  356, 357, 371
  picture **Vol. 7:** 6, 262, 304
  Vol. 12: 42–44, 46, 47, 54, 61, 68, 69,
  209, 213, 221, 222, 252, 256, 260, 364
  strange (*see also* elementary particle)
  Vol. 12: 84
particle/wave, *see* wave/particle
partition between subject and object, *see*
  subject/object distinction/relation
Pascal, B. **Vol. 12:** 371
Paschen series **Vol. 2:** 169, 285
  Vol. 3: 207

Paschen–Back effect **Vol. 2:** 362, 366–367
  Vol. 4: 216, 629, 773
  Vol. 5: 289, 448, 452
  Vol. 6: 391
  Vol. 10: 339
Paschen, F. **Vol. 1:** xxxi
  Vol. 2: 169, 285, 335, 362, 365, 366,
  409, 445, 591, 603
  Vol. 3: 7, 52, 130, 134, 135, 164, 177,
  184, 207, 280, 281, 314, 316, 330, 331,
  332, 424, 445, 510, 526, 539, 554, 583,
  589, 591, 600, 653, 654, 687, 690, 691
  Vol. 4: 34, 93, 112, 119, 132, 197, 216,
  271, 333, 346, 369, 374, 379, 440, 472,
  565, 573, 576, 589, 605, 622, 629, 631,
  640, 651, 702, 712
  Vol. 5: 278, 408f
  Vol. 6: 386
  Vol. 8: 673
  Vol. 10: 338, 375, 378
  Vol. 12: 115, 340–342, 342p, 525
Pasteur, L. **Vol. 7:** 20
Patterson, R.A. **Vol. 4:** 493, 498, 530, 531
"Pauli effect" **Vol. 5:** 367f
Pauli exclusion principle (*see also* closing of
  electron groups; exclusion principle)
  Vol. 7: 176, 187, 220, 226
  Vol. 11: 567
Pauli, F. **Vol. 10:** 491, 543, 547, 550–552,
  556, 559, 561–563, 567, 568, 572
  Vol. 12: 366
Pauli, W. **Vol. 1:** xxvii, xxxii, xxxvi, xxxvii,
  xxxviii
  Vol. 3: 41, 42, 467, 469, 516, 531, 545,
  558, 649, 656, 657, 658, 659, 661,
  673–675
  Vol. 4: 23, 38, 39, 42, 601, 607, 649,
  653, 669, 670, 671, 672, 673, 675, 676,
  688, 700, 716
  Vol. 5: 3, 6, 19f, 23, 29, 31–37, 41f, 55,
  57–61, 63, 74–75, 79ff, 84, 86–88, 91,
  105, 116, 216, 220, 223–224, 225, 228ff,
  235, 238, 256, 260, 262, 266, 279f, 288,
  293, 296, 297, 329, 332, 336, 338, 340,
  343, 351, 354ff, 358ff, 370ff, 400,
  408–482, 498, 501, 503, 507f
  Vol. 6: vi, 3, 8ff, 16f, 19, 24, 30, 31,
  32–35, 36, 37, 41ff, 46, 53, 96, 125,
  136, 144, 146, 153, 158, 162, 164, 166,
  177, 181, 191ff, 304, 305, 306, 307,
  308, 309f, 318, 326–330, 337–347,

391*f*, 402, 411, 413, 415, 422, 425, 431, 432–456, 467*f*, 472
Vol. 7: 4, 8–11, 14, 17, 18*n*, 23, 31, 32*p*, 35, 38, 39, 49, 50, 64, 69, 82, 121, 126, 138, 165, 166*n*, 176, 182, 187, 191, 230, 234, 250, 252, 253*p*, 267–269, 277, 279, 328, 329, 348, 396*p*, 435, 441, 445, 449, 465, 469, 471, 472, 474, 477, 481, 485–487, 490, 495*n*, 496, 499, 501, 502, 514
Vol. 8: 211, 215, 231, 264, 304, 662, 677*f*, 692, 699, 701*ff*, 737*f*, 754, 812
Vol. 9: 5–6, 8, 9, 13, 17, 28, 36, 37, 44*f*, 46, 78, 79, 132, 134, 135–136, 138*f*, 275, 280, 307, 321, 533, 536, 541*ff*, 570*f*, 606–608, 644*ff*, 649, 651
Vol. 10: xxiv, xxxv*n*, xxxvi, xli, 13, 20, 173, 275*n*, 339, 379, 394, 395, 400, 403, 404, 412, 419, 424, 426, 442, 446, 447, 451, 454, 462, 478–481, 489, 491, 531, 533, 557*n*, 564*n*, 572*n*, 575*n*
Vol. 11: 567
Vol. 12: 11, 38*p*, 48, 116, 117, 210, 215, 348, 351, 362*p*, 363–366, 500, 502, 533
Pauli, W. Sr.   **Vol. 5:** 75, 435, 437*f*
Pauling, L.   **Vol. 8:** 698
Paulsen, J.   **Vol. 12:** 410*p*, 423*n*
Pauly, A.   **Vol. 10:** 543*n*, 566
Pavlov, I.P.   **Vol. 11:** 198, 201, 213
Peaceful Uses of Atomic Energy
   International Conference on, *see* Geneva Conference
   New York Conference on, 1954
      Vol. 11: 359, 361, 362*p*, 563, 717–719, 734
Peaslee, D.C.   **Vol. 9:** 508
Pedagogical Society (Denmark)
      Vol. 12: 400, 404
Pedersen, H.   **Vol. 11:** 406*p*
Pedersen, J.   **Vol. 10:** 221, 274*p*, 275, 546, 550
      Vol. 11: 343, 345*n*, 346, 400, 401, 406*p*, 410, 411, 440, 441, 465*n*, 525*p*
      Vol. 12: 100, 107, 126, 127, 248, 249, 471*n*
Pedersen, P.O.   **Vol. 1:** 6, 8, 30, 44, 50, 64
      Vol. 5: *137*
      Vol. 12: 184, 193
Pegram, G.B.   **Vol. 9:** 61, 550*ff*

Peierls–Kapur method, *see* Kapur–Peierls method
Peierls, R.E.   **Vol. 5:** vi, 94, 216
      Vol. 6: 330
      Vol. 7: 7–12, 19, 20, 28, 32*p*, 51*p*, 59, 80, 82, 101, 121, 123, 137, 139, 151, 165, 172, 173, 177, 179, 184, 185, 188, 189, 199, 229–235, 237, 287*n*, 441, 442, 464, 466, 468
      Vol. 9: 28, 32, 38, 45, 46, 47, 48, 49*ff*, 76, 80*f*, 134, 261, 332, 340, 377, 385, 391–393, 487–519, 537, 539, 570*f*, 609–621, 644*f*, 659, 663, 668, 670, 678*f*
      Vol. 10: 407, 455*p*, 531, 532
      Vol. 11: 328, 331
      Vol. 12: 8, 31, 34, 38*p*
Peltier coefficient   **Vol. 1:** 147, 350*ff*, 483, 490, 543
pendulum orbit   **Vol. 5:** 223, 279, 434, 436, 446, 450–451
penetrating orbits   **Vol. 4:** 8, 78, 147, 454, 478
penetration of charged particles   **Vol. 6:** 76, 82, 131, 156, 375
penetration of outer electron into atomic core
      Vol. 3: 24, 43
perception   **Vol. 7:** 259, 322
      Vol. 10: xxx, xxxv, xxxvi, xlii, xlvi, xlvii, 8, 45, 60, 62, 70, 91, 104, 169, 231, 235, 242, 405, 495
Perelman, C.   **Vol. 7:** 328
Perfilov, N.A.   **Vol. 8:** 331, 541, 568, 591
Pericles   **Vol. 12:** 152, 164
periodic alternation of properties
      Vol. 4: 587, 638
periodic system of elements (*see also* atomic number)   **Vol. 1:** xxii, xxxiv–xxxv
      Vol. 2: 117–120, 133, 136*f*, 146, 148, 189, 198–199, 207–210, 236, 289–290, 314, 393, 419, 429, 440, 530, 551, 552, 625, 632
      Vol. 4: 199
      Vol. 5: 131, 138, 168, 171, 242, 244, 246, 248, 251*f*, 256, 260, 262, 266, 272, 276, 278*f*, 282, 284, 398, 410, 413, 498, 499
      Vol. 6: 136, 158, 224, 228, 237, 243, 333, 375, 378*f*, 383, 386*f*, 390*ff*
      Vol. 9: 113, 156, 164, 173, 267, 269, 279, 304, 319, 407*f*, 417, 446, 593*f*

diagram of   **Vol. 4:** 272, 412, 432, 468, 489, 522, 584, 636, 658
hafnium, *see* hafnium
on the explanation of the   **Vol. 4:** 421*ff*
periodic system, perturbed
possessing two degrees of freedom **Vol. 3:** 116
possessing more than two degrees of freedom   **Vol. 3:** 117*ff*
periodic table of elements   **Vol. 8:** 24, 169, 172, 175, 177, 179, 780, 782, 786
Vol. 10: 373, 376, 379, 385, 387, 393, 394, 396, 400, 403, 437, 438, 446, 496*n*, 556
Vol. 11: 514, 522, 575, 579, 590, 598, 613, 614, 621, 623
Vol. 12: 10, 42, 43, 45, 46, 48, 51, 61, 67, 210, 215, 252, 288
general interpretation of   **Vol. 4:** 153–174
survey of   **Vol. 4:** 315*ff*
periodicity   **Vol. 12:** 42, 252, 357, 364
degree of   **Vol. 3:** 503, 533
systems, *see* multiply periodic systems
systems, mechanical   **Vol. 3:** 507, 516, 524, 528, 536, 546, 554, 558
systems, quantum theory of   **Vol. 4:** 554*ff*, 614*ff*
periods of vibration, long   **Vol. 1:** 358
permanent state (*see also* normal state; stationary state)   **Vol. 2:** 9, 107–109, 117, 164, 175, 180–185, 200*ff*, 233, 258, 296, 313, 353, 396, 475, 481*ff*, 500, 579*f*
Vol. 6: 177*ff*, 311*f*, 383*f*, 386, 395
Perrin, F.   **Vol. 7:** 32*p*, 302
Vol. 8: 269, 278
Vol. 9: 100, 134, 156
Vol. 10: 455*p*
Vol. 11: 563
Perrin, J.B.   **Vol. 1:** 455
Vol. 3: 361
Vol. 5: 274
Vol. 10: 435
Vol. 12: 125, 372
personality, integrity/dissolution of
Vol. 10: 71, 93, 201, 216
perturbation (*see also* Born approximation; disturbance of object by measurement)
Vol. 5: 64, 166, 168*f*, 180*f*, 196, 237, 243–244, 247, 331, 358*f*, 362, 477, 478, 480, 481, 493

Vol. 8: 8, 35, 205, 208*f*, 225, 256, 272, 275*f*, 281, 284*f*, 289, 291*ff*, 297*f*, 456, 492, 499, 504*f*, 511, 607, 629, 635*ff*, 772, 775, 798
of electron orbit   **Vol. 2:** 46–51, 338, 633
theory   **Vol. 9:** 108*f*, 275, 646*f*, 669
perturbations   **Vol. 3:** 55*ff*
central   **Vol. 3:** 266*ff*
secular   **Vol. 3:** 111*ff*, 153, 463, 503*ff*, 533*ff*
perturbing field, small   **Vol. 3:** 101
Peshkov, V.P.   **Vol. 11:** 298
Peters, B.   **Vol. 8:** 624*f*
Petersen   **Vol. 11:** 518, 525
Petersen, A.   **Vol. 7:** 56, 165*n*, 523
Vol. 10: xxxix*n*, xl, 110, 172, 480, 482, 484, 588–593
Vol. 11: 361, 734, 735
Vol. 12: 14, 16, 94, 96, 530–532
Petzoldt, J.   **Vol. 6:** 192
Pflüger, E.   **Vol. 10:** 5
Pfund series   **Vol. 2:** 116, 169
pH concept   **Vol. 12:** 124
pharmacology/pharmacy   **Vol. 10:** 148, 150, 153, 345, 352, 359, 365, 592
phase   **Vol. 6:** 20, 37, 49, 60, 62, 64, 79, 86, 98, 106, 118, 132, 133, 149*f*, 156, 164*ff*, 170*ff*, 179*ff*, 186, 193, 270, 295, 439*ff*, 443, 466
Vol. 8: 218, 271, 280, 337, 345*f*, 350*f*, 381, 449, 457, 497, 634, 638, 760
Vol. 9: 49, 374, 377, 379, 392*f*, 492, 494, 496*ff*, 501, 514*ff*, 614*f*
difference, between current and field **Vol. 1:** 363, 386
function   **Vol. 7:** 22
relations (*see also* coherence)   **Vol. 5:** 44, 52, 61, 109, 123, 146, 150, 309, 390, 393, 395*f*, 449, 453
space   **Vol. 6:** 143, 163, 172, 320, 467, 470
Vol. 7: 226
velocity   **Vol. 6:** 29, 31, 76*ff*, 84*f*, 118, 149
wave (*see also* de Broglie wave)
Vol. 6: 77*f*, 84*f*, 92, 96, 117, 126, 129, 132, 149*f*, 153, 155*f*
Phelps, E.R.   **Vol. 5:** 387
phenomenon   **Vol. 7:** 254, 258, 260, 262, 264–266, 268, 272, 275, 276, 285, 293, 295, 296, 298, 316, 321, 335, 341, 349, 362, 364, 368–370, 372, 376, 378, 391, 393, 488, 491, 510, 511

Vol. 10: xxxvii, xlvii, 89, 129, 178, 179, 183, 187, 209, 212, 213, 284, 286, 290, 291, 428, 444, 539, 541, 544, 545, 548, 549, 564, 574
Vol. 11: 370, 540, 549, 670
Vol. 12: 223
phenotype **Vol. 10:** 13, 246, 544, 549
Philadelphia Symposium, 1946, *see* National Academy of Sciences (U.S.)
Phillips, M. **Vol. 9:** 257, 387*f*, 484
philology **Vol. 12:** 248, 249
philosophers **Vol. 7:** 339, 340, 380
Philosophical Magazine (journal)
Vol. 12: 145
philosophy **Vol. 7:** 38, 249, 250, 326, 358, 372, 375, 379, 385–394, 407, 419, 464, 466, 502, 503, 524
Vol. 10: v, vi, ix, xxix, xxxi, xxxii, xxxix, xl, xliii–xlv, xlix, 7, 9, 16, 20, 67, 84, 97, 111, 117, 119, 127, 141, 159, 160, 172, 174, 182, 188, 202, 208, 212, 229, 234, 276, 285, 291, 297, 298, 300, 301, 303, 306–309, 313–322, 345, 346, 351, 358, 359, 364, 415, 425, 448, 484, 486, 509–515, 529, 530, 545–547, 549–551, 566, 573, 582, 584, 591
Vol. 11: vi, 26, 196, 200
Vol. 12: 3, 12, 118, 124, 136*n*, 154, 166, 201, 205, 225
photo-effect, nuclear **Vol. 9:** 30, 42, 43–52, 261–262, 263, 297–299, 306, 316, 320, 328, 331–332, 384, 393, 493, 519, 542*f*, 574*f*, 585*ff*, 606–608, 611, 635, 637, 678
photo-fission **Vol. 9:** 388–389
photobiology **Vol. 10:** 483
photochemistry **Vol. 10:** 386, 496*n*
photoelectric effect **Vol. 2:** 117, 124*f*, 162, 176–177, 289
Vol. 3: 244, 256
Vol. 5: 3, 14–16, 103, 315, 319, 323, 448, 452
Vol. 6: 10, 23, 69, 115, 148, 162–163, 227, 241, 270, 295, 420
Vol. 7: 279, 342, 401, 415
Vol. 8: 20*f*, 684, 690*f*
Vol. 9: 27, 304, 315, 318, 327, 603*f*
Vol. 10: 58, 120, 338, 376, 397, 432, 434, 437, 438
Vol. 11: 567
Vol. 12: 44, 356

photon
absorption/emission **Vol. 7:** 213, 305, 306, 342, 488
Vol. 12: 61, 69, 357, 358
concept **Vol. 7:** 3, 23, 71, 73, 96, 99, 131, 132, 148, 150, 180, 190, 315, 343
constituting field/radiation **Vol. 7:** 29, 72, 73, 84, 87, 103, 104, 115, 116, 132, 140, 142, 153, 154, 161, 162, 200, 334, 448, 450
deflection **Vol. 7:** 87, 142, 347, 356
density **Vol. 7:** 72, 97, 132, 149, 201, 238
distribution **Vol. 7:** 30, 72, 132
Einstein's idea of **Vol. 7:** 304, 344
emission **Vol. 7:** 11, 213, 305, 306, 342, 344, 488
energy **Vol. 7:** 6, 342, 343, 368, 369
localization **Vol. 7:** 491
momentum, *see* momentum of photon
polarization **Vol. 7:** 73, 132
produced by measurement **Vol. 7:** 207
propagation **Vol. 7:** 352, 358, 361, 362, 365, 369
properties of **Vol. 7:** 332, 350
uncertainty relation for **Vol. 7:** 310
wavelength of **Vol. 7:** 79, 136
Vol. 10: 29, 30, 211, 213, 376, 388, 402, 403, 426, 432, 440, 441, 443, 445, 450
Vol. 11: 660, 665
Vol. 12: 6*n*, 47, 51, 61, 69, 157, 170, 282, 356, 378
photon-free space **Vol. 7:** 220
phycomyces **Vol. 10:** 24, 476, 477
physical
chemistry **Vol. 3:** 225
Vol. 10: 149, 400
Vol. 12: 438, 440, 446, 447
experience, *see* experience
reality **Vol. 7:** vii, viii, 219, 254, 257*n*, 267, 274, 282, 284, 289–298, 317, 333, 371–375, 408, 420, 425–430
Physical and Chemical Societies of Copenhagen
lecture before **Vol. 3:** 9, 21, 39
lecture before (18 Oct 21) **Vol. 4:** 20*ff*, 187–256, 257–340
Physical Science Study Committee (U.S.)
Vol. 12: 15, 93, 95
Physical Society (Copenhagen) (*see also* Danish Physical Society)
Vol. 11: 479, 483, 488, 490

Vol. 12: 111*n*, 305–311, 511, 512, 524–526, 529, 530

A. Landé's lecture before   **Vol. 3:** 23, 24

N. Bohr's lecture before, 1920   **Vol. 3:** 3, 24, 30

   Vol. 4: 7*ff*, 43–69

Physical Society (London)   **Vol. 12:** 111

Physical Society (The Netherlands)
   Vol. 12: 116, 355

physicalism   **Vol. 10:** 15, 16

physiology   **Vol. 10:** xliii, xlv, 5, 8, 13, 24, 33, 34, 54, 67, 91, 104, 118, 119, 130, 135, 150, 152, 167, 175, 298, 468, 469, 471, 472, 500, 534, 535, 574

   Vol. 11: 213, 348

   Vol. 12: 329*n*

Piaget, J.   **Vol. 7:** 328

Piccard, A.   **Vol. 5:** 94
   Vol. 7: 32*p*

Piccard, F.   **Vol. 1:** 30

Pickering lines/series   **Vol. 2:** 116, 121–122, 123, 170–171, 200, 202, 274–276, 285, 297–300, 333, 352, 385–388, 504, 508–509, 532*f*, 558*f*, 561*f*, 582*f*
   Vol. 3: 588, 591
   Vol. 10: 378, 390, 392

Pickering, E.C.   **Vol. 2:** 121, 170, 200, 274, 276, 285, 352
   Vol. 3: 214, 216, 588, 589
   Vol. 6: 385
   Vol. 10: 378, 390

Pickup, E.   **Vol. 8:** 810*f*

pictorial description/concepts/interpretation/ representation   **Vol. 7:** 400, 401, 406, 414, 415, 419
   Vol. 10: xl, 42, 44, 70, 76, 87, 93, 103, 127, 136, 150, 158, 165, 166, 179, 208, 209, 211, 213–215, 277, 301, 379, 403, 404, 406, 440, 442, 443, 453, 500, 512–514, 582

Pihl, M.   **Vol. 5:** 137
   Vol. 10: 220*n*, 283, 289
   Vol. 11: 496, 499, 729

pion   **Vol. 10:** 453

Pippard, A.B.   **Vol. 12:** 38*p*

Placzek, G.   **Vol. 7:** 520
   Vol. 8: 224, 714, 716
   Vol. 9: 28, 29, 47, 48, 49*ff*, 53, 62, 63–64, 66*f*, 71, 80, 156, 209, 248, 292, 312, 325, 332, 340, 377*f*, 391–393, 398, 487–519, 539*f*, 544,

554, 609, 611*ff*, 615*f*, 644*f*, 659, 663, 678*f*

Plancherel, M.   **Vol. 7:** 328

Planck
   Festschrift, 1938   **Vol. 9:** 42, 46, 301–329
   Jubilee Issue, 1929   **Vol. 6:** 189–195, 201–217, 268, 293, 309, 441*ff*, 445, 447*f*, 456, 479
   Medal   **Vol. 6:** 317, 480

Planck harmonic oscillator   **Vol. 3:** 49, 78, 79, 105, 223, 243, 247, 250, 251, 262, 265, 310, 327, 369, 429, 437, 569, 586, 594
   Vol. 4: 211, 285, 344, 348, 354, 366, 617

Planck–Einstein relations, *see* quantum relations

Planck's constant, *see* quantum of action

Planck's formula/law (*see also* black-body radiation)   **Vol. 6:** 10, 178, 227, 241, 363, 382, 398, 423
   Vol. 9: 241, 276, 303, 308, 316, 318, 322, 427, 454
   Vol. 10: 400, 439, 442

Planck's theory   **Vol. 2:** 9, 34, 63, 88, 92, 108*ff*, 137*ff*, 162*ff*, 172, 180, 185, 214, 232, 258, 274, 280, 288–289, 292*ff*, 308, 312–313, 350*ff*, 393–395, 434, 443, 451*ff*, 474, 500, 553*ff*, 572, 579, 582, 588, 600, 603*f*, 630*ff*
   Vol. 5: 14, 103*f*, 132, 139, 221, 243, 247, 274, 275, 277, 315, 513
   Vol. 8: 10*f*, 19, 66, 88, 104, 158, 675*f*

Planck, M.   **Vol. 1:** xxiii, xxiv, xxviii, xxix, xxxiii, xxxviii, 101, 125, 127, 144, 174, 273, 299, 415, 448, 466, 467, 484
   Vol. 2: 104, 114, 115, 123, 124, 137, 147, 163, 164, 165, 167, 168, 232, 280, 288, 292, 293, 296, 308, 312–313, 322, 338, 340, 341, 342, 350, 351, 393, 394, 421, 434, 435, 443, 451, 452, 453, 454, 455, 456, 458, 459, 460, 474, 554, 555, 560, 563, 565, 567, 569, 571, 583, 600
   Vol. 3: 7, 20, 22, 69, 72, 73, 78, 79, 80, 82, 84, 105, 110, 139, 190, 208, 209, 213, 214, 223, 232, 233, 234, 243, 244, 250, 258, 274, 285, 350, 372, 381, 398, 402, 406, 426, 430, 433, 458, 460, 461, 470, 569, 586, 593, 627, 638, 640, 677, 686
   Vol. 4: 27, 28, 87, 190, 191, 193, 212, 264, 267, 285, 344, 345, 351, 354, 423,

435, 436, 447, 469, 470, 475, 551, 557, 558, 612, 616, 617, 659, 685
Vol. 5: 103, 132, 139, 221, 238, 274
Vol. 6: 21, 36, 37, 39, 47, 50, 69, 113, 178, 189, 192, 193, 194, 203*ff*, 208*ff*, 214, 217, 227, 233, 241*f*, 250, 261, 268, 282, 293, 359, 374, 380*f*, 413, 415, 442*f*, 456–457, 463, 467
Vol. 7: 279, 303, 309, 341, 342, 386, 389, 400, 414
Vol. 8: 10
Vol. 9: 42, 111, 215, 218, 303, 317*f*, 329
Vol. 10: xxxix, 23, 57, 69, 76, 87, 103, 127, 142, 149, 158, 176, 195, 209, 211, 338, 351, 364, 376, 384, 386, 389, 391, 400, 402, 406, 427, 432, 433, 435, 437, 518, 519, 521
Vol. 11: 6, 511, 512, 520, 521, 529–531, 567, 659, 664, 670
Vol. 12: 20, 43, 51, 61, 68, 90, 108, 125, 145, 149, 150, 156, 168, 199, 203, 209, 214, 222, 281, 282, 356, 358, 363, 378, 439, 441, 524
platinum, absorption in **Vol. 2:** 37, 83
platinum metals **Vol. 4:** 81, 243, 313, 404
Plato **Vol. 10:** 318, 322
Vol. 11: 669
Vol. 12: 155, 166
Platzman, E. (R.L. Platzman's wife)
Vol. 8: 763, 766
Platzman, R.L. **Vol. 8:** 248*ff*, 252, 428, 646, 759–766
Plesset, M.S. **Vol. 8:** 224, 795, 804
Vol. 11: 7*p*
Ploug, C.P. **Vol. 12:** 152, 164
Plum, H. **Vol. 3:** 286, 295
Plum, N.M. **Vol. 3:** 286, 295
plutonium **Vol. 9:** 71
Vol. 11: 579, 580, 582, 583, 597, 600, 603, 604, 615, 627, 628
Pockets, A. **Vol. 1:** 62
Podolsky, B. **Vol. 7:** 250, 251, 254, 256, 277, 290, 292, 295, 296, 372–374
poetry **Vol. 10:** 95, 141, 160, 318, 322, 511, 512, 545*n*
Vol. 12: 127, 159, 172
Pohl, R. **Vol. 1:** 158
Vol. 4: 24, 696
Poincaré, H. **Vol. 1:** 290, 353, 394
Vol. 2: 351

Vol. 5: 273, 314*f*
Vol. 10: 509, 510
Poisson bracket (*see also* commutation relation) **Vol. 5:** 501
Poisson distribution/statistics **Vol. 7:** 23, 238
Vol. 8: 325, 466
polar research **Vol. 10:** 259, 270
polarization (*see also* vacuum polarization; photon polarization) **Vol. 7:** 44, 207–209, 213, 215, 216, 484
Vol. 8: 216, 393, 395, 398, 400, 494–497, 611, 633, 662, 699*f*, 708
of electrons **Vol. 6:** 309, 347, 349
of fluorescent light, *see* fluorescent light, polarization of
of ions **Vol. 5:** 355*f*
of radiation **Vol. 3:** 11, 13, 28, 123, 142, 151, 263, 339–356
of spectral lines **Vol. 2:** 326, 345, 356, 360*ff*, 400*ff*, 468, 511
Vol. 5: 105, 107, 272, 277
Police (Danish), Language Club of **Vol. 11:** 729
Polish Intellectual Co-operation Committee **Vol. 7:** 260
Politiken (Danish newspaper) **Vol. 11:** 9, 54, 55, 61, 65, 206–209, 211, 338, 349
Vol. 12: 114, 525, 529, 531, 532
Pollard, E.C. **Vol. 9:** 619
polonium **Vol. 2:** 13, 78
Vol. 8: 65, 86, 102, 148, 212, 678, 752
Vol. 11: 574, 575, 587, 590
polyhedrons **Vol. 11:** 669
Polytechnical Institute (Denmark), *see* Technical University of Denmark
Popper, K.R. **Vol. 7:** 328
Vol. 10: 36*p*, 126
Pos, H.J. **Vol. 7:** 328
Pose, H. **Vol. 9:** 185*f*
Vol. 10: 412
positive rays **Vol. 1:** 525, 531
Vol. 2: 168, 171, 228, 402, 476, 479, 483, 499, 627
Vol. 5: 188, 202
Vol. 8: 86, 101, 184, 189, 671, 757*f*
positon (positron) **Vol. 12:** 11, 51
positron (*see also* pair production)
Vol. 7: 165*n*, 170, 175, 176, 181, 183, 186, 187, 190, 241, 242, 451, 453, 493, 520
Vol. 8: 339, 802, 811

Vol. 9: 11f, 14, 125f, 135ff, 140, 166, 174f, 185, 188, 216, 219, 274f, 280, 305f, 319, 321, 415, 420, 444, 448, 652
Vol. 10: 404, 407, 450

postulates
  Bohr's (*see also* assumptions, Bohr's principal; frequency condition, Bohr's)
    Vol. 5: 18, 45f, 102, 103–104, 113, 121, 255f, 258, 261ff, 265, 272, 275–279, 282, 284
    Vol. 6: 70, 80, 88, 143, 162, 178, 263, 285, 381–382, 383f, 393
  of quantum mechanics (*see also* quantum postulate/principle) **Vol. 6:** 14f, 70f, 423f

potassium, dispersion in (*see also* alkali metals, spectra of) **Vol. 2:** 337, 448–450, 564, 565–566, 604

potency **Vol. 10:** 91, 104, 174, 187

potential energy/barriers **Vol. 7:** 206, 371

potential scattering **Vol. 9:** 50, 392, 490ff, 496ff, 501f, 506, 508, 512, 515ff, 614f

Potter, R.D. **Vol. 9:** 60

Poulsen, F. **Vol. 11:** 406p

Poulsen, V. **Vol. 5:** 129, 137
    Vol. 11: 729
    Vol. 12: 123, 527

Powell, C.F. **Vol. 7:** 522
    Vol. 8: 554, 567, 570ff
    Vol. 10: 452
    Vol. 11: 569

power law of force of nucleus **Vol. 2:** 96–99

power sources **Vol. 11:** 246, 346, 573, 585, 587, 611, 618, 635, 636, 647, 651

Poynting's theorem/vector **Vol. 5:** 7ff, 12, 492–493

pragmatism **Vol. 10:** 160

Prankl, F. **Vol. 8:** 405, 555, 567
    Vol. 9: 365

Pre, Huntley du **Vol. 10:** 590

precession
  of electron orbit (*see also* Larmor's theorem)
    Vol. 3: 518ff, 528, 547ff, 559
    Vol. 5: 166, 168f, 171, 239, 288f, 331, 384, 386, 390–391, 393, 406, 425, 428, 467ff, 471, 473, 500
  of electron spin (*see also* Thomas precession) **Vol. 5:** 239, 288f, 406, 467ff, 471, 473, 500
  of orbital plane **Vol. 4:** 570, 626

predestination **Vol. 10:** 97, 279, 555

Preissmann, A. **Vol. 7:** 328

Preiswerk, P. **Vol. 9:** 652

prejudice **Vol. 10:** 64, 187, 188, 230, 235, 248, 249, 286, 292, 415
    Vol. 11: 94, 137, 141, 531, 538, 548, 690, 694
    Vol. 12: 61, 68, 119, 156, 168, 201, 205, 372, 378, 402, 404, 434, 435, 462, 464

Present, R.D. **Vol. 8:** 239, 750

Preston, T. **Vol. 3:** 421, 446, 527, 555
    Vol. 10: 338

Preußische Akademie, Berlin
    Vol. 6: xii, 21f, 317, 419f, 480

Price **Vol. 8:** 667

Prigogine, I. **Vol. 10:** 455p, 492

Princeton (*see also* lectures, Bohr's)
    Vol. 8: 232, 252, 260, 571, 739, 788ff
    Vol. 9: 39, 51, 56ff, 63f, 68ff, 74, 345, 355f, 361, 383, 402, 404, 550, 552, 554, 556, 557, 558, 564, 590, 591, 621, 633, 635, 658, 660f, 664, 679
    Vol. 10: 409, 476, 547, 550, 558, 561, 562

Princeton University **Vol. 11:** 12, 327, 330, 370, 372, 529, 564, 569, 711, 712
    Vol. 12: 4, 20

principal series **Vol. 2:** 170, 177, 274f, 356, 361, 366, 370, 385, 407, 449–450, 509, 564, 566
    Vol. 5: 122, 496

principle of mechanical transformability, *see* adiabatic principle/transformation

Pring, J.N. **Vol. 2:** 501, 597
    Vol. 8: 672, 676

Pringle, G.E. **Vol. 8:** 710

Pringsheim, E. **Vol. 1:** 273, 378
    Vol. 2: 124

Pringsheim, P. **Vol. 5:** 145, 148

prize essay **Vol. 1:** xix
  addendum to **Vol. 1:** 4
  facsimile of **Vol. 1:** 7, 8, 22–23
  judgement of **Vol. 1:** 5ff
  problem **Vol. 1:** 4
  Royal Danish Academy **Vol. 1:** 3ff, 21
  translation of addendum **Vol. 1:** 67–78

probability (*see also* causality, conservation of energy and momentum; transition probability) **Vol. 5:** 8f, 16, 29f, 42, 71f, 75f, 85, 89, 94, 98, 104f, 109,

$115f$, 178, $183ff$, 194, $198ff$, 209, 211, 302, $304f$, $310f$, 349, $362f$, $387f$, 402, 448, 453, 490
    Vol. 7: 73, 132, 207, 260, 331, 332, 343–345, 371, 403, 416, 428
amplitude/function **Vol. 6:** 34, 105, 139, 142, $164ff$, 168, 171, $175f$, 178, 181, 270, 295, 423, 434
distribution of energy loss
    Vol. 2: 9, 59–60, 64–70, 77–78, 81–83, 90, $493ff$, 515, 517
of capture/loss (*see also* cross section, capture/loss) **Vol. 8:** 36, 241, 329, 362–364, 381, 509, 529, 536, 538, 613, 620, 623, 750
of stationary state, *see* statistics of stationary states
of transition, *see* transition probability
Proca, M.A. **Vol. 12:** $38p$
professorship (teaching post) at University of Copenhagen **Vol. 2:** $329ff$, 496, 520–524, 535, $554ff$, 573, $591ff$
progress
  of cultures **Vol. 10:** 205, 261, 272
  of quantum physics **Vol. 10:** 149, 283, 289, 439
  of science **Vol. 10:** 35, 96, 102, 153, 157, 160, 205, 275, 353, 367, 415, 512, 514
Prometheus **Vol. 11:** 346
proper vibration **Vol. 6:** 12, 13, 33, 34, 49, $70f$, $96f$, 126–127, 129, 132, $134f$, $154ff$, $171f$, 174, $177f$, 181, 326, 423, 433, 434, $439f$, 450, 460, 466
    Vol. 7: 5, 242, 428–430
protactinium **Vol. 9:** 72, 372, 399–404, $407f$, 469, 473, 479, 484, 661, 679
proteins, cyclol theory of **Vol. 12:** 12, 54
protofluorine **Vol. 2:** 107
proton **Vol. 7:** 180, 190, 222
    Vol. 10: 398, 403, 409–413, 443, 449, 451
    Vol. 11: 568, 569, 577, 580, 593, 594, 599, 613, 614, 622–624, 646, 649
    Vol. 12: 48, 51, 54
  negative **Vol. 9:** 12–13, $126f$, $578f$
  structure of **Vol. 9:** 11, 13, 87, 103, 125, 131, 415, 445
Proust, J.L. **Vol. 6:** 378
providence (*see also* God)
  divine **Vol. 10:** 97, 276
  personified **Vol. 10:** 546, 550, 555

Providence, University of **Vol. 11:** 564
Prytz, K. **Vol. 1:** xix, 5, 7, 15, 534, 535
    Vol. 2: 520, 522
    Vol. 12: 326, 328
Przibram, K. **Vol. 6:** 50
pseudo-classical collisions, *see* collisions, pseudo-classical
Pseudo-Longinus, *see* Longinus, D.
psychiatry **Vol. 10:** 71, 93, 201
psychical experience/phenomena **Vol. 10:** 47, 62, 94, 143, 152, 159, 168, 174, 175, 199, 200, 215, 244, 245, 285, 286, 291, 540, 542, 543, 545, 549
psychoanalysis **Vol. 10:** 93, 566
psychology **Vol. 6:** 27, 189, 194, 204–206, 212–217, 220–221, 233–234, 250–251, 259, 269, 272, 275–276, 279, 293, 298, 301–302, 352, 445, 447
    Vol. 7: 259, 322, 336, 364, 376, 408, 420
    Vol. 10: v, ix, xxiii, xxiv, xxvi–xxx, xxxii, xl, xli, xliv, xlix, 7, 14, 16, 34, 45, 47, 53, 60, 62, 64, 71, 93, 98, 105, 123, 152, 157, 160, 184, 185, 188, 203, 204, 216, 244, 245, 248, $275n$, 279, 297, 298, 300, 314–317, 320–322, 324, 325, 405, 480, 481, 483, 502, 504, 518, 519, 521, 536, 538, 540, 542, 543, 547, 548, 561, 562, 565, 567, 573, 582, 589, 592
    Vol. 11: 134
    Vol. 12: 157, 158, 170, 171, 201, 205, 212, 217, 366
psychophysical parallelism **Vol. 7:** 376
    Vol. 10: xliii–xlv, xlvii, xlviii, 17, 34, 35, 47, 175, 518, 519, 521, 536, 537, 540, 542
public opinion **Vol. 7:** 252, 480
Public Records Office (London) **Vol. 11:** 5
Pupin, M.I. **Vol. 12:** 228, 229
purpose **Vol. 10:** 4, 34, 70, 104, 117, 118, 122, 151, 174
Pürschel, V. **Vol. 11:** 206, 207, 210
Pyenson, L. **Vol. 5:** 62
Pythagoras/Pythagoreans **Vol. 10:** 58, 177, 551
    Vol. 11: 669

quadrupole **Vol. 9:** 27, 44, 82, 194, $238f$, 315, 327, 544, $607ff$, $668ff$
quantization, sharp **Vol. 5:** 56, 72, 147, 153, 344

quantized energy transfer, *see* energy transfer, quantized

quantum **Vol. 2:** —

concept **Vol. 12:** 125, 126

conditions (*see also* commutation relation)
    Vol. 5: 45, 145, 149, 158, 160, 162, 164, 255, 257*ff*, 262, 264*f*, 271, 277, 279*f*, 446, 449*f*, 453, 500

conditions in the old quantum theory
    Vol. 2: 451–452, 464*ff*

domain **Vol. 7:** 71, 74, 101, 131, 133, 152

effects **Vol. 7:** 219, 258, 309, 342, 349, 350, 358, 368, 375
    Vol. 11: 659, 660, 664, 665
    Vol. 12: 6, 9, 14, 222, 223

electrodynamics **Vol. 5:** 215*f*
    Vol. 6: 4, 95, 146, 155, 156, 359, 402, 439*f*, 446*f*
    Vol. 7: —
    Vol. 8: 214*f*, 221, 655*f*, 714, 716
    Vol. 9: 110, 570*f*, 617, 621, 644, 671
    Vol. 10: 178, 212, 213, 403, 407, 452, 453, 531, 532
    Vol. 12: 348, 351, 359

field theory **Vol. 12:** 6

mechanics **Vol. 1:** xxxvii–xxxviii
    Vol. 5: 27, 35, 46*f*, 57, 61, 62, 88, 215, 220–223, 225, 230, 238*f*, 258, 261, 264*f*, 267, 271, 272, 280, 283*f*, 289, 312, 330, 337*f*, 366*ff*, 374*f*, 377*f*, 405*f*, 434, 436, 445–453, 454, 456, 457–461, 463, 465*ff*, 469, 478, 480*f*, 484*f*, 497, 498, 501
    Vols. 6–9: —
    Vol. 11: 11, 52, 515, 523, 531, 659, 661, 664, 665, 671, 713*n*
    Vol. 12: 7, 10, 22, 25, 26, 31, 34, 39, 44, 47, 51, 79, 81, 116, 118, 200, 204, 220, 223, 224, 281, 340, 341, 344, 345, 348, 351, 358, 364, 365, 439, 441, 499–502, 528, 534

mechanics, relativistic (*see also* Dirac equation; electron theory, relativistic; relativistic quantum mechanics)
    Vol. 8: 221, 273, 275, 282, 284, 435, 502, 516–517, 518, 522, 637*f*, 715
    Vol. 9: 6*f*, 88, 106–111, 112, 126, 547, 555, 605–606

number **Vol. 5:** 35, 38, 45, 52, 58, 158, 162*f*, 166–171, 234, 251*f*, 255, 260, 262, 267, 271, 272, 277*f*, 280, 288*f*,
362*f*, 395, 411, 413, 417, 420, 424, 426, 428*f*, 431*ff*, 436, 445–453, 457*ff*, 490*f*, 500*f*, 503, 505*f*
    Vol. 6: 26, 35, 49, 126, 134*f*, 154*f*, 157, 173, 228, 230, 243, 245, 388*f*, 391, 394, 397, 435, 466
    Vol. 8: 264, 386, 389, 506, 525, 541, 604, 609, 613, 616, 618, 621, 623, 680
    Vol. 9: 79, 138, 145, 153, 176, 193, 279, 304, 317, 319, 376*f*, 420, 449, 524
    Vol. 12: 45–47, 51, 344, 345

of action **Vols. 2–5:** —
    Vol. 11: 52, 133, 512, 515, 520, 523, 530, 567, 659–661, 664, 665, 670, 671
    Vol. 12: 9, 14, 22, 26, 43, 44, 47, 51, 61, 62, 68, 69, 90, 94, 96, 108, 156, 157, 168, 170, 199, 200, 203, 204, 209, 214, 222, 223, 233, 281, 356, 357, 363, 364, 374, 375, 378, 439, 441, 533

physics **Vol. 12:** 10, 12, 13, 90, 94, 96, 115, 224, 363–365, 372

postulate/principle **Vol. 1:** xxv–xxx
    Vol. 6: 7, 30, 44, 48, 52, 70, 75, 81, 89, 91, 93*ff*, 107, 108, 113, 114, 116, 119*f*, 122*ff*, 130*f*, 134*f*, 147, 148*ff*, 157*f*, 190, 200, 229, 232, 244, 248, 265, 288, 393, 431, 465, 478
    Vol. 7: 305, 306, 402, 415, 416
    Vol. 9: 87, 281, 304, 309, 318, 323
    Vol. 10: xxiv, xxxi, 331, 376, 377, 389, 391, 396, 463
    Vol. 12: 47, 357, 358

processes **Vol. 12:** 156, 168, 222

processes, radiative **Vol. 12:** 357, 358, 378

relations **Vol. 6:** 21*f*, 39, 59, 62, 64, 69*f*, 76*ff*, 83*f*, 86, 92, 117, 119, 126, 129, 132, 149*f*, 154*ff*, 363, 382, 388, 392*f*, 416, 419

rules **Vol. 8:** 33*f*, 659*f*, 662

states (*see also* stationary states)
    Vol. 12: 51, 340, 341, 357, 358

statistics (*see also* Bose–Einstein statistics; Fermi–Dirac statistics) **Vol. 6:** 50, 144*f*, 349, 397*f*, 404*f*, 466
    Vol. 8: 715*f*
    Vol. 10: 179, 403, 447, 452, 453, 519, 522
    Vol. 12: 22, 25, 344, 346, 364

quantum numbers **Vol. 4:** 193, 486, 520

coupling **Vol. 3:** 508

[127]

effective **Vol. 4:** 300, 501, 533, 581, 607, 634, 653
invariance and permanence of
    Vol. 4: 578, 632
principal **Vol. 3:** 514, 517
    Vol. 4: 196, 373, 455, 475, 486, 520, 561, 619
principle of existence and permanence of
    Vol. 3: 41, 473, 508, 538
proper **Vol. 3:** 508, 518, 537
second subordinate **Vol. 4:** 509, 541
subordinate **Vol. 3:** 515, 517, 544, 546
    Vol. 4: 448, 455, 476, 561, 619, 660
Quantum Theory of Line-Spectra, On the
    Vol. 3: 3*ff*, 67–200
correspondence related to **Vol. 3:** 10*ff*
German translation of **Vol. 3:** 9
Part I **Vol. 3:** 71–102
Part II **Vol. 3:** 103–166
Part III **Vol. 3:** 167–184
Part IV Fragmentary Drafts
    Vol. 3: 185–200
quantum theory (*see also* Planck's theory *or more specific subjects, e.g.*, angular momentum, quantization of; assumptions, Bohr's principal; frequency condition, Bohr's; stationary state) **Vol. 1:** xxiii–xxiv, 571, 581
    Vol. 12: 4*n*, 5, 10, 43–45, 47, 51, 54, 145, 156, 168, 210, 214, 344–346, 348, 351, 356, 357, 363, 365, 368, 439, 441, 508
formal development of **Vol. 3:** 593*ff*
fundamental postulates of **Vol. 4:** 437
origin of **Vol. 4:** 435*ff*, 469*ff*
postulates of **Vol. 4:** 264
principles of **Vol. 4:** 210
"quasi-molecule" **Vol. 5:** 189, 203, 362*f*
Quebec Agreement/meetings **Vol. 11:** 15, 35–37, 39, 255*n*, 275, 280, 284, 285, 290, 292
quenching **Vol. 9:** 332
Quensel **Vol. 3:** 652, 654
Quine, W. **Vol. 10:** 82

rabbi, teachings of the **Vol. 7:** 284
Rabi, I.I. **Vol. 9:** 31, 544, 634, 637
    Vol. 11: 367, 721, 722
    Vol. 12: 378
racial ideology/theories **Vol. 11:** 11, 199, 202

radiation **Vol. 12:** 10, 51, 277, 356, 358, 500, 501
absorption/emission **Vol. 12:** 42, 44, 45, 47, 288, 357, 439, 441
beam **Vol. 7:** 15, 16
black-body, *see* black-body radiation
classical, *see* electrodynamics, classical/ordinary
damping/reaction/resistance **Vol. 5:** 22, 37–38, 43, 53, 64, 108, 200–201, 358*f*, 396, 401, 491, 504*f*
    Vol. 6: 13, 133, 138, 142–143, 156, 177, 401*f*, 460
    Vol. 7: 60, 124, 171, 180, 183, 189, 232, 235–237, 402, 415
    Vol. 9: 107*f*, 502, 607*f*
effects **Vol. 7:** 353, 368, 371
energy **Vol. 7:** 10, 82, 138, 234
equilibrium **Vol. 7:** 334, 344, 400, 414
field **Vol. 7:** 3, 8, 64, 82, 89, 116, 126, 138, 143, 162, 204, 207, 235, 345
in free space **Vol. 7:** 304
interacting with particles **Vol. 7:** 47, 230, 232
momentum **Vol. 7:** 82, 138, 520
packet **Vol. 7:** 84, 87–89, 111, 140–143, 158
phenomena/processes **Vol. 7:** 80, 88, 137, 143, 176, 187, 220, 342, 343, 345–347
pressure **Vol. 5:** 116–117
quantum **Vol. 7:** 401, 415
quantum theory of **Vol. 7:** 64, 126, 179, 189, 345
scattering of, *see* scattering of radiation
secondary **Vol. 2:** 238
theory **Vol. 10:** 210, 211, 403, 432, 433, 443
theory, Einstein's **Vol. 2:** 633
treatment **Vol. 12:** 327, 329, 330, 454, 459
virtual, *see* virtual oscillator/radiation
radiation, from non-isolated systems
    Vol. 3: 487*ff*
problem in the quantum theory
    Vol. 3: 372*ff*, 410*ff*, 476*ff*
radiative collisions, *see* collisions, radiative
radio transmission **Vol. 10:** 29, 209, 350, 363
radioactive
isotopes/substances (*see also* isotopes)
    Vol. 11: 363, 373, 574–577, 579, 580, 583, 587, 589, 591, 592, 594, 597, 599, 600, 604, 606, 614, 616, 617, 623, 627, 630, 646, 649, 718

[128]

Vol. 12: 42, 60, 66, 79, 81, 84, 211, 215n, 216, 255–257, 259–261, 265, 277, 287

methods, Bohr's experiments on
Vol. 2: 3–4, 11–14, 576, 624

transformation/transition, *see* disintegration; transformation/transmutation

transformations, Bohr's lecture on
Vol. 2: 4, 624

radioactivity (*see also* β-decay; nuclear transmutations)
Vol. 2: 117–120, 189, 190, 212–214, 236–238, 418, 423–425, 428, 440–441, 528, 530, 551, 624
Vol. 5: 95, 130, 138, 274
Vol. 6: 224f, 228, 230, 237ff, 243, 246, 375–376, 377ff, 404, 405–407
Vol. 7: 345, 371
Vol. 8: 4, 17, 75, 90, 149, 171ff, 175ff, 179, 329, 459, 467, 469, 473, 562f, 670, 680, 712f, 719, 734, 736, 755, 779, 781, 787, 793
Vol. 10: 56, 75, 86, 103, 104, 179, 210, 338, 384, 385, 391, 392, 408, 409, 411–413, 438, 447, 450
Vol. 11: 530, 565, 578, 583, 595, 596, 606, 615, 627, 659, 664, 671, 726
Vol. 12: 20, 42, 51, 145, 252, 265, 277, 287, 502, 504

absorption of α-rays **Vol. 1:** xxiv, 555, 559, 573

artificial **Vol. 11:** 515, 516, 523, 531, 569, 578, 595
Vol. 12: 51, 84, 109

displacement laws **Vol. 1:** xxii–xxiii

isotopes **Vol. 1:** xxii, 581

Radioisotopes in Physical Sciences and Industry, Copenhagen conference on, 1960 **Vol. 11:** 736

radiology **Vol. 10:** 75, 78

Radiology, 7th International Congress of, 1953 **Vol. 10:** 74, 75, 78, 588

Radiology, 8th International Congress of, 1955 **Vol. 10:** 478

radium
Vol. 11: 90, 102, 515, 523, 565, 568, 574–576, 578, 587, 589, 591, 596, 614, 624, 682, 685
Vol. 12: 60, 61, 66, 67, 255, 259, 287, 524

C **Vol. 2:** 29, 74ff, 85ff

preparation **Vol. 11:** 517, 524, 574, 578, 587, 596, 683, 686

stations **Vol. 11:** viii, 373, 375, 516n, 673p, 682–687, 726
Vol. 12: 327, 330, 332, 333

Radium Foundation **Vol. 11:** 373, 682, 683, 685, 686

Radium Gift **Vol. 11:** 517n, 525p

radius of atom, *see* dimensions, atomic

Rahbek, K. **Vol. 12:** 152, 163

Rahbek, K.L. **Vol. 12:** 152, 163

Rahmy, R. **Vol. 10:** 575n

Raman, C.V. **Vol. 5:** 59f

Raman effect (*see also* Smekal–Raman effect)
Vol. 6: 393
Vol. 10: 401

Ramanujan, S. **Vol. 9:** 258

Ramsauer effect **Vol. 5:** 31, 79, 80–81, 84, 90, 108, 192–193, 206, 350, 353, 364f
Vol. 6: 163
Vol. 8: 29f, 226, 254, 269, 278, 289f, 297, 303f, 457, 689f, 692ff

Ramsauer, C. **Vol. 3:** 508, 537, 631
Vol. 5: 79ff, 90, 108, 192, 206
Vol. 8: 29, 457, 568, 689f, 692ff

Ramsay, W. **Vol. 1:** 29
Vol. 6: 376
Vol. 11: 445

Ramsey, W.H. **Vol. 9:** 649, 650

Randall, A. **Vol. 11:** 55p

Randall, H.M. **Vol. 4:** 590, 641
Vol. 5: 383ff
Vol. 8: 664

Randall, J.H. **Vol. 10:** 82

Randers, G. **Vol. 11:** 365, 618n, 628n

range of charged particles **Vol. 2:** 29f, 64–65, 68, 74ff, 493, 495, 499–500, 581

range–velocity relation, *see* velocity–range relation

Rankine, A.O. **Vol. 3:** 45

rare earth atoms, structure of **Vol. 2:** 210

rare earths **Vol. 4:** 81, 172ff, 242ff, 312, 405, 459, 480, 512, 543, 662
Vol. 12: 46, 288

Rashevsky, N. **Vol. 10:** 36p

Rask–Ørsted Foundation **Vol. 5:** 134, 141
Vol. 8: 659, 690, 775, 778f, 781
Vol. 10: 260, 272

Vol. 11: 7, 343, 403, 413, 448, 451,
456, 461, 513, 514, 522, 546, 551
Vol. 12: 160, 173, 183, 191
Rask, R. **Vol. 10:** 259, 269
Vol. 11: 544, 551
Rasmussen, E.K. **Vol. 8:** 205
Vol. 9: 28, 58, 68, 537, 590f, 621–640
Vol. 11: 358p, 518, 526, 653p
Vol. 12: 13, 71p, 119, 380p,
381–384, 526
Rasmussen, K. **Vol. 10:** 250p, 259,
270, 271p
Vol. 12: 106, 128, 240, 244, 245, 478,
479, 524
Rasmussen, R.E.H. **Vol. 9:** 28, 630f
Vol. 12: 305, 308
Rasmussen, S.V. **Vol. 1:** xx
Rathenau, G.W. **Vol. 12:** 38p
ratio between thermal and electric
conductivities **Vol. 1:** 337ff, 341, 342
rationalism **Vol. 12:** 224
Ratner, S. **Vol. 2:** 597
Rau, H. **Vol. 2:** 331, 333, 388, 399, 405, 407,
559, 563
Vol. 8: 20
Rausch von Traubenberg, H. **Vol. 3:** 434, 485
Vol. 8: 25, 646, 766–769
Rawlinson, W.F. **Vol. 2:** 500
Vol. 8: 111, 114, 669, 671
Ray, B.B. **Vol. 5:** 328, 483, 485
Vol. 11: 514, 522
Rayleigh–Jeans' law **Vol. 2:** 288
Vol. 5: 303
Rayleigh, Lord **Vol. 1:** 4, 5, 6, 7, 13, 14, 15,
29, 30, 31, 36, 43, 44, 62, 65, 67, 74,
76, 77, 81, 82, 88, 147, 231, 262, 272,
273, 274, 300, 344, 370, 378, 379, 419,
474, 556, 557
Vol. 2: 24, 125, 288
Vol. 3: 343, 352, 591, 595
Vol. 5: 84, 137, 274, 350
Vol. 6: 76, 82, 111, 150, 396
Vol. 8: 56, 163f
Vol. 9: 70
Vol. 10: 375, 388, 393, 406, 432, 433
Vol. 11: 214, 445
Rayleigh's formula **Vol. 8:** 164f, 170, 173,
177, 347, 352
Rayton, W.M. **Vol. 8:** 554, 568
reactors **Vol. 11:** 360, 363–365, 578,
581–583, 596, 601–606, 614–617,

626–630, 646, 650, 718
Vol. 12: 359
readership at University of Manchester
Vol. 2: 8, 330ff, 594–595
realism **Vol. 10:** 95
reality (*see also* physical reality)
Vol. 10: xxxii, 14, 15, 95, 503,
504, 564
reality of stationary states **Vol. 6:** 14f, 32,
35, 49, 98, 130–135, 155–157, 267,
290, 423, 424, 434
reason **Vol. 10:** 47, 92, 105, 185, 245, 246,
536, 537, 545, 549
Rebbe, O. **Vol. 11:** 358p
Rebild Bakker **Vol. 12:** 240, 244
Rechenberg, H. **Vol. 9:** 578
reciprocity **Vol. 6:** 192–194, 202, 204f,
210ff, 214f, 271, 296–297, 441,
443ff, 479
between emission and absorption
Vol. 5: 111, 113
of collisions **Vol. 5:** 70–71, 178ff,
194ff, 455f
of language **Vol. 5:** 75, 438f
recognition **Vol. 10:** 575, 576
recoil **Vol. 8:** 204, 257, 271, 280, 372, 459,
468f, 473, 549, 563ff, 717f, 803,
807, 819
Red Cross **Vol. 12:** 126, 332, 334
red-shift **Vol. 7:** 366
Reddemann, H. **Vol. 9:** 22, 24, 544, 546
reduced mass **Vol. 2:** 122, 275, 299, 301,
326, 333, 356, 404, 416, 419, 440, 588
Vol. 8: 253, 310, 430, 457, 511
Vol. 9: 73f, 376, 658
reduction of wave packet **Vol. 6:** 33, 94, 142,
174, 433
reference body/system **Vol. 7:** 90, 144, 298,
333, 364, 366, 389, 504, 507
reflection **Vol. 3:** 495ff
of electrons (*see also* Davisson–Germer
experiment; diffraction of electrons;
interference of electrons) **Vol. 6:** 70,
77, 84, 92, 116, 149, 176–177, 230,
245, 265, 288, 334
of light **Vol. 5:** 7, 85, 113ff, 273, 311, 400
of wave **Vol. 9:** 154, 245, 497ff
of X-rays (*see also* diffraction of X-rays;
X-rays, reflection of) **Vol. 6:** 77, 84
Reformation **Vol. 10:** 254, 263

[130]

refraction **Vol. 2:** 31, 32, 90
Vol. 5: 18, 85, 102, 113, 115, 273, 311, 317*f*
Vol. 8: 63*f*, 84*f*, 99*f*, 117, 160, 292*f*, 665
refractive index **Vol. 8:** 10, 292, 666
refugees
Vol. 11: 11, 20, 61, 137, 141, 217, 228, 239, 338, 601*n*
Vol. 12: 124, 402, 404, 434, 435
regeneration **Vol. 10:** 4, 117, 137, 215
Regensen (student hall of residence)
Vol. 12: 182, 191
regulation mechanisms/processes
Vol. 10: 35, 55, 60, 68, 117, 150, 166, 167
Reiche, F. **Vol. 3:** 180
Vol. 4: 53, 65, 132
Vol. 5: 21, 42, 106, 400*f*, 403
Reichenbach, H. **Vol. 7:** 329
Vol. 10: 16, 525, 526
Reidemeister, K. **Vol. 7:** 328
Reinganum, M. **Vol. 1:** 298, 338, 339, 341, 403, 404, 405, 406, 409, 446, 534, 535
relativistic
argument **Vol. 7:** 330, 368, 378, 379
covariance/invariance **Vol. 7:** 86, 141, 212, 225, 297, 309
effects **Vol. 7:** 173, 185, 226, 227
electron theory, *see* electron theory, relativistic
many-body problem **Vol. 6:** 35, 435
mass **Vol. 2:** 9, 28, 37, 60, 70–73, 79–84, 89*f*, 115, 191–192, 326–329, 374, 379–382, 441–442, 466–467, 469, 504*ff*, 515, 517, 604, 629
quantum mechanics/theory (*see also* electron theory, relativistic; quantum mechanics, relativistic) **Vol. 6:** 233, 249, 292, 314, 333, 339, 402, 426*f*, 430, 432
Vol. 7: vii, 49, 63, 121, 126, 165, 171, 184, 231, 309, 320, 463, 466
Vol. 10: 87, 407, 450, 515, 516, 531, 533
stopping formula, *see* stopping formula, Bohr's relativistic
wave mechanics **Vol. 7:** 442
relativity **Vol. 5:** 7, 34, 35, 37, 41, 216, 229, 233, 235*ff*, 239, 274, 277, 288*f*, 312,

330, 339, 374, 376*f*, 406, 422–425, 427–429, 445*f*, 448, 450*ff*, 457*f*, 462*f*, 467–474, 478–482, 499, 504, 506
and complementarity **Vol. 7:** 266, 394
effect on hydrogen **Vol. 3:** 119*ff*
equivalence in theory of **Vol. 10:** 90, 103, 409, 448, 449
general **Vol. 6:** 136, 157, 260, 281, 328, 368*f*, 453
Vol. 7: 182, 191, 257*n*, 281, 297, 298, 320, 367, 378, 476, 479
Vol. 9: 8, 132, 568*ff*
Vol. 10: 31, 44, 85, 142, 340
Vol. 11: 134, 670
notion of **Vol. 7:** 276, 285, 287, 388
of human concepts/judgement
Vol. 10: 241, 306, 308
of simultaneity **Vol. 7:** 3
postulate/principle/theory
Vol. 10: xxxviii, 40–42, 59, 85, 90, 103–105, 158, 176, 187, 204, 209, 230, 235, 242, 243, 247, 277–279, 284, 290, 350, 363, 409, 425, 433, 448, 518, 521, 568, 570, 573
theory of (*see also* relativistic mass)
Vol. 7: 8, 12, 181, 182, 191, 199, 224, 225, 257*n*, 258, 266, 297, 298, 309, 317–319, 330, 342, 351, 358, 364, 365, 375, 377–379, 389, 401, 402, 415, 416, 502, 503, 505, 508, 522
Vol. 11: 530, 540, 549, 568, 576, 591, 613, 622, 658, 659, 663, 664, 670
Vol. 12: 20, 47, 156, 168, 221, 363, 365, 372, 374, 375, 378, 533
religion **Vol. 10:** xxxix, xl, 96, 97, 275, 276, 278–280, 298, 483, 511, 512, 546, 550, 555, 567, 588
Renaissance **Vol. 7:** 399, 412, 502, 503
Vol. 10: 54, 67, 91, 97, 102, 104, 117, 130, 135, 157, 165, 174, 194, 208, 257, 267, 276, 343, 357
Vol. 11: 123, 536, 547, 612, 620, 669
Vol. 12: 371
replication **Vol. 10:** 8, 13, 22–24, 477
reproduction **Vol. 10:** 4, 575, 576
residual excitation (*see also* excitation, residual) **Vol. 9:** 68, 242–243, 276*f*, 337, 351, 353, 355–357, 375*ff*, 382, 388, 393, 424, 452, 477*f*, 484, 679
resistance, *see* electric resistance

resistance, Danish **Vol. 11:** 14, 47, 56, 61, 137, 141, 288, 297, 372, 702, 703
    Vol. 12: 211, 216, 454, 460
resolving power **Vol. 6:** 121, 150*f*
resonance (*see also* Heisenberg resonance)
    Vol. 2: 72
    Vol. 5: 51, 55, 146, 151, 273, 279, 318, 394
    Vol. 6: 13, 97, 127, 154, 460
    Vol. 8: 19*f*, 142, 208, 235, 254, 256, 303, 355, 457, 459, 464, 499*f*, 504–517, 525, 527, 532, 562, 592, 683, 685, 688, 691
    fluorescence/radiation/spectra **Vol. 3:** 236, 254*ff*
    Vol. 5: 17, 39, 48, 50, 57*f*, 60, 65*ff*, 146, 273, 317, 335, 346, 401, 416, 420, 495
    Vol. 6: 20, 133, 156, 179–181, 186
    sharpness of, *see* sharpness of resonance
respiration **Vol. 10:** 10, 33, 54, 91, 104, 119, 137, 151, 166, 169
responsibility **Vol. 10:** 95, 105, 160, 185, 202, 205, 287, 292, 354, 367, 415
    Vol. 11: 6, 22, 47, 54, 61, 74, 80, 89, 118, 124, 128, 133, 137, 140, 141, 143, 146, 148, 156, 159, 166, 184, 297, 331, 332, 369, 570, 611, 615, 619, 627, 643, 646, 650, 702, 703
    Vol. 12: 84, 454, 460
rest
    energy **Vol. 7:** 227
    mass **Vol. 7:** 60, 124, 216, 225–227
retardation effects **Vol. 7:** 3, 102, 152
retardation, relativistic **Vol. 8:** 214, 215, 275, 284, 435, 497, 516, 712*f*
Retherford, R.C. **Vol. 7:** 48
Reventlow, E. **Vol. 11:** 327, 330
reversibility (*see also* irreversibility)
    Vol. 5: 185, 200, 305
Reymond, A. **Vol. 7:** 328
Ricci-Curbastro, G. **Vol. 12:** 372
Rice University **Vol. 12:** 532
Richardson, Mrs O.W. **Vol. 3:** 14, 15
Richardson, O.W. **Vol. 1:** 111, 112, 113, 114, 115, 116, 117, 118, 119, 234, 347, 440, 441, 442, 443, 447, 448, 460, 462, 482, 484, 486, 488, 489, 540, 541, 542, 543, 562, 563, 578, 579
    Vol. 2: 71, 263, 374, 395, 616, 618, 619
    Vol. 3: 13–15, 631, 657, 658

    Vol. 5: 30, 106, 239, 407, 490
    Vol. 6: 36, 37, 311, 318, 349, 415
    Vol. 7: 32*p*, 172, 184
    Vol. 8: 141
    Vol. 10: 339, 440
    Vol. 12: 4, 5, 9, 20, 376*p*
Richter, S. **Vol. 5:** x, 80, 238
Richtmyer, R.D. **Vol. 8:** 765
Riecke, E. **Vol. 1:** 97, 124, 132, 133, 134, 143, 144, 149, 172, 173, 223, 224, 225, 242, 297, 298, 337, 338, 354, 413, 446
Rieff, P. **Vol. 7:** 411
Riemann surfaces **Vol. 3:** 22, 634
    Vol. 6: xxvi
    Vol. 10: xxviii
Riemann, G.F.B. **Vol. 5:** 274
    Vol. 6: xxiv
    Vol. 10: xxix
    Vol. 12: 372
Riesz, M. **Vol. 3:** 653, 654
    Vol. 10: 558
rigid body/frame **Vol. 7:** 83, 84, 109, 110, 117, 139, 157, 158, 163
    Vol. 10: 30, 128, 158, 165, 184, 277
Rindal, E. **Vol. 12:** 17*p*
Risø **Vol. 11:** vii, 365, 368, 369, 617, 629*n*, 630*n*, 646, 648, 653*p*
Ritter, J.W. **Vol. 10:** 345–347, 359, 361
Ritz
    formula (*see also* Rydberg–Ritz formula)
        Vol. 3: 43
    law (*see also* Rydberg law)
        Vol. 2: 171–172, 284–285, 293–294, 300, 325, 351, 357, 368, 386, 395, 403, 412, 445, 470, 508
    principle, *see* combination principle
    series **Vol. 2:** 386
Ritz, W. **Vol. 2:** 169, 171, 284, 351, 357, 368, 395, 403
    Vol. 3: 71, 169, 172, 207, 251, 321, 336, 373, 381, 411, 423, 514, 543, 567, 568, 583, 587
    Vol. 4: 113, 206, 279, 346, 434, 438, 469, 471, 553, 559, 613, 618
    Vol. 5: 157, 161, 274
    Vol. 6: 228, 242, 382
    Vol. 10: 58, 331, 332, 338, 375, 464
Ritzau Press Agency **Vol. 11:** 9, 207, 209, 211, 294, 295, 301, 302
Roberts, G.E. **Vol. 8:** 813

Roberts, R.B.   **Vol. 9:** 31, 56, 357, 365, 383f, 389
Robertson, P.   **Vol. 5:** x, 6, 93
   Vol. 8: 25
Robertson, W.A.   **Vol. 2:** 501
   Vol. 8: 672
Robinson, H.R.   **Vol. 2:** 74, 428, 580, 597
   Vol. 8: 144, 667, 676
   Vol. 10: 393
   Vol. 12: 287
Rochester, University of   **Vol. 9:** 39
Rockefeller Foundation   **Vol. 5:** 6, 240, 313, 387, 389, 407, 410, 413
   Vol. 8: 221, 223, 232, 663f, 753, 791–792, 807, 809
   Vol. 9: 217, 221, 540f
   Vol. 10: 23, 473
   Vol. 11: 7, 512n, 514, 516, 522, 524, 529, 531, 727
   Vol. 12: 160, 173, 183, 192, 211, 216, 327, 329
Rockefeller philanthropy
   Vol. 11: 371, 514n, 727
Rogers, C.   **Vol. 11:** 563
Romanticism   **Vol. 10:** 256, 266
Rome Conference, 1931   **Vol. 6:** 319
   Vol. 7: 192, 193, 463, 466
   Vol. 8: 701, 703
   Vol. 9: 8, 99–114
   Vol. 12: 500
Rømer, O.   **Vol. 5:** 49, 141
   Vol. 10: 258, 268, 350, 354, 355, 363, 368
   Vol. 11: 544, 551
   Vol. 12: 122, 428, 430, 431
Röntgen rays, *see* X-rays
Röntgen spectra (*see also* X-ray spectra)
   Vol. 3: 199
Röntgen, W.C.   **Vol. 10:** 75, 102
   Vol. 11: 530, 565
Roosevelt University   **Vol. 12:** 532
Roosevelt, F.D.   **Vol. 1:** xlv
   Vol. 11: 4, 15, 19–22, 24–27, 30–39, 39n, 40–43, 45–48, 89, 98, 101, 109, 110, 156, 176, 235, 237, 240–242, 245, 246, 249, 251–253, 255n, 257–260, 262, 267–269, 271–275, 277, 278, 280, 283–293, 328n
Rosanes, J.   **Vol. 5:** 312
Rose, D.C.   **Vol. 8:** 632
Rose, M.E.   **Vol. 7:** 485

Rose, W.   **Vol. 11:** 514n
Rosen, N.   **Vol. 7:** 250, 251, 254, 256, 277, 290, 292, 295, 296, 372–374, 480
Rosenblum, M.S.   **Vol. 7:** 32p
   Vol. 10: 447
Rosenfeld, L. (*see also* Bohr–Rosenfeld paper)
   Vol. 1: xvii, xxi, xxvi, xli, xlii, 96
   Vol. 2: 110, 114, 116, 119
   Vol. 3: 46
   Vol. 5: v, vi, 7, 216, 318
   Vol. 6: v, xvii, xxvi, 27f, 30, 304, 305, 306, 307, 309, 310, 320
   Vol. 7: 3, 7, 9, 10, 12–15, 17, 18, 20n, 26, 33–35, 37, 38, 40n, 40p, 41, 43–47, 49, 50, 53–166, 169, 178, 189, 211–216, 235, 242, 252, 256, 261p, 302, 435, 445, 450, 453, 457–459, 465, 467, 469, 471, 472, 474, 477, 480, 482, 485, 487–491, 493, 495–498, 500–502, 511, 513, 514, 520–522
   Vol. 8: 5, 220f, 224f, 268, 336, 342, 655f, 704f, 711ff, 716, 730, 732f, 735, 804, 821, 823
   Vol. 9: 10, 19, 22, 24, 28, 50, 54ff, 60, 64–65, 68, 80f, 116, 130, 134, 144, 192, 196, 272f, 284, 292, 348, 356, 468, 529, 533f, 537, 544, 545–547, 557f, 564, 570f, 617, 620ff, 627, 629ff, 637, 641–651, 661, 666, 668, 671, 676ff
   Vol. 10: xxvii$n$, xxxii$n$, 110, 330, 332, 333, 407, 420, 455p, 460, 492, 493, 531, 532, 534, 535, 583, 584, 587, 589, 591–593
   Vol. 11: 7p, 731
   Vol. 12: vi, 6, 7, 9, 17p, 24, 28, 30, 33, 38p, 105, 499, 501, 517, 525, 526, 528, 532–534
Rosenfeld, Y.   **Vol. 7:** 522
Rosenhain, W.   **Vol. 10:** 440
Rosenkjær, J.   **Vol. 9:** 626, 628
   Vol. 12: 128, 474, 475, 530
Ross, P.A.   **Vol. 5:** 319
Rossel, J.   **Vol. 7:** 328
Rosseland, S.   **Vol. 3:** 404, 469, 489, 510, 540, 592, 657, 658, 660, 662
   Vol. 4: 411, 417, 547, 588, 639, 698, 699, 700
   Vol. 5: 70, 89, 179, 195, 275, 298, 383, 385, 387, 389, 482–486
   Vol. 6: 304, 383, 405
   Vol. 8: 26, 27, 28, 262, 647, 683, 692,

769–782, 783*ff*
    Vol. 9: 95
    Vol. 10: 400
    Vol. 11: 354, 513, 522
Rossi, B.   **Vol. 8:** 699, 701
Rossi, R.   **Vol. 2:** 213, 238
    Vol. 6: 377
    Vol. 10: 385, 489
Rossier, P.H.   **Vol. 7:** 328
rotational state of nucleus   **Vol. 9:** 27, 30, 38,
    80, 236, 294, 312, 325, 522, 524, 656*f*
rotator, quantized   **Vol. 6:** 182–183,
    396*f*, 426*f*
Rothe, O.   **Vol. 10:** 588
Rousseau, J.J.   **Vol. 10:** 298
Rovsing, T.   **Vol. 12:** 452, 457, 503, 504
Rowland, H.A.   **Vol. 3:** 289, 297
    Vol. 10: 388
Royal Danish Academy of Sciences and
    Letters   **Vol. 2:** 345
    Vol. 5: vi, xi, 69, 95, 173*f*, 219, 224,
    439, 442, 454*f*
    Vol. 6: viii, xxi, 4, 41, 55–56, 107–108,
    199–200, 317, 353–354
    Vol. 7: 22, 53, 287, 323, 523
    Vol. 8: vii, xix, 226*f*, 235, 237*ff*, 244,
    250, 261, 313–318, 423, 573–581, 593,
    650, 688, 727, 729, 733, 735, 749, 783,
    798, 822, 824
    Vol. 9: vii, xviii, 12, 15, 17, 39, 41, 47,
    49, 75, 119–121, 149–150, 151–156,
    211, 223, 287–289, 299, 306, 320, 332,
    339–340, 393, 405–408, 409–410, 413,
    443, 469, 481–482, 484, 485–486, 505,
    539, 612, 621, 662, 679
    Vol. 10: 141, 192, 221, 297, 298, 302*n*,
    303, 313, 314, 317–319, 322, 324, 325,
    353, 367, 369*n*, 585, 589
    Vol. 11: vi, 52, 58, 63*n*, 303, 307, 308,
    310–319, 337–352, 356, 357, 380–383,
    384*p*, 386, 387, 390, 391, 394–396,
    398–403, 405, 406*p*, 407–414, 416,
    417, 421–424, 426–432, 434, 435, 438,
    440, 441, 444, 445, 454–461, 464–466,
    471, 474, 563, 564, 669, 726, 728,
    729, 732
    Vol. 12: 103, 105, 107, 113, 115*n*, 123,
    124, 127, 138, 140, 144, 180, 182–184,
    188, 190, 192, 193, 212, 217, 248, 249,
    277, 291, 294, 296, 297, 300, 302, 315,

    317–320, 322–324, 356, 357, 382, 384,
    527, 528, 531, 537
  Collections for the History of
    Vol. 12: 319, 323
  Communications   **Vol. 11:** 339, 340
    Vol. 12: 6*n*, 317, 322
  gold medal   **Vol. 1:** xix, 7
    Vol. 12: 122, 135, 136, 138, 140,
    209, 213
  Historical–Philosophical Class
    Vol. 11: 339, 345, 346
  Proceedings   **Vol. 11:** 340, 345, 347
    Vol. 12: 6*n*, 113, 114, 320*n*, 537
  Scientific–Mathematical Class
    Vol. 11: 339, 345*n*, 348
Royal Danish Geographical Society
    Vol. 12: 106, 241, 245
Royal Danish School of Educational Studies
    Vol. 12: 16
Royal Danish Society for National History
    Vol. 11: 399, 400, 409, 410
Royal Engineers   **Vol. 12:** 285
Royal Festival Hall (London)   **Vol. 12:** 120
Royal Institution (London)   **Vol. 11:** 352
Royal Mint (Copenhagen)   **Vol. 12:** 242, 246
Royal Society (Edinburgh)   **Vol. 6:** 317,
    351–352, 480
Royal Society (London)   **Vol. 5:** 488*f*
    Vol. 8: 231, 648*f*, 678*f*, 787, 806–807
    Vol. 9: 39, 539, 642*f*
    Vol. 11: 10, 25, 26, 212, 213, 215, 216,
    218–220, 241, 339, 347, 348, 444,
    445, 730
    Vol. 12: 103, 135, 136, 144, 220, 286,
    318, 323
  first paper   **Vol. 1:** 8*ff*, 25–65
  second paper   **Vol. 1:** 11, 79–89
Royal Technical University (Stockholm)
    Vol. 11: 628*n*
Royal Theatre (Denmark)   **Vol. 11:** 373
Royal Veterinary and Agricultural College
    (Copenhagen)   **Vol. 11:** 402, 412, 516,
    518*n*, 524
    Vol. 12: 119, 125, 184, 193, 382, 383
Royds, T.   **Vol. 10:** 387, 417
Rozental, S.   **Vol. 1:** xvii
    Vol. 5: 92, 354
    Vol. 6: xix
    Vol. 7: 38*n*, 252*n*, 270, 522
    Vol. 8: 240, 247*f*, 250, 262, 336,
    358*f*, 370, 384, 428, 647, 822*f*

Vol. 9: 52, 680
Vol. 10: xxxi*n*, xxxii*n*, 110, 172*n*, 220*n*, 221, 477*n*, 584–587, 589
Vol. 11: 328, 331, 375, 699*n*, 731
Vol. 12: 528–530
Rozhdestvenskii, D.S.   **Vol. 3:** 9, 178, 518, 547, 617
   Vol. 4: 115, 230, 298, 394
   Vol. 11: 198*n*
Ruark, A.E.   **Vol. 5:** 145, 148
Rubens, H.   **Vol. 1:** 101, 125, 147, 150, 156, 250, 258, 360, 366, 408, 447, 464
   Vol. 2: 238
   Vol. 3: 190
   Vol. 4: 24, 696, 714
   Vol. 10: 433
Rubin, E.   **Vol. 1:** 498, 499, 506, 507, 512, 513, 514, 515, 586
   Vol. 2: 617
   Vol. 6: xxvi
   Vol. 10: xxix, xxx, xxxii, xliv, xlvi, 16, 36*p*, 299*n*, 300, 462
Rubinowicz, A.   **Vol. 3:** 12, 13, 16, 27, 28, 37, 126, 181, 274, 317, 332, 341, 344, 345, 346, 347, 350, 352, 354, 355, 412, 438, 498, 629, 680, 686, 688
   Vol. 4: 12, 23, 28, 110, 217, 368, 573, 628, 733–735
   Vol. 7: 302
   Vol. 10: 396, 419
Rubow, P.   **Vol. 11:** 406*n*
Rud Nielsen, J.   **Vol. 2:** 110
   Vol. 4: 690
   Vol. 5: v, ix, x, *318*
   Vol. 6: 195, 278
   Vol. 10: 173*n*, 192, 193, 534, 535
   Vol. 12: 17*p*
Rüdinger, E.   **Vol. 7:** 169, 524
   Vol. 10: 592–594
   Vol. 12: vi, vii, 534
Rudorf, G.   **Vol. 2:** 613, 614
Rumer, Y.B.   **Vol. 11:** 7*p*
Runge, C.   **Vol. 1:** 566, 567
   Vol. 2: 127, 362
   Vol. 3: 446, 527, 555
   Vol. 4: 34, 374, 702
   Vol. 10: 338, 375, 378
   Vol. 12: 340, 341
Russell, A.S.   **Vol. 2:** 119, 213, 214, 238
   Vol. 6: 377*f*
   Vol. 10: 385

Russell, B.   **Vol. 12:** 119, 120, 390
Russell, H.N.   **Vol. 5:** 256, 262, 279
Russian Academy of Sciences, *see* Academy of Sciences of the U.S.S.R.
Rutherford, E.   **Vol. 1:** xxi, xxii, xxiii, xxiv, xxv, xxvii, xxviii, xxx, 106, 108, 109, 538, 539, 548, 549, 550, 551, 554, 555, 558, 559, 560, 561, 572, 573, 574, 575, 576, 577, 580, 581
   Vol. 2: 3, 4, 5, 6, 7, 8, 18, 29, 32, 39, 59, 60, 73, 74, 75, 90, 96, 97, 98, 103, 104, 106, 107, 108, 111, 112, 113, 115, 117, 119, 120, 121, 122, 123, 125, 126, 127, 129, 133, 136, 145, 161, 162, 163, 179, 188, 189, 211, 212, 214, 215, 232, 280, 289, 290, 297, 308, 312, 319, 320, 325, 329, 330, 331, 332, 342, 343, 344, 350, 368, 392, 393, 418, 420, 424, 428, 433, 440, 447, 473, 474, 493, 495, 501, 515, 516, 517, 520, 522, 523, 556, 557, 558, 561, 575–598, 603, 604, 613, 614, 615, 616, 617, 619, 620, 624
   Vol. 3: 11, 14, 21, 29, 32, 34, 69, 103, 104, 159, 167, 191, 203, 204, 215, 242, 291, 299, 300, 361, 363, 364, 381, 392, 398, 422, 424, 584, 610, 615, 626, 629, 631, 667, 682–684
   Vol. 4: 22, 26, 28, 30, 31, 32, 45, 52, 58, 85, 87, 190, 259, 263, 343, 346, 422, 430, 431, 432, 437, 467, 468, 470, 551, 558, 612, 617, 658, 659, 675, 676, 677, 678, 685
   Vol. 5: 6, 10*f*, 70, 88–89, 130, 138, 178, 194, 238, 274, 298, 335, 337*f*, 486–489
   Vol. 6: 8, 52, 76, 82, 138, 225, 238, 373, 375*ff*, 379, 381, 387, 405, 413, 422, 424, 457–458
   Vol. 7: 32*p*, 304, 343, 401, 415
   Vol. 8: 4*ff*, 16, 21, 24, 28*f*, 50, 61, 71, 129, 144, 160, 203*ff*, 210, 223, 233, 256, 427, 529, 535, 539, 568, 647, 667, 672, 677*f*, 702*f*, 717*f*, 752, 783–788
   Vol. 9: ii, 3, 6, 14, 20*f*, 95, 117, 134, 152, 163, 165, 172*f*, 185, 201, 215*f*, 218*ff*, 279, 303, 308, 310, 317*f*, 322, 324, 329, 407*f*, 413, 443, 537, 555*f*, 572*f*, 576*f*, 580*f*, 630, 651–652
   Vol. 10: 23–25, 56, 57, 86, 102, 168, 177, 210, 329–333, 338, 376, 377, 380*p*, 383–385, 387, 390–394,

396–400, 403, 404, 408–415, 417, 420, 435, 436, 438, 443, 449, 451, 462, 463, 474, 583, 592, 594
  Vol. 11: 6, 9, 10, 54, 87n, 138, 142, 214–216, 220–222, 231, 241, 261, 282, 296, 311, 313, 348, 445, 511–513, 515, 520–523, 529–531, 563, 566–568, 575–577, 589, 592, 593, 646, 649, 659, 664, 692, 697
  Vol. 12: 42, 43, 51, 61, 67, 84, 99, 108–110, 110p, 111, 112, 114, 117, 119, 138, 140, 144, 145, 149, 150, 160, 173, 198, 199, 202, 203, 209–211, 213–215, 221, 254–261, 265, 265p, 266, 269, 272, 274, 277, 278, 281, 282, 285, 286, 326, 329, 356, 502n, 525, 526, 529, 533
Rutherford, M.   Vol. 10: 387, 414, 415
  Vol. 11: 220
  Vol. 12: 110, 274
Rutherford Memorandum   Vol. 2: 103–105, 114, 135–158, 624
Rutherford Memorial Lecture   Vol. 10: 329, 330, 333, 383, 387, 392, 408, 413, 463n, 464, 591, 593
  Vol. 12: 109, 111, 112, 281
Rutherford's atomic model, see atomic model, Rutherford's
Rutherford scattering   Vol. 8: 6, 77, 92, 143, 160, 163, 170f, 173f, 177ff, 207f, 210f, 253f, 269ff, 277ff, 286, 289, 298f, 305f, 337, 343, 348, 407, 417, 426, 435–450, 454, 463ff, 467, 480, 482, 484f, 510, 535, 631f, 634, 636, 663, 813
Ryckmans, P.   Vol. 11: 563
Rydberg
  constant   Vol. 1: xxvii–xxix, 571
    Vol. 2: 111, 116, 120, 122, 123, 126, 131, 169ff, 199, 202, 275, 284f, 294–295, 300–301, 333, 353, 356, 374, 395, 397, 404, 409, 410, 508, 539–540, 544, 552, 553, 559, 562, 582, 588, 603, 628
    Vol. 4: —
    Vol. 5: 157f, 161, 163, 259, 261, 265, 267, 276, 280, 407
    Vol. 10: 331, 332, 374, 377, 378, 388, 389, 392, 437, 464
    Vol. 12: 43, 44
  formula   Vol. 10: 332, 375

Rydberg–Ritz formula (see also Ritz formula)
    Vol. 5: 34, 63, 156–171, 422, 438, 512
  law (see also Ritz law)   Vol. 2: 116, 122, 167, 171–172, 233, 270, 274f, 280, 284–285, 297–298, 327–328, 351, 368, 374, 385–388, 403, 505, 506
Rydberg, J.R.   Vol. 1: xxvii, xxviii, xxix
    Vol. 2: 167, 233, 270, 274, 280, 284, 285, 297, 298, 328, 334, 351, 368, 385, 386, 387, 403, 509
    Vol. 3: 71, 169, 172, 251, 336, 373, 411, 423, 511, 512, 540, 541, 543, 545, 552, 583, 587, 588, 589, 690
    Vol. 4: 27, 52, 434, 442, 445, 469, 473, 474, 572, 628, 659
    Vol. 5: 157, 161, 274, 277
    Vol. 6: 228, 242, 382, 385, 387
    Vol. 10: 58, 329, 332, 338, 372p, 373–376, 378, 379, 388, 390, 436, 556, 558, 589
    Vol. 11: 567
    Vol. 12: 43, 45, 100, 114
Ryle, G.   Vol. 10: xli

Sachs, M.   Vol. 12: 400, 404
Saha, M.N.   Vol. 3: 591
    Vol. 6: 16
    Vol. 8: 248, 647, 788–789
    Vol. 10: 400
Sahner, B.   Vol. 5: 35, 411
St Olav (Norwegian order)   Vol. 12: 138, 140
Sakata, S.   Vol. 7: 496
Salam, A.   Vol. 10: 455p
Salmonsen, see Salomonsen, C.J.
Salomon, H.   Vol. 12: 242, 246
Salomonsen, C.J.   Vol. 2: 523, 524
    Vol. 8: 707
Sampson, R.A.   Vol. 6: 312
Sander, F.   Vol. 11: 525p
Sandersen, M.   Vol. 11: 358p
Sandfeld, K.   Vol. 11: 406p
Sandler, R.   Vol. 11: 365, 618n
Sargent curve   Vol. 9: 385f
Saturnian atom   Vol. 2: 127, 549
Saunders, F.A.   Vol. 5: 256, 262, 279
Saunderson, J.L.   Vol. 8: 633
Savart, F.   Vol. 10: 349, 362
Savitch, P.   Vol. 9: 344, 365, 407f
Saxo   Vol. 10: 263n
Saxtorph, I.   Vol. 12: 400, 404
scalar potential   Vol. 7: 94, 147

Scandinavian Mathematics Congress, 1925
  Vol. 5: 220, 225f, 238, 269f, 273,
    338, 405
  Vol. 6: 3, 7, 263, 285
Scandinavian Meeting of Natural Scientists
  Vol. 11: 354, 356, 502p, 503, 505, 727
  Vol. 12: 318, 322
  1839 **Vol. 10:** 367n
  1847 **Vol. 10:** 366p, 367n
  1929 **Vol. 6:** xxii, 195f, 198, 219–253,
    256, 274, 299, 428f, 479
  Vol. 10: xxv, 515, 519, 522, 528, 529
  1936, *see* Helsinki conference, 1936
Scandinavian scientific cooperation
  Vol. 11: 354, 365, 707, 708
Scandinavian Students' Meeting **Vol. 11:** 727
scattering (*see also* Compton scattering; Møller
    scattering; Mott scattering; potential
    scattering; Rutherford scattering)
  Vol. 7: 33n, 173, 175, 185, 186, 315
  anomalous **Vol. 8:** 214, 457, 709, 795, 804
  coherent **Vol. 9:** 12, 492–493, 497, 501,
    515–516, 574f, 577
  elastic **Vol. 9:** 14, 50, 336, 490, 494ff, 506,
    510, 513ff, 614
  inelastic **Vol. 9:** 14, 349, 351, 382, 397,
    506, 515ff, 614
  multiple **Vol. 8:** 16, 75, 91, 161–165, 184,
    189, 244, 253, 255, 257, 417, 426, 465,
    469, 480–486, 546f, 630, 632–634,
    636f, 669, 813
  of α-rays **Vol. 6:** 376, 379, 398, 403, 405
    Vol. 9: 93, 95, 188, 256
  of charged particles **Vol. 2:** 6–7, 10, 18,
    19, 29, 44, 46, 60, 82–83, 90, 92,
    95–99, 126, 136, 145, 161, 189, 343,
    374, 588, 596, 624, 627, 628
  of electrons/protons (*see also* reflection of
    electrons) **Vol. 6:** 105, 309
    Vol. 9: 11, 109–110, 529
    Vol. 12: 24, 28, 39, 54, 386, 387
  of fission fragments **Vol. 2:** 10
    Vol. 9: 432, 458
  of γ-rays **Vol. 9:** 12, 573–577
  of light/radiation (*see also* Compton
    effect/recoil; dispersion; dispersion
    theory, Kramers' (and Heisenberg's))
    Vol. 2: 178, 183–184, 270, 584, 597
    Vol. 5: 42, 44, 62, 65, 67, 81, 102, 113,
    115f, 145f, 149, 151, 256, f., 263,

    267, 274, 279f, 309, 311, 356f, 398ff,
    416, 420
    Vol. 6: 17, 19, 104, 119ff, 140, 150f,
    162, 165
    Vol. 9: 108–109, 615
  of neutrons **Vol. 9:** 14f, 29, 146, 152, 154,
    181, 245ff, 249, 312, 325, 336, 349,
    351, 382, 397, 560, 566
  of X-rays (*see also* X-rays, scattering of)
    Vol. 6: 359, 375, 379
scepticism **Vol. 10:** 35
Schaefer, C. **Vol. 4:** 12, 737
Schaposchnikow, K. **Vol. 3:** 101
Scharff, M. **Vol. 8:** 264, *265*, 590, 607, 625
    Vol. 9: 528
Scharff, W. **Vol. 12:** 419p, 423n
Scheel, K. **Vol. 2:** 458, 459
    Vol. 5: 62, 79f, 298, 350f, 357, 360f,
    408f
Scheffers, H. **Vol. 8:** 679
Schelling, F.W.J. von **Vol. 10:** 345, 358
Schenck, R. **Vol. 1:** 147, 148, 231, 232,
    344, 345
Scherrer, P. **Vol. 6:** 439f
Schidlof, A. **Vol. 2:** 166
Schiff, L.I. **Vol. 9:** 83
Schiller, F. **Vol. 10:** 545, 550
Schilling, K. **Vol. 1:** 230, 343
Schilpp, P.A. **Vol. 6:** 37, 452
    Vol. 7: 250n, 339, 340
    Vol. 10: xxxiii$n$, 477n
Schimmelmann, E.H. **Vol. 11:** 340n
Schlick, M. **Vol. 10:** 15, 19
Schmidt-Nielsen, S. **Vol. 11:** 399, 408
Schöning, G. **Vol. 11:** 399, 408
Schonland, B.F.J. **Vol. 8:** 720, 725
School of Dentistry (Copenhagen)
    Vol. 11: 516, 524
Schott, G.A. **Vol. 3:** 152
Schottky, W. **Vol. 1:** 116, 117, 482, 483,
    484, 485, 487, 488, 490
Schrieffer, J. **Vol. 12:** 9
Schrödinger
  equation (*see also* wave equation)
    Vol. 6: 9, 70, 96f, 126, 128f, 154f,
    172, 264, 287
    Vol. 7: 261, 307, 429
    Vol. 12: 47, 358, 499, 501
  function, *see* wave function
  state function **Vol. 7:** 392
  wave, *see* wave function

[137]

Schrödinger, E. **Vol. 1:** xxxvii
  Vol. 3: 9, 179, 485
  Vol. 4: 12, 13, 230, 298, 394,
    587, 639, 737, 738
  Vol. 5: 29–31, 41, 298, 490–492
  Vol. 6: 3f, 7ff, 10–15, 29, 32, 36, 37,
    39, 46–51, 52, 70f, 80, 88, 96ff, 126ff,
    134, 137, 143, 153ff, 169, 172f, 184,
    191f, 229, 245, 264f, 287f, 359,
    393f, 413, 415, 417, 422ff, 439f,
    459–467
  Vol. 7: 3, 32p, 255n, 278, 307, 347,
    348, 435, 469, 472, 506, 509, 511
  Vol. 8: 36, 207f, 270, 279
  Vol. 9: 9, 134, 305, 319
  Vol. 10: 212, 402, 426, 427,
    440–442, 454
  Vol. 11: 567
  Vol. 12: 10, 11, 47, 358
Schroeter, J.F.W. **Vol. 5:** 483, 485
Schrum, G.M. **Vol. 3:** 596
Schultz, B. **Vol. 3:** 42, 659, 661
  Vol. 5: 156, *364*, 365
  Vol. 6: 220, 480
  Vol. 8: 182, 711f, 803, 820
  Vol. 9: 57ff, 124, 557f, 564, 567, 588,
    597f, 622, 624, 626ff, 633, 635, 676
  Vol. 10: 582, 586
  Vol. 11: 358p
  Vol. 12: 529
Schulze, F.A. **Vol. 1:** 148
Schuster, A. **Vol. 10:** 375, 388, 390
Schwarzschild, K. **Vol. 2:** 321, 323, 361,
    474, 592, 599–601, 616, 617, 628
  Vol. 3: 61, 69, 84, 87, 96, 106, 124,
    136, 137, 139, 140, 142, 191, 222, 262,
    314, 315, 330, 372, 383, 409, 430, 463,
    505, 534, 598, 638, 640
  Vol. 4: 213, 358, 368, 369, 449, 453,
    476, 478
  Vol. 5: 277
Schweitzer, A. **Vol. 7:** ii
Schwinger, J. **Vol. 7:** 37n, 49, 212, 482, 485,
    494n, 495
  Vol. 9: 617
  Vol. 10: 453, 455p
science
  activity to the benefit of **Vol. 12:** 102, 111,
    212, 217, 316, 319, 320, 324, 326, 328,
    429, 431, 439, 441, 512, 513

advancement/dissemination of
  Vol. 12: 305, 306, 308, 309, 429, 431
Science (journal) **Vol. 11:** 51, 54, 67, 81,
    297, 300, 324, 329, 372, 713, 731
Science and Society (study group)
  Vol. 11: 297n, 348, 465n
Science Commission of 1946 (Denmark)
  Vol. 11: 464, 465
Science Society of Denmark **Vol. 11:** 308,
    479, 483, 488, 490
  Vol. 12: 109, 113, 254, 258, 268, 291,
    294, 300, 302, 318, 322, 524, 525, 527
Scientific American (journal) **Vol. 12:** 119
Scientific Monthly (journal) **Vol. 11:** 372
Scientific Revolution **Vol. 10:** 150, 157
scientific schools, so-called **Vol. 12:** 161, 174
scintilloscope **Vol. 12:** 61, 67
Scipio **Vol. 12:** 152, 164
Scott Prize **Vol. 6:** 317, 351, 480
screening **Vol. 8:** 238f, 253ff, 264, 320, 328,
    330, 356, 366, 368, 372, 378–380, 417,
    426, 438, 443–462, 466, 471, 481,
    484f, 490, 510, 521, 533–534, 563,
    592, 609, 613, 632, 636, 723, 739
Seaborg, G. **Vol. 12:** 84
Sealey **Vol. 1:** 532, 533
Sears, P.B. **Vol. 7:** 411
second external field, small compared to first
  Vol. 3: 126
second law of thermodynamics (*see also*
  entropy; irreversibility) **Vol. 5:** 273f
  Vol. 6: 226, 240, 320ff, 326, 470, 473
secondary electrons **Vol. 8:** 156f, 226,
    289f, 756ff
Secret Service
  British **Vol. 11:** 24, 25, 29, 32, 54,
    104, 227
  Danish **Vol. 11:** 56, 57, 227, 333
  Soviet **Vol. 11:** 57
security (*see also* world security)
  Vol. 11: 18, 32, 36, 44–46, 50, 57, 65,
    66, 71, 72, 74, 78, 93, 106, 108–110,
    133, 157, 160, 164–167, 175, 177, 178,
    180, 182, 185, 192, 249, 266, 269, 274,
    280, 309, 320, 321, 331, 332, 611, 619
Seelig, C. **Vol. 5:** 27
Segrè, E. **Vol. 9:** 55, 100
selection rule **Vol. 3:** 38, 274, 447–452
  Vol. 5: 166, 168, 169f, 251f, 272, 278,
    288f, 331, 347, 424, 428, 445, 450

Vol. 6: 389, 396
  Vol. 9: 261, 317, 329, 493, 513
selective reflection of X-rays, *see* X-rays,
  reflection of
selectivity **Vol. 9:** 15, 32, 45, 47, 50, 154*ff*,
  158, 179–181, 189, 209, 211, 244, 256,
  278, 294, 298*f*, 311, 316, 324*f*, 328,
  332, 335, 428–430, 438, 440, 454–457,
  463*f*, 470, 472, 585, 587, 611, 663, 677
self (*see also* ego) **Vol. 10:** 160, 200, 201
self-awareness **Vol. 10:** xlii, 160, 168, 215
self-charge **Vol. 7:** 50, 484–486
self-energy **Vol. 7:** 20, 33, 35*n*, 46, 64, 126,
  220, 221, 226, 228, 449, 451, 454, 473
  of electron **Vol. 6:** 446, 448
self-preservation/regulation **Vol. 10:** 34, 118,
  151, 160, 187, 199, 204, 215, 527, 528
Sellmeier formula **Vol. 2:** 449
semiconductors **Vol. 12:** 24, 28
Semyonov, N.N. **Vol. 11:** 230
sense impression/perceptions **Vol. 6:** 45, 75,
  82, 115, 148, 152, 158, 203, 205, 209,
  214, 215, 220*f*, 223, 224, 233, 234*f*,
  236, 250, 251*f*, 259, 262, 269, 272,
  279, 283, 294, 299, 327, 453
sense organs **Vol. 7:** 399, 412
  Vol. 10: 86, 102, 180, 404, 564
sentiment **Vol. 10:** xlv, 143, 159, 168, 184,
  199, 215, 279
Serber, R. **Vol. 1:** xliii
  Vol. 7: 522
  Vol. 8: 570*f*
  Vol. 9: 30, 48, 211, 261, 263, 493,
  499, 510
  Vol. 12: 386, 388
series
  diagrams **Vol. 4:** 280*ff*, 299, 331*ff*, 412*ff*,
    563, 608*ff*, 621, 654*ff*
  diffuse subordinate **Vol. 3:** 170, 253
  fundamental (Bergmann) **Vol. 3:** 170, 253
  principal **Vol. 3:** 170, 253
  sharp **Vol. 2:** 170, 275, 356, 361, 366, 370,
    385, 387, 509
  sharp subordinate **Vol. 3:** 170, 253
series spectra **Vol. 3:** 41*ff*, 379
    Vol. 4: 106*ff*, 205*ff*, 212, 329*ff*
  and atomic structure **Vol. 4:** 576*ff*, 631*ff*
  effect of electric fields on **Vol. 3:** 174, 175
  effect of external fields on
    Vol. 3: 419–446, 521*ff*, 549*ff*

effect of magnetic fields on
    Vol. 3: 176, 177
  general structure of **Vol. 3:** 167*ff*
  laws of **Vol. 3:** 560*ff*, 563*ff*
  of individual elements **Vol. 3:** 171*ff*
  of the elements **Vol. 3:** 24*ff*, 241–282
    Vol. 4: 100*ff*
  of the third kind **Vol. 4:** 589, 640
  physical basis of **Vol. 3:** 591*ff*
  structure of **Vol. 3:** 598*ff*
  theory of **Vol. 3:** 41, 270*ff*, 502–558, 578*ff*
series terms
  interpretation of the empirical laws of
    Vol. 3: 511*ff*, 540*ff*
Serwer, D. **Vol. 5:** 61–62, 229
Shadow, M. **Vol. 11:** 328, 331
Shakespeare, W. **Vol. 10:** 96, 189, 318, 322
    Vol. 11: 26
Shankland, R.S. **Vol. 5:** 94, 215
    Vol. 8: 807
    Vol. 9: 13, 602, 604
Shapley, H. **Vol. 7:** 411
sharpness
  of resonance (*see also* level width)
    Vol. 9: 152, 154*f*, 169*f*, 177*f*, 187,
    209, 246, 255*f*, 299, 311, 315, 324,
    327, 337
  of spectral line, *see* line width
  of spectral lines, limit of **Vol. 3:** 133, 223
  of stationary state **Vol. 5:** 33, 416,
    419, 491
Shaw, B. **Vol. 11:** 206, 208
shell model **Vol. 9:** 74, 76–83, 145, 529,
    618*ff*, 669*f*
shell structure **Vol. 5:** 168, 171
Shelter Island conference **Vol. 7:** 48
Sherrington, C.S. **Vol. 11:** 445
Shimizu, T. **Vol. 4:** 493, 530
    Vol. 5: 235
Shire, E.S. **Vol. 9:** 201
Shoenberg, D. **Vol. 12:** 38*p*
Shuttleworth, R.G. **Vol. 11:** 563
Sidgwick, N.V. **Vol. 1:** 121, 123
Siegbahn, M. **Vol. 2:** 618, 619
    Vol. 3: 17, 361, 601, 688
    Vol. 4: 31, 32, 34, 201, 253, 324, 460,
    461, 481, 493, 515, 530, 546, 677,
    678, 702
    Vol. 5: 278
    Vol. 6: 385, 390
    Vol. 10: 379, 396, 438

Siegumfeldt, H.   **Vol. 3:** 286, 295
Sigurgeirsson, T.   **Vol. 8:** 560, 566, *734*, 736
Sikjær, S.   **Vol. 12:** 16, 94, 96
Silberstein, L.   **Vol. 5:** 7
Silliman Lecture, Bohr's   **Vol. 5:** 410, 413
  notes to   **Vol. 3:** 582–601
silver   **Vol. 11:** 578, 579, 595–597
  absorption in   **Vol. 2:** 83
Simon, A.W. (*see also* Compton–Simon
    experiment)   **Vol. 5:** 78, 210*f*,
    215, 221
  Vol. 6: 139
Simons, L.   **Vol. 9:** 560, 566, 638, 640
simple harmonic motion   **Vol. 6:** 134, 157,
    173, 393, 396
simultaneity   **Vol. 6:** 34, 38, 49, 96, 120, 141,
    167, 205, 214, 260, 281, 314, 369,
    416, 434
simultaneous influence of electric and
    magnetic fields   **Vol. 3:** 155*ff*
Sinclair, J.S.   **Vol. 11:** 361, 562, 563, 717,
    719, 720
single-mode excitation   **Vol. 9:** 293*f*
singular motion, *see* crossed electric and
    magnetic fields; degeneracy; pendulum
    orbit
Sitte, K.   **Vol. 9:** 9
Skobeltzyn, D.   **Vol. 8:** 802
Skot-Hansen, J.P.   **Vol. 11:** 373
Skov, P.   **Vol. 10:** 299, 300
Slack, F.G.   **Vol. 9:** 61, 365, 380, 385, 469
Slater, J.C. (*see also* Bohr–Kramers–Slater
    theory)   **Vol. 1:** xxxv, xxxvii
  Vol. 3: 44, 46
  Vol. 5: 5–13, 19*ff*, 40, 43, 52, 65–69,
    86, 93, 95, 97*ff*, 102, 183, 199, 229,
    256, 263, 279, 288, 296, 298, 324, 326,
    381*f*, 395, 404*f*, 487, 492–502
  Vol. 6: 37, 39, 139, 417
  Vol. 8: 31, 36, 206, 737*f*
  Vol. 9: 599, 601
  Vol. 10: 419, 425
  Vol. 12: 6*n*
Sleator, W.W.   **Vol. 5:** 387*f*
Slebsager, E.   **Vol. 11:** 328, 331
Slomann, V.   **Vol. 6:** xxiv
  Vol. 10: 250*p*, 299*n*, 300
Smekal, A.   **Vol. 3:** 12, 472
  Vol. 4: 253, 254, 324, 409, 410
  Vol. 5: 59–60, 146, 151, 257, 263, 279
  Vol. 6: 393

Smekal–Raman effect   **Vol. 5:** 59–60, 146,
    151, 257, 263, 279, 318, 358*f*
Smit, J.   **Vol. 12:** 38*p*
Smith, E.   **Vol. 10:** 397
Smith, J.D. Main, *see* Main Smith, J.D.
Smoluchowski, M. von   **Vol. 2:** 64
  Vol. 5: 82*f*
  Vol. 8: 134
  Vol. 12: 372
Smoluchowski, R.   **Vol. 7:** 261*p*
Smuts, J.C.   **Vol. 11:** 26, 27, 33, 34, 78, 246,
    277, 324, 327, 329, 330
Smyth, H.D.   **Vol. 8:** 681
  Vol. 11: 563
Smyth Report   **Vol. 11:** 56, 57, 139*n*, 324,
    329, 333
Snorri Sturlason   **Vol. 10:** 265*n*
Snyder, H.S.   **Vol. 6:** 450
social
  impact on/of science
    Vol. 11: 183, 207, 210, 347, 349, 350,
      367, 402, 404, 412, 414, 428, 432,
      448–451, 455, 459
    Vol. 12: 3, 84, 94–96, 103, 158, 162,
      171, 175, 180–184, 189–194, 201, 205,
      228, 229, 290, 292, 306, 309, 332–334
  problems   **Vol. 10:** 185, 203, 405, 415
  sciences   **Vol. 10:** 157, 300
society, activity to the benefit of   **Vol. 12:** 59,
    63, 123, 182, 190, 212, 217, 300–302,
    315–317, 319–321, 324, 326–328, 330,
    374, 375, 446, 447, 454, 459, 469–472,
    474, 475, 495
Society for Philosophy and Psychology
    (Copenhagen)   **Vol. 6:** 196, 428*f*
Society for the Dissemination of Natural
    Science   **Vol. 5:** 47–49, 125*ff*, 132,
    134–135, 136, 140, 142
  Vol. 9: 75, 411, 443, 627*f*
  Vol. 10: 354, 367, 368
  Vol. 11: 12*n*, 337, 352–354, 470–475,
    478–485, 488–491, 494, 495, 497–499,
    726, 729, 733
  Vol. 12: 107, 113, 123, 127, 291, 293,
    300, 302, 305–310, 317, 321, 322, 337,
    338, 381, 383, 527, 529, 530
Society for the Protection of Science and
    Learning (London)   **Vol. 9:** 542*f*
  Vol. 11: 291
Society of Randrusians   **Vol. 11:** 394, 395

sociology **Vol. 7:** 336
    Vol. 10: 474
Socrates **Vol. 6:** xx, 24, 424
    Vol. 10: 188
    Vol. 12: 155, 162, 166, 175
Soddy's rule **Vol. 8:** 171, 175, 179
Soddy, F. **Vol. 2:** 4, 119, 214, 418, 419, 420,
    440, 594, 597
    Vol. 4: 431, 467, 658
    Vol. 6: 375*ff*
    Vol. 8: 171, 175, 179
    Vol. 9: 308, 322
    Vol. 10: 385, 391, 417
    Vol. 11: 566
    Vol. 12: 42
sodium **Vol. 12:** 45
    absorption and dispersion in (*see also* alkali
        metals, spectra of)
    Vol. 2: 116–117, 176, 177–178, 337,
        448–450, 564, 566–567, 604
    spectrum **Vol. 6:** 429*f*
solar system **Vol. 12:** 61, 67, 68
solid state physics **Vol. 11:** 196, 199
solipsism **Vol. 10:** 15
Solomon, J. **Vol. 7:** 166*n*
    Vol. 9: 10, 592
    Vol. 12: 17*p*
Solvay Conference in Physics **Vol. 10:** 207,
    330, 431, 437, 451
    Vol. 11: 8, 76, 326, 330
    Vol. 12: 118, 294*n*, 317, 322, 376*p*
    1st, 1911 **Vol. 5:** 27
        Vol. 10: 384, 432, 433, 435
    2nd, 1913 **Vol. 10:** 435, 437
    3rd, 1921 **Vol. 10:** 437, 439
    4th, 1924 **Vol. 5:** 93
        Vol. 10: 439, 440, 448
    5th, 1927 **Vol. 5:** 93, 207
        Vol. 6: 4, 8, 30, 32, 35–41, 44, 46,
        99–106, 415–418, 478–479
        Vol. 7: vii, 255, 351, 363, 364, 506
        Vol. 10: 406, 441, 448, 453, 496
    6th, 1930 **Vol. 6:** 20, 307, 310, 318, 319,
        337–349, 481
        Vol. 7: 319, 368
        Vol. 8: 665
        Vol. 10: 406, 446–448
    7th, 1933 **Vol. 7:** 31, 32*p*, 167–194, 370,
        446, 448, 474, 477, 520
        Vol. 8: 702*ff*

        Vol. 9: 8–10, 12, 37, 129–141, 596*f*
        Vol. 10: 448–451
    8th, 1948 **Vol. 7:** 50, 223–228, 278, 326,
        488, 492, 495, 497, 522
        Vol. 8: 257, 569–572
        Vol. 10: 452, 454
    12th, 1961 **Vol. 10:** 454
    discussions with Einstein at
        Vol. 7: 250, 351
Solvay Conference, report to third
    Vol. 3: 28*ff*, 357–395
    discussion of report **Vol. 3:** 388–395
    extract of second half
        Vol. 4: 14*ff*, 92–97
Solvay Institute for Physics (Brussels)
    Vol. 10: 216, 431, 446, 454,
        455*p*, 492*n*
    Vol. 12: 291, 294, 300, 302, 317, 322
Solvay, E. **Vol. 3:** 362, 363
    Vol. 10: 431, 446
    Vol. 12: 294*n*
Sommerfeld, A. **Vol. 1:** xxx, xxxi, xxxii,
    122, 566, 567, 572, 573, 580, 581, 584,
    585, 586
    Vol. 2: 123, 127, 166, 256, 331, 336,
    337, 338, 340, 342, 343, 363, 367, 448,
    464, 465, 466, 467, 468, 469, 474, 558,
    561, 564, 565, 566, 567, 568, 570, 571,
    572, 595, 602–604, 615, 618, 619, 632
    Vol. 3: 3, 7, 8, 9, 10, 12, 13, 16, 17, 18,
    23, 27, 37, 39, 40, 50, 52, 55, 56, 58,
    61, 69, 82, 83, 84, 91, 92, 96, 98, 102,
    106, 107, 122, 123, 129, 130, 131, 132,
    133, 144, 145, 149, 150, 158, 169, 172,
    178, 179, 181, 183, 191, 193, 199, 217,
    222, 224, 258, 259, 264, 265, 267, 270,
    271, 272, 273, 307, 308, 309, 314, 315,
    316, 319, 320, 322, 325, 330, 331, 333,
    334, 335, 336, 337, 372, 383, 409, 430,
    431, 440, 463, 486, 504, 505, 506, 517,
    528, 530, 534, 535, 546, 556, 558, 595,
    596, 617, 619, 628, 637, 638, 647,
    686–692
    Vol. 4: 7, 11, 12, 13, 20, 48, 49, 53, 55,
    56, 58, 68, 75, 76, 78, 93, 96, 109, 111,
    112, 113, 116, 118, 120, 145, 157, 159,
    195, 196, 197, 200, 201, 207, 209, 213,
    214, 215, 217, 231, 232, 249, 250, 252,
    253, 269, 270, 271, 273, 274, 280, 281,
    282, 300, 320, 321, 322, 323, 324, 356,
    360, 366, 370, 374, 375, 379, 392, 398,

408, 409, 423, 424, 436, 447, 448, 449, 452, 457, 460, 470, 475, 476, 477, 479, 481, 487, 492, 504, 520, 525, 536, 563, 565, 567, 568, 570, 574, 599, 600, 602, 606, 621, 623, 624, 626, 629, 647, 648, 649, 652, 660, 707, 737, 740–742

Vol. 5: 4, 36–38, 57, 88, 92, 157, 162, 167, 169–170, 251f, 277ff, 288, 298, 304, 314, 424, 428, 435, 441, 446, 450, 500, 502–507

Vol. 6: 10, 41, 52, 126, 144, 153, 228, 243, 318, 325, 349, 389, 398

Vol. 7: 346

Vol. 8: 19, 184, 189, 664, 673, 688

Vol. 9: 100, 110, 304, 319

Vol. 10: 339, 379, 395, 398, 400, 419, 424, 434, 438, 446, 448

Vol. 11: 567

Vol. 12: 22, 25, 45, 115, 344–346, 346p, 357, 363, 524

Sommerfeld, speech of appreciation to
    Vol. 3: 18, 217ff

  translation of speech **Vol. 3:** 19

Sonning, L. **Vol. 10:** 288p

Sonning Prize **Vol. 10:** 221, 283, 289, 594

Sophocles **Vol. 10:** 189

Sørensen, H. **Vol. 11:** 346

Sørensen, S.P.L. **Vol. 9:** 626, 628
    Vol. 11: 341, 353, 380–382, 471, 474, 480, 484, 525p
    Vol. 12: 100, 124, 127, 307, 310

soul **Vol. 10:** 7, 117, 478, 483, 485, 486, 536, 537

soul and body, distinction between
    Vol. 7: 407, 420

space curvature, *see* curvature of space
  quantization (*see also* Stern–Gerlach experiment) **Vol. 5:** 50, 54, 153, 237, 288, 358f, 504
    Vol. 6: 345, 391

space–time
  absolute **Vol. 10:** 85, 103, 104, 142, 157, 158, 176
    Vol. 12: 221, 371, 378
  concepts **Vol. 7:** 34, 172, 184, 209, 321, 351, 362, 392, 510, 511, 520, 521
    Vol. 10: 85, 89, 158, 189, 209, 241, 277, 278, 284, 290, 340, 404, 454, 518, 521, 545n
    Vol. 11: 404, 414, 540, 549, 670, 671

Vol. 12: 79, 81, 119, 156, 168, 234, 371, 374, 375

coordinates/description/pictures **Vol. 5:** 13, 21, 59, 79ff, 85, 90ff, 96, 106f, 115, 190–193, 204–206, 209–212, 216, 221, 224, 258f, 264, 276, 280, 309, 311, 350, 353–354, 360f, 446, 448, 451, 453, 489, 497

    Vol. 7: 20, 61, 124, 213, 405, 418

    Vol. 8: 30, 36, 207, 228, 231, 288, 290, 306, 381, 411f, 499, 627–638, 694, 805

    Vol. 9: 87, 93, 104f, 110f

    Vol. 10: xxxv–xxxvii, xli, 40–42, 46, 104, 148, 166, 178, 183, 197, 203, 213, 215, 230, 235, 496, 512, 513, 531–533

coordination/localization **Vol. 7:** 31, 173, 178, 185, 188, 198, 225, 271, 314, 315, 343, 364, 371, 389, 392, 406, 419

    Vol. 10: 30, 88, 104, 131, 137, 158, 175, 196, 277, 284, 290, 453, 573

    Vol. 11: 657, 663, 670, 671

    Vol. 12: 220, 221, 223

course **Vol. 7:** 86, 96, 141, 148, 294

dimensions/scale **Vol. 7:** 38, 60, 92, 124, 145, 176, 178, 187, 189, 225

integral **Vol. 7:** 22, 96, 148

quantities/variables **Vol. 7:** 37, 225, 264

reference frame **Vol. 7:** 15, 84, 140, 314, 350, 355, 365, 513

separation between **Vol. 7:** 317, 505, 508

Spallanzani, L. **Vol. 10:** 55

Spärck, R. **Vol. 10:** 36p

spark spectra **Vol. 1:** 571
    Vol. 2: 275, 326, 386–388, 405, 469, 503, 559, 562, 603
    Vol. 3: 254, 590
    Vol. 4: 206ff, 278, 445, 474
    Vol. 5: 251f, 288, 407

specialization **Vol. 10:** 64, 141, 306, 308

specific heat **Vol. 2:** 126, 162, 224, 261, 289, 312, 338, 342, 416, 421–422, 434, 452, 455f, 458–459, 565, 567, 589, 604
    Vol. 6: 145, 158, 396ff
    Vol. 7: 279
    Vol. 9: 209, 233, 293, 313, 326, 422, 450
    Vol. 10: 376, 386, 427, 432, 433, 443
  of metals **Vol. 1:** 108, 553

Specker, E. **Vol. 7:** 328

spectra (*see also* band spectra; line spectra; resonance radiation/spectra; spark

spectra; X-rays)   **Vol. 8:** 20, 31*f*, 169, 172, 177, 184*f*, 189*f*, 198, 200, 658, 663, 672*f*, 683, 689, 691, 695*ff*, 769, 771, 780, 782, 785, 786, 788
  Vol. 10: 57, 58, 121, 183, 211, 331–333, 337–340, 373–379, 385–396, 400, 401, 409, 412, 434, 436–440, 447, 451, 463, 464, 510, 511, 589
atomic (*see also the element in question*)
  Vol. 7: 76, 171, 183, 187, 348
  Vol. 11: 512, 521, 671
  Vol. 12: 44, 45, 209, 214, 326, 329, 340, 341, 356
continuous   **Vol. 7:** 429
coronal and stellar (*see also* Pickering lines/series)   **Vol. 2:** 107*ff*, 121*f*, 166, 169, 183–184, 194, 200, 202, 270–271, 274–276, 285, 296–298, 313, 315, 354, 385, 503, 579, 582
excitation of   **Vol. 7:** 344
high-frequency (*see also* X-ray spectra)
  Vol. 11: 567
  Vol. 12: 51, 288
infrared   **Vol. 2:** 117, 139, 169*f*, 224, 230–231, 238, 285, 338–339, 386, 420, 446–448, 457, 565, 567, 631
line   **Vol. 11:** 567
molecular   **Vol. 2:** 117, 139, 224*f*, 230–232, 237, 338–339, 341, 420, 437, 446–448, 452, 457, 565, 567, 631
  Vol. 11: 671
  Vol. 12: 439, 441
multiplet   **Vol. 12:** 45, 47
of alkali metals, *see* alkali spectra
of alkaline earths, *see* alkaline earths, spectra of
of atoms with one electron
  Vol. 3: 502*ff*, 532*ff*
of atoms with several electrons
  Vol. 3: 509*ff*, 538*ff*
of elements of higher atomic number
  Vol. 3: 167*ff*
of helium, *see* helium spectrum
of hydrogen, *see* hydrogen spectrum
of many-atomic gases   **Vol. 12:** 125
of mercury, *see* mercury spectrum
of sodium, *see* sodium spectrum
of sun   **Vol. 7:** 366
optical (*see also* band spectra)   **Vol. 9:** 94*f*, 108*f*, 111, 113, 127, 164, 172, 197, 204, 209, 231, 304*ff*, 309, 318*f*,

321*f*, 621
  Vol. 12: 45, 51
physical basis of the theory of   **Vol. 3:** 591
ultraviolet   **Vol. 2:** 108, 117, 169*f*, 177, 201, 203, 222, 337, 386, 399, 420
X-ray   **Vol. 12:** 43–46, 112, 288
spectral
absorption/emission   **Vol. 11:** 566
  Vol. 12: 45, 358
analysis   **Vol. 7:** 88, 143
  Vol. 11: 658, 663
distribution   **Vol. 7:** 389
doublet, *see* doublet, spectral
laws (*see also* combination principle)
  Vol. 6: 136, 158, 228, 242*f*, 382, 391
  Vol. 7: 227, 279, 344
  Vol. 12: 10
laws, empirical   **Vol. 3:** 511*ff*, 540*ff*
laws, general   **Vol. 3:** 251, 254
lines (*see also* line spectra)   **Vol. 7:** 176, 187, 472, 473
  Vol. 12: 42, 44–47, 114, 288, 340–342, 348, 351, 356, 358, 368, 381, 383
lines, effect of electric and magnetic fields on   **Vol. 3:** 419–446
lines, width of (*see also* line width)   **Vol. 6:** 13, 133, 138, 143, 156, 173, 396, 460
structure   **Vol. 11:** 568
spectroscopy   **Vol. 1:** xxvi–xxviii, xxx–xxxvi, 519
  Vol. 10: 29, 30, 58, 179, 210, 214, 332, 373, 385, 392, 406, 446
  Vol. 11: 566
  Vol. 12: 20, 112, 114, 115, 117, 119, 326, 329, 368
combination principle
  Vol. 1: xxvii–xxviii, 571
dispersion   **Vol. 1:** xxxv–xxxvi, 581
He$^+$ spectrum   **Vol. 1:** xxix, 567, 571, 573
hydrogen spectrum
  Vol. 1: xxvii–xxviii, 567
Nicholson's theory   **Vol. 1:** 563
Rydberg constant   **Vol. 1:** xxvii–xxix, 571
spark spectra   **Vol. 1:** 571
Stark effect   **Vol. 1:** 569, 573
X-rays   **Vol. 1:** xxxi
Zeeman effect   **Vol. 1:** xxxvi
Speziali, P.   **Vol. 6:** 8
spheres, elastic   **Vol. 1:** 515
spin   **Vol. 7:** 175, 186, 220, 242, 524
  Vol. 8: 210*f*, 442, 449, 502

[143]

Vol. 10: 339, 395, 396, 403, 404, 406, 418, 427, 438, 443, 447, 451

exchange (*see also* exchange forces)
    Vol. 9: 195–198, 204, 677

of electron (*see also* electron spin; magnetic moment of electron) **Vol. 5:** 226–239, 283, 285, 287–289, 330, 332, 339, 372, 376*f*, 406, 457–482, 497, 499–500, 507–508
    Vol. 9: 36, 94, 105*ff*, 125, 275, 304, 319, 523, 583, 585

of neutrino **Vol. 9:** 275

of nucleon **Vol. 9:** 30, 36, 197, 202, 204, 237, 261, 274, 280, 307, 321, 376, 583, 585, 679

of nucleus (*see also* angular momentum)
    Vol. 9: 5, 7, 29*f*, 36, 38, 42, 79, 80, 153, 187, 195–198, 202*ff*, 211, 231, 236*f*, 248, 256, 261, 274, 280, 294, 306, 321, 356, 376, 378, 381, 392, 491–492, 494, 496, 500, 507, 509, 511, 513, 518*f*, 524, 583*f*, 614*f*, 646*f*, 654*ff*, 679

of positron **Vol. 9:** 11, 135*f*

of proton (*see also* magnetic moment of proton) **Vol. 6:** 145, 158, 398, 404

spin–orbit coupling **Vol. 5:** 229, 234–235, 236–237, 467*ff*, 471, 473*f*, 499*f*, 507
    Vol. 9: 30, 78*f*

Spinoza, B. **Vol. 6:** xx
    Vol. 7: 376, 377
    Vol. 10: xliii, xliv, 7, 175, 298, 318, 322, 546, 550
    Vol. 11: 196, 200

spiritual
    influence on material phenomena
        Vol. 10: 35, 536, 537
    traditions **Vol. 10:** 247

spiritualism **Vol. 10:** 20, 175

Sponer, H. **Vol. 8:** 685, 687

spontaneous transition **Vol. 3:** 73
    probability of **Vol. 3:** 6, 81, 98, 375

sport **Vol. 11:** 677, 679

Springer, J. **Vol. 5:** 402

Squire, C.F. **Vol. 9:** 56
    Vol. 12: 38*p*

stability **Vol. 8:** 9, 169, 172, 177, 184, 189, 197*ff*, 269, 278, 686*f*, 786
    Vol. 10: 30, 33, 34, 44, 59, 69, 70, 77, 87, 103, 128, 130, 136, 142, 149, 158, 165, 177, 180, 209–211, 213, 242, 244,

284, 290, 333, 376, 386, 388, 389, 405, 407, 435, 436, 498, 499, 519, 522, 539, 541, 544, 548
    of atom **Vol. 4:** 200*ff*, 246, 247, 273, 434, 468
        Vol. 5: 131, 139, 242, 244, 246*ff*, 275, 277–278
    of atoms and molecules **Vol. 2:** 46, 49–51, 104*ff*, 127, 134, 136*ff*, 162, 167, 180, 183, 185, 189*ff*, 216*ff*, 291, 338, 396–397, 442–443, 460, 466, 473*ff*, 482*ff*, 528, 529*f*, 577, 587, 625, 633
    of atoms/stationary states **Vol. 6:** 70–71, 130–131, 155–156, 225, 238, 275, 301, 317, 319, 355–356, 374, 376, 379–380, 381, 383*f*, 387, 393, 397, 399, 401, 403, 406*f*
    of cells **Vol. 10:** 91, 104, 122, 151, 214, 544, 548
    of elementary particles **Vol. 6:** 401–402, 403, 407
    of genes **Vol. 10:** 22, 131, 168, 180
    of living organisms **Vol. 10:** 136, 166, 214, 497, 499, 501
    of measuring bodies **Vol. 10:** 184, 444
    of molecule **Vol. 5:** 189, 203
        Vol. 10: 180, 198, 214, 471, 472, 497
    of nuclei **Vol. 6:** 145, 225, 238, 403
        Vol. 10: 121, 408, 412, 451
    of organisms **Vol. 6:** 275, 301
        Vol. 10: 485, 486
    of stationary state **Vol. 5:** 90, 103, 108, 111, 113, 167, 169, 179, 182*f*, 187*ff*, 191*f*, 194, 197*f*, 201*ff*, 205*f*, 224, 245, 248, 255, 262, 275, 277–278, 305, 454*f*
    properties **Vol. 10:** 57, 61

Stachel, J.J. **Vol. 5:** 27
    Vol. 7: 55, 56
    Vol. 9: 55

Stäckel, P.G. **Vol. 5:** 277

Stahel, E. **Vol. 5:** 94
    Vol. 7: 32*p*

Stahl, G.E. **Vol. 10:** 4*n*

Stainer, W.W. **Vol. 8:** 675

Stalin, J. **Vol. 11:** 10, 34, 35, 53*n*, 198, 201, 213*n*, 223, 225, 246, 289, 290
    Vol. 12: 506*n*

Stanford University **Vol. 9:** 540*f*, 544

Stanley, C.W. **Vol. 8:** 562

Stanley, W.M. **Vol. 10:** 23

Stansfield, H.   **Vol. 8:** 696f
Stark effect   **Vol. 1:** 569, 573
    Vol. 2: 304, 319–326, 335, 344, 345,
    350, 354, 356–362, 370–371, 400–403,
    406–407, 439, 446, 467ff, 511, 554f,
    559, 562, 589ff, 596f, 600–601, 606,
    608, 627, 628
    Vol. 3: —
    Vol. 4: —
    Vol. 5: 63, 92, 112, 247, 277, 280,
    390–391, 393, 407, 445, 448, 450, 452
    Vol. 6: 311–312, 315f
    Vol. 8: 226, 672, 673, 687, 795
    Vol. 12: 368
Stark, J.   **Vol. 1:** xxxi, 108, 109, 115, 127,
    437, 438, 447, 448, 552, 553, 558, 559,
    568, 569
    Vol. 2: 110, 111, 166, 319, 320, 321,
    322, 323, 324, 335, 350, 356, 357, 360,
    361, 362, 365, 368, 370, 371, 400, 401,
    402, 406, 407, 409, 446, 468, 511, 554,
    555, 589, 590, 591, 592, 593, 600, 601,
    605–606, 608, 615, 616, 628
    Vol. 3: 9, 51, 52, 69, 87, 135, 136, 137,
    141, 142, 144, 145, 164, 166, 174, 175,
    176, 182, 275, 276, 313, 315, 316, 322,
    329, 330, 331, 336, 419, 421, 422, 438,
    444, 597, 641
    Vol. 4: 114, 213, 286, 376, 436, 449,
    451, 452, 470, 476, 477
    Vol. 5: 112, 277
    Vol. 10: 394, 395
Starling, E.H.   **Vol. 11:** 445
Stason, D.E.B.   **Vol. 11:** 563
state relations   **Vol. 3:** 366, 372, 405ff
  for periodic and multiply periodic motions
    Vol. 3: 426ff
stationary state (*see also* normal state;
    permanent state; quantum states)
    Vol. 2: 112, 128, 164ff, 190ff, 232, 243,
    264f, 293ff, 313–316, 326, 351ff, 378,
    380, 394ff, 409, 434ff, 464ff, 474ff, 559,
    562, 590, 604
    Vol. 3: 210, 245, 366ff, 402ff, 422, 459
    Vol. 5: —
    Vol. 8: 17, 18ff, 27f, 30, 32f, 169,
    172, 177, 185, 190, 198, 200, 207, 263,
    306, 366, 368, 376, 381, 411f, 498f,
    542f, 598, 602, 604, 608, 613, 619,
    621ff, 637f, 657f, 675, 683, 687, 706

    Vol. 10: 30, 58, 177, 376, 377,
    388–390, 392, 395, 398, 401–403,
    437–439, 443, 450, 552
    Vol. 12: 10, 43–45, 47, 51, 157, 170,
    210, 214, 357
assumption of   **Vol. 3:** 71
classification of   **Vol. 3:** 574ff
description and classification of
    Vol. 3: 572ff
fixation for simply and multiply periodic
    systems   **Vol. 3:** 461ff
for systems in presence of external
    conservative field   **Vol. 3:** 463ff
general properties of   **Vol. 3:** 402ff
of hydrogen atom, taking relativity
    modifications into account
    Vol. 3: 119ff
of perturbed periodic system
    Vol. 3: 107ff, 115
reality of, *see* reality of stationary states
stability of, *see* stability of atoms/stationary
    states
stability of, in presence of external varying
    fields of force   **Vol. 3:** 467ff
statistical weights of   **Vol. 3:** 473ff
statistical
  atom mechanics   **Vol. 10:** 9, 58, 439, 496
  character of quantum processes
    Vol. 10: 30, 33, 35, 70, 87, 129, 158,
    166, 195, 389, 405, 426, 432, 442, 445,
    499, 564, 568
  description   **Vol. 12:** 156, 168
  description in quantum physics   **Vol. 7:** 35,
    249, 262, 264, 265, 303–309, 315, 345,
    363, 371, 377, 391, 392, 403, 406, 417,
    419, 471
  distribution of energy loss, *see* probability
    distribution of energy loss
  distribution of fission fragments (*see also*
    fission, symmetrical/asymmetrical)
    Vol. 9: 66, 72–74, 344, 350, 352–353,
    361, 366, 386–387, 467–473, 553,
    662–663, 679
  equilibrium   **Vol. 9:** 54, 392, 477, 492,
    499ff, 510ff
  interpretation   **Vol. 8:** 36, 207
  mechanics (*see also* classical statistics;
    kinetic theory; Maxwell distribution;
    mechanics, statistical; microcanonical
    ensemble)   **Vol. 1:** 304ff

Vol. 9: 209, 259, 262, 314, 326, 344, 366, 373, 386, 433, 459, 499, 511, 555
Vol. 10: 177, 208, 433, 442
Vol. 12: 9, 39
mechanics, classical (*see also* thermodynamics) **Vol. 6:** 143, 204, 209, 226*f*, 240*f*, 316, 319–330, 359, 374, 398, 400–401, 469–473, 477*f*
Vol. 7: 241, 295, 331, 343, 353, 400, 413
mechanics, quantum **Vol. 6:** 50, 144*f*, 349, 397*f*, 404*f*, 466
thermodynamics (*see also* mechanical theory of heat) **Vol. 10:** 103, 399, 432, 500
weight **Vol. 5:** 44, 255, 278, 354, 356, 411, 413, 446, 451, 478, 481
Vol. 9: 392, 500, 509, 511, 663
weights, invariance of
Vol. 3: 529*ff*, 559*ff*
weights of stationary states
Vol. 4: 646*ff*
statistics, *see* probability
statistics of nuclei **Vol. 6:** 330, 404, 406
Vol. 9: 5, 7, 94, 96, 111, 113, 117, 593*f*
statistics of stationary states **Vol. 2:** 338, 341–342, 450–461, 565, 567, 604, 631
Stauning, T. **Vol. 12:** 241, 245
Steensen, N. **Vol. 11:** 517, 525
Steensgaard, E. **Vol. 12:** 240, 245
Steensholt, G. **Vol. 9:** 638, 640
Vol. 12: 17*p*
Steenstrup, J. **Vol. 12:** 152, 164
Stefan, J. **Vol. 12:** 228, 229
stellar spectra, *see* spectra, coronal and stellar
Steno Lecture **Vol. 10:** x, 116, 546*n*, 561*n*, 586, 590
Steno, N. **Vol. 11:** 544, 551
Stensen, N. (Steno) **Vol. 10:** 116, 117, 123, 258, 268, 354, 355, 368
Stenström, W. **Vol. 4:** 493, 512, 530, 543
Stepanowa, E. **Vol. 8:** 802
Stern–Gerlach effect **Vol. 7:** 346, 524
Vol. 10: 426
Stern–Gerlach experiment **Vol. 5:** 255, 262, 278, 328, 446–447, 451, 504
Vol. 6: 10, 32, 37, 132, 156, 164, 166, 170–172, 177–178, 179–181, 307, 309*ff*, 333–336, 338, 341–343, 348*f*, 392
Stern, O. **Vol. 3:** 23, 484, 664, 673
Vol. 4: 51, 367, 449, 476, 596, 645, 695
Vol. 5: 228*f*, 255, 262, 278, 439, 442,

447, 449, 451, 453, 455*f*, 504, 507
Vol. 6: 35, 41, 43, 318, 326*f*, 329*f*, 333*f*, 392, 414, 432, 435, 437, 449*ff*, 456, 467–473
Vol. 7: 181, 190
Vol. 8: 212, 647
Vol. 9: 28, 100, 126, 131
Vol. 10: 339, 558, 559, 561
Stevenson, E.C. **Vol. 8:** 810
Stevin, S. **Vol. 10:** 208
Stifler, W.W. **Vol. 5:** 323
Stimson, H. **Vol. 11:** 47–49, 69, 293
Stintzing, H. **Vol. 3:** 27, 307, 323, 337
Vol. 4: 49
Vol. 5: 398
Stocker, H. **Vol. 2:** 616
Stockholm appeal, 1950 **Vol. 11:** 80, 189
Stokes, G.G. **Vol. 1:** 37, 67, 76, 77
Stolzenburg, K. **Vol. 6:** vii, 8, 37, 51
Vol. 8: vi, 31*f*, 203
Stone, S. **Vol. 11:** 715
Stoner, E.C. **Vol. 1:** xxxvi
Vol. 4: 40, 41, 42, 679, 680, 681
Vol. 5: 81, 251*f*, 256, 260, 262, 266, 279, 335, 424, 428, 431*f*
Vol. 6: 391
Vol. 10: 394, 419
Vol. 12: 48
Stoney, G.J. **Vol. 4:** 658
Vol. 6: 374
Vol. 12: 42
stopping
of charged particles (*see also* absorption of charged particles) **Vol. 5:** 69–75, 79*f*, 89, 178–206, 334, 337, 345, 347, 350, 360*ff*
Vol. 6: 76, 82, 131, 156, 375
Vol. 9: 26, 75, 431*f*, 457*f*, 524, 559, 603*f*
of fission fragments **Vol. 2:** 10
power **Vol. 2:** 93, 374, 500
Vol. 5: 71, 82, 89, 181*f*, 186, 195, 197, 200, 362*f*
stopping formula
Bethe's **Vol. 8:** 15, 208*f*, 214*ff*, 226, 232, 239, 241, 256, 269, 273*ff*, 282*ff*, 289*f*, 305, 330*f*, 339, 372, 379*f*, 407, 500*ff*, 505, 512, 516*ff*, 549, 635*ff*, 654, 737*f*
Bloch's **Vol. 8:** 219, 372, 380, 407, 503, 515*f*, 518*f*, 524

Bohr's non-relativistic **Vol. 8:** 9, 59*f*, 81*ff*,
    97*f*, 133*f*, 139*f*, 170*f*, 174*f*, 178*f*,
    208*f*, 213, 218*f*, 226, 232, 236, 241,
    260*f*, 269, 273, 275, 278, 282, 284,
    289*f*, 305, 320, 330, 338, 371, 379,
    407, 492, 497, 500, 515, 608, 635*ff*,
    774, 778, 791
Bohr's relativistic **Vol. 8:** 15, 142*f*, 214,
    268*f*, 275, 278, 284, 290, 724*f*
Gaunt's **Vol. 8:** 208*f*, 215*f*, 225, 241,
    371, 499, 791
Henderson's **Vol. 8:** 28*f*, 209, 241,
    339, 355, 371, 395, 399, 498, 512,
    657, 737*f*
Møller's **Vol. 8:** 214, 275, 284, 502, 517*f*,
    522, 792
Thomson's (*see also* Thomson's theory)
    Vol. 8: 59*f*, 68*f*, 76, 83, 92, 98, 154*f*,
    276, 285, 487*f*, 520, 522
Storm, K. **Vol. 10:** 74
Störmer, C. **Vol. 1:** 204, 322
    Vol. 12: 524
straggling **Vol. 2:** 66, 77–78
    Vol. 8: 14*ff*, 28, 89, 105, 111, 114, 121,
    125, 134–140, 147–148, 151–153, 160,
    213, 221*ff*, 225, 255, 257, 274, 283,
    325, 328, 332–333, 473–479, 486,
    518–519, 527–528, 547, 551–555,
    560–562, 564, 571, 665, 668*ff*, 678,
    683, 706, 783, 791, 793
"strain", *see* "Zwang"
Strassmann, F. **Vol. 1:** xliv
    Vol. 8: 405
    Vol. 9: 52, 56, 63, 67, 342, 344*f*,
    365*f*, 378*f*, 382, 407*f*, 431, 457, 469,
    553, 555, 557*f*, 565, 660
    Vol. 10: 412
    Vol. 11: 12, 569, 579, 597
    Vol. 12: 84
Strauss, L. **Vol. 11:** 325–327, 329, 330, 363,
    368*p*, 713
Street, J.C. **Vol. 8:** 810
Strelitz, P.
    Vol. 11: 328, 331
    Vol. 12: 400, 404
stripping **Vol. 8:** 329, 541, 623, 765
Strömgren, B. **Vol. 11:** 406*p*
    Vol. 12: 17*p*
Strömgren, E. **Vol. 1:** 168, 294
    Vol. 5: 483
strontium **Vol. 11:** 580, 599

structure
    of atoms, *see* atomic constitution/structure
    of molecules, *see* molecular structure
    of nuclei, *see* nuclear constitution/structure
Strutt, R.J. **Vol. 2:** 238, 239, 307, 308, 627
    Vol. 3: 21, 237, 255, 256, 275
Strutt, W., *see* Rayleigh, Lord
Stuewer, R.H. **Vol. 5:** 3, 7, 30, 38, 490
    Vol. 8: 225
    Vol. 9: vi, 16
Stüssi, F. **Vol. 7:** 328
subconsciousness **Vol. 10:** 93, 566
subject/object (*see also* ultimate subject)
    Vol. 10: xxv, xxviii, xxxi, xxxiv,
    xxxvi–xxxviii, xli–xlvi, xlviii, xlix,
    6–8, 15, 17, 47, 62, 71, 93, 95, 97, 105,
    123, 142, 157, 159, 160, 185, 203, 215,
    216, 248, 276, 279, 280, 286, 292, 476,
    479–481, 483, 485, 486, 567, 570
    distinction/relation (*see also* objectivity)
    Vol. 6: xxiv,
    xxvi, 141, 158, 189*f*, 192, 204–205,
    212, 234, 251, 269, 293, 367, 431
    separation **Vol. 11:** 134
subjectivity (*see also* objectivity; subject/object
    distinction/relation) **Vol. 6:** 46, 190,
    205, 213, 220, 233, 250, 259, 279, 431
    Vol. 10: 95, 284, 290, 496
successive disintegrations **Vol. 9:** 55, 74, 210,
    244, 277, 295*f*, 351, 353, 368*f*, 384,
    425, 431, 437, 452–453, 457, 462,
    475–479, 484
Sudoplatov, P. **Vol. 11:** 332
Suess, H.E. **Vol. 9:** 78
    Vol. 11: 243, 244, 264, 265
Suhm, P.F. **Vol. 11:** 399, 408
sulphur **Vol. 12:** 45
sum rules **Vol. 5:** 168, 171, 272, 423, 427,
    448, 452
Supek, I. **Vol. 12:** 104*p*, 105, 516, 517
superconductivity **Vol. 1:** xlv
    Vol. 5: 440, 443, 479, 482
    Vol. 6: 25, 425
    Vol. 8: 219, 651–652, 654*f*, 679,
    702*ff*, 714*ff*
    Vol. 9: 25, 600*f*
    Vol. 10: 434, 448, 534, 535
    Vol. 12: v–vii, 3, 4*n*, 5–9, 17*p*, 22–28,
    30–34, 39, 51, 54, 498–501, 525, 530,
    532–534

transition temperature **Vol. 12:** 8, 9, 24, 28, 30–34, 39, 499, 501
transition to **Vol. 12:** 6–8, 24, 28, 30, 33, 498–501
superfluidity **Vol. 11:** 57, 298, 302
  Vol. 12: 12, 51, 54
superlattice **Vol. 12:** 54
superposition **Vol. 7:** 30, 36, 99, 150, 177, 180, 188, 190, 221, 262, 304, 306–308, 315, 347, 348
  Vol. 8: 292, 343, 348, 445
  Vol. 9: 48, 108, 386, 392, 492, 494, 514, 669
superposition (principle) **Vol. 6:** 27, 48f, 59ff, 69ff, 76f, 79f, 83f, 87f, 92, 96f, 115ff, 126f, 129, 132, 135, 142, 148f, 153ff, 193, 206, 216, 441, 443, 465f
surface tension, curves **Vol. 1:** 17ff
surface tension/energy (*see also* capillarity)
  Vol. 9: 34, 47, 52f, 56, 194, 234, 336, 365, 369ff, 421, 432, 436, 449, 458, 461, 561f, 582, 584, 666
surface tension, water **Vol. 1:** xix, 8, 12
  Vol. 12: 135, 136, 138, 140, 144, 149, 150
  by method of jet vibration **Vol. 1:** 25ff
  discussion of results **Vol. 1:** 64–65
  effect of surrounding air **Vol. 1:** 44ff
  in recently formed surface **Vol. 1:** 81ff
  possible variations of **Vol. 1:** 83, 87
  results **Vol. 1:** 57–64
surgery **Vol. 12:** 454, 459
Suter, P. **Vol. 1:** 119, 120, 121, 122
Sutherland, W. **Vol. 1:** 175, 300, 413
Swan, C.P. **Vol. 1:** 478
Swann, W.F.G. **Vol. 8:** 496, 568
Swinne, J.R. **Vol. 4:** 251, 322, 410
Sylvestri, M. **Vol. 11:** 563
Sylvius, J. **Vol. 10:** 54
symbolic character of quantum theory
  Vol. 6: 12, 32, 52, 60, 62, 65, 71, 75, 80, 82, 88, 93ff, 115f, 118f, 124ff, 132, 135, 149f, 152ff, 203, 208, 244, 246, 266, 270, 290, 295, 317, 360, 393ff, 398, 459
symbolic nature of description/model
  Vol. 5: 90, 191, 205, 258, 265, 276, 280, 311
symmetry **Vol. 7:** 68, 129
  Vol. 8: 148, 211f, 215, 226, 306, 386ff, 433, 450, 458, 462, 477, 495, 543, 678ff, 712f, 795

Vol. 9: 12f, 125, 202, 211, 274, 280, 306f, 315, 321, 328, 336, 392, 549, 578f
sympathy **Vol. 10:** 188, 202
synthesis, *see* analysis; analysis/synthesis
systems consisting of nucleus and single electron **Vol. 4:** 48ff, 100ff, 345ff, 360
systems of one degree of freedom
  Vol. 3: 76ff, 369
Szczeniowski, S. **Vol. 7:** 302
Szent-Györgyi, A. **Vol. 10:** 134
Szilard, G. **Vol. 6:** 330
Szilard, L. **Vol. 6:** 330, 470, 473
  Vol. 9: 384, 386
Szilard machine **Vol. 6:** vii, 330, 468, 470–471, 473

Taconis, K.W. **Vol. 12:** 38p
Tagore, R. **Vol. 5:** 328
"tail" of energy distribution **Vol. 8:** 15, 111, 114, 152, 255, 475, 477ff, 483, 520, 559
Tait, P.G. **Vol. 3:** 625
Takamine, S. **Vol. 12:** 368
Takamine, T. **Vol. 3:** 9, 182, 524, 553, 599
  Vol. 4: 604, 650
  Vol. 5: 87
  Vol. 8: 687
  Vol. 10: 593
  Vol. 12: 117, 117p, 118, 368, 533
Tamm, I. **Vol. 8:** 497, 567f
  Vol. 11: 7p, 230
Tänzler, P. **Vol. 2:** 363
Tata Institute of Fundamental Research (Bombay) **Vol. 12:** 534
Tate, J.T. **Vol. 8:** 749, 803
  Vol. 9: 73, 664
Taylor, G.I. **Vol. 3:** 497
Taylor, T.S. **Vol. 2:** 28, 29, 30, 74, 75, 76, 78, 87, 88, 581
  Vol. 8: 60ff, 75, 91, 110, 113f, 144ff, 148, 157f, 668, 752
Tchaikovsky, P. **Vol. 11:** 376
Technical Museum of Denmark
  Vol. 12: 310n
Technical University of Denmark **Vol. 11:** 352, 354, 355, 402, 412, 470, 473, 478, 480, 482, 483, 488, 490, 513n, 518n
  Vol. 12: 112, 184, 193, 307, 310, 348, 350
  physical laboratory of **Vol. 12:** 326, 328

Technical–Scientific Research Council
  (Denmark)   **Vol. 11:** 351, 459
  Vol. 12: 183, 192
technology   **Vol. 10:** 141, 148, 151, 153, 157,
  160, 182, 183, 193, 260, 272, 283, 286,
  289, 292, 344, 358, 415
  Vol. 11: 74, 79, 81, 82, 133, 139, 143,
  151, 153, 165, 166, 175, 185, 188, 198,
  201, 203, 204, 351, 368, 455, 459, 538,
  548, 556, 559, 565, 578, 596, 615–617,
  627, 630, 635, 636, 643, 672, 736
  Vol. 12: 75, 84, 181, 182, 189, 190,
  236, 327, 330
teleology   **Vol. 6:** 275, 301
  Vol. 10: 4, 9, 18, 34, 53, 122, 167, 276,
  519, 522, 524, 526, 534, 535, 565
Teller, E.   **Vol. 1:** xliii
  Vol. 6: 326f, 449, 453
  Vol. 8: 224, 239f, 248, 252, 407, 537,
  540f, 556, 566f, 572, 647, 709ff, 750,
  753, 759f, 762, 765, 789
  Vol. 9: 28, 31, 56
  Vol. 12: 386, 387
temperature   **Vol. 6:** 26, 145, 325, 364f, 397,
  400f, 469ff
  concept of   **Vol. 10:** xxxvii, 10, 119
    Vol. 12: 122, 428, 430
  of nucleus   **Vol. 9:** 27, 29, 209ff, 233, 241ff,
    252, 259f, 262f, 268f, 276f, 293, 298,
    312ff, 325ff, 336, 342, 380, 419–423,
    424ff, 434, 448–450, 452ff, 459f, 477
  radiation, *see* black-body radiation
Templeton, D.H.   **Vol. 6:** 414
Terletzkii, Y.   **Vol. 11:** 57, 58, 299, 302–304,
  325, 329, 332, 333, 703n
terminology
  physical   **Vol. 7:** 273, 335, 390
  questions of   **Vol. 7:** 377
terms
  singlet   **Vol. 4:** 601, 649
  triplet   **Vol. 4:** 600, 648
Tesla Centenary Celebrations, 1956
    Vol. 11: 722
    Vol. 12: 105, 236, 531
Tesla, N.   **Vol. 7:** 397, 523
    Vol. 10: 590
    Vol. 11: 722n
    Vol. 12: 105, 106, 228, 229, 236, 531
Thanatological Argument/Principle
    Vol. 10: 11, 12, 19

thermal
  conductivity, non-metallic   **Vol. 1:** 340
  energy   **Vol. 9:** 209f, 259f, 277, 313, 316,
    326, 328, 335, 342, 344, 366, 404,
    419ff, 424ff, 432ff, 441, 448ff, 452ff,
    458f, 461f, 465
  equilibrium   **Vol. 10:** 400, 413
  fluctuations   **Vol. 10:** 136
  motion   **Vol. 11:** 568
  radiation (*see also* black-body radiation)
    Vol. 10: 87, 103, 149, 376, 391, 400,
    402, 427, 432
thermodynamics (*see also* kinetic theory;
    mechanical theory of heat; statistical
    mechanics)   **Vol. 5:** 31, 33, 70, 81, 84,
    179, 185, 191, 195, 199f, 205, 224, 255,
    262, 273f, 278, 350, 415, 419, 454f
    Vol. 6: 226, 240, 316, 319–330, 374,
    383, 400–401, 423, 449, 453–454,
    469–473, 477f
    Vol. 7: 303, 342, 345, 400, 414
    Vol. 8: 27, 36, 530, 718
    Vol. 10: 10, 25, 47, 57, 86, 91, 104,
    119, 131, 136, 137, 149, 150, 165, 166,
    176, 208, 214, 363n, 395, 439, 492,
    493, 540, 542, 582
    Vol. 11: 657, 662, 670
    Vol. 12: 221, 291, 293, 348, 351, 438,
    440, 533, 534
  Bohr's lectures on
    Vol. 2: 106, 515, 517
thermoelectric, open circuit   **Vol. 1:** 346ff
  all parts of same metal   **Vol. 1:** 347
  different metals, non-uniform temperature
    Vol. 1: 348
  power   **Vol. 1:** 349
  uniform temperature   **Vol. 1:** 347
thermoelectricity   **Vol. 1:** 159, 345ff
  discussion with O.W. Richardson
    Vol. 1: 116, 481ff
  rejoinder to paper by O.W. Richardson
    Vol. 1: 443–444
  thermodynamic theory   **Vol. 1:** 352ff
thermomagnetic phenomena   **Vol. 1:** 383ff
thermometer   **Vol. 12:** 428, 430
Thiele, T.N.   **Vol. 1:** 508, 509, 542, 543
Thirring, W.E.   **Vol. 10:** 556
Thomas B. Thrige Factories   **Vol. 11:** 359,
    518, 525, 557–560, 573, 585
Thomas B. Thrige Foundation
    Vol. 9: 217, 221

Vol. 11: 359, 360, 363, 517, 524, 530, 557–560, 573, 585, 647, 650
Vol. 12: 182, 190, 211, 216
Thomas precession (*see also* precession of electron spin) **Vol. 5:** 231–232, 235, 339, 372–377, 462–482, 508
Vol. 6: 3, 7, 158, 422
Thomas–Fermi model **Vol. 8:** 220*f*, 238*f*, 255, 330, 461, 524*f*, 540, 653, 750
Thomas, L.H. **Vol. 5:** 225, 231*f*, 235*f*, 337*ff*, 372–377, 462–482, 508
Vol. 6: 3, 158, 422, 458
Vol. 8: 203*ff*, 245, 363*f*, 372, 383, 385*ff*, 530*f*, 535, 537, 568, 677
Thompson, G.C. **Vol. 11:** 563
Thomsen, A. **Vol. 10:** xliv
Thomsen, C.J. **Vol. 10:** 269*n*
Thomsen, J. **Vol. 1:** 7, 15
Vol. 3: 582, 601
Vol. 4: 22, 272, 316, 388, 405, 431, 463, 468, 482, 658
Vol. 6: 387, 390
Vol. 12: 42, 438, 440
Thomsen, M. **Vol. 11:** 406*p*
Thomsen, V. **Vol. 6:** xx*f*
Vol. 10: 250*p*, 298, 312*p*, 313, 319
Vol. 11: 340, 342, 392*p*, 394–396
Vol. 12: 100, 123, 124*n*, 153, 166, 198, 202, 527
Thomson
  coefficient **Vol. 1:** 350*ff*
  effect **Vol. 1:** 483, 489*ff*, 543
Vol. 12: 22, 25
Thomson, G.P. **Vol. 6:** 7, 394, 445*f*
Vol. 9: 5, 87
Vol. 10: 402, 440
Thomson, J.J. **Vol. 1:** xxi, xxiii, 97, 102, 104, 105, 106, 124, 132, 133, 149, 150, 152, 155, 156, 157, 158, 173, 175, 181, 204, 228, 238, 239, 240, 241, 244, 247, 255, 273, 276, 278, 298, 300, 305, 322, 341, 350, 351, 352, 355, 357, 358, 364, 365, 368, 379, 381, 383, 398, 407, 412, 413, 416, 417, 422, 423, 424, 427, 428, 429, 431, 432, 433, 440, 442, 446, 447, 448, 460, 462, 463, 464, 476, 479, 482, 483, 484, 485, 488, 489, 490, 518, 519, 520, 521, 522, 523, 524, 525, 526, 527, 528, 529, 530, 531, 532, 533, 538, 539, 548, 549, 558, 559

Vol. 2: 3, 18, 19, 20, 22, 28, 60, 84, 85, 90, 103, 125, 126, 136, 139, 145, 152, 162, 166, 168, 178, 183, 200, 210, 224, 228, 239, 267, 268, 278, 290, 344, 423, 476, 483, 486, 576, 588, 589, 625, 627
Vol. 3: 203, 584
Vol. 4: 26, 44, 104, 198, 200, 202, 273, 274, 388, 422, 429, 432, 457, 467, 468, 479, 658, 659, 661
Vol. 5: 3, 130, 138, 178, 184, 186, 194, 199*f*, 274, 278, 280
Vol. 6: 76, 82, 359, 373, 375, 378*f*, 387
Vol. 7: 4
Vol. 8: 4*f*, 12, 44, 50*ff*, 54, 60, 76*ff*, 86, 91*ff*, 101, 109, 113, 130, 154, 183, 185, 188, 191, 239, 255, 270, 279, 343, 347, 427, 487*f*, 568, 780, 782, 818
Vol. 9: 109
Vol. 10: 56, 102, 337, 383, 384, 386, 387, 393, 398, 399, 434, 436
Vol. 11: 241, 445, 530, 563, 566, 568
Vol. 12: 42, 43, 45, 108, 109, 138, 140, 144, 149, 150, 160, 173, 198, 199, 202, 203, 221, 252, 255, 259, 265, 266, 326, 329, 531
Thomson, Mrs G.P. **Vol. 8:** 750
Thomson, W. *see* Kelvin, Lord
Thomson's atomic model, *see* atomic model, Thomson's
Thomson's theory (*see also* atomic model, Thomson's; stopping formula, Thomson's) **Vol. 8:** 5, 11*f*, 16, 45, 50*f*, 76, 78*f*, 82, 91*ff*, 98, 130, 160, 163, 169*ff*, 173*f*, 177, 179, 239, 255, 270, 276, 279, 285, 365*f*, 396, 400, 430, 453, 487*ff*, 757*f*, 818
thorium **Vol. 9:** 56, 57*f*, 64*ff*, 75, 343*ff*, 349*ff*, 353, 357, 361, 366*f*, 370, 372, 374, 377, 381, 383*f*, 388*f*, 401, 404, 407*f*, 431, 433, 437, 439*ff*, 457, 459, 462, 464*f*, 469, 473, 478, 484, 551*ff*, 559*f*, 566, 588*f*, 625, 627, 629, 631, 634, 636, 638*f*, 659, 662
Vol. 11: 579, 583, 597, 604, 616, 627
Thorsen, V. **Vol. 3:** 659, 661
Vol. 8: 786
Thorvaldsen, B. **Vol. 10:** 258, 269
Vol. 12: 152, 164
thought **Vol. 10:** xli, xliii–xlv, xlvii, 8, 47, 62, 143, 159, 168, 184, 199, 215, 230,

231, 235, 245, 279, 285, 286, 291, 317, 321

thought experiment (*see also* double-slit experiment; Einstein weighing experiment) **Vol. 6:** 15, 17, 160, 170–172, 177–183

Bohr's **Vol. 5:** 58, 60–61, 416–417, 420

Thrige, T.B.
Vol. 11: 359, 517, 524, 556–560
Vol. 12: 182, 190, 216*n*

Thule expeditions **Vol. 12:** 128

Thullie, Z. **Vol. 1:** 276, 277, 382

Tigerstedt, R. **Vol. 10:** 5*n*

Tillich, P. **Vol. 7:** 411

time
concept **Vol. 7:** 209
direction of **Vol. 10:** 12, 104, 122, 496
in biology **Vol. 10:** 554, 565
interval/region **Vol. 7:** 75, 86, 134, 141, 201
measurement **Vol. 7:** 296
of collision (*see also* collision, time of)
Vol. 8: 9, 15, 33, 52*ff*, 80*f*, 96, 109, 113, 118, 131*f*, 142, 170, 174, 178, 183, 189, 198*f*, 206, 272, 281, 291, 378, 394, 398, 434*f*, 490*f*, 494, 498, 504, 507*f*, 510, 516, 604, 635*f*, 657*ff*, 672

Times (London newspaper) **Vol. 11:** 50–52, 54, 61, 67, 296, 300, 324, 326, 328, 329

Timoféeff-Ressovsky, N.W. **Vol. 10:** 22, 23, 168, 471, 472, 552

tin, absorption in **Vol. 2:** 30, 35, 37

Tisdale, W. **Vol. 8:** 753

Tisza, L. **Vol. 11:** 7*p*

Titeica **Vol. 9:** 519, 568

Tizard, H.T. **Vol. 11:** 277

Toll, J. **Vol. 9:** 667

Tolman, R.C. **Vol. 2:** 441
Vol. 6: 181
Vol. 11: 325, 329

Tomonaga, S. **Vol. 7:** 49, 212, 485
Vol. 10: 453, 455*p*
Vol. 12: 38*p*

Tönnies, F. **Vol. 10:** 301

Topsøe, H. **Vol. 11:** 583, 606

Toronto, University of **Vol. 9:** 39

total charge, *see* effective charge

Tout, T.F. **Vol. 10:** 397

Townsend, J.S. **Vol. 1:** 442
Vol. 2: 631
Vol. 12: 285

Toyoda, T. **Vol. 9:** 76

Trabacchi, G.C. **Vol. 9:** 100, 478

tracer **Vol. 9:** 167, 175, 216, 220

tracer method, *see* isotopic indicators

tracks, *see* cloud-chamber tracks; fission-fragment tracks

transfer of electricity and energy
Vol. 1: 331*ff*

transformation
adiabatic, *see* adiabatic principle/process/transformation
principle, *see* adiabatic principle/process/transformation
theorem **Vol. 7:** 292, 297
theory **Vol. 6:** 3*f*, 23, 46, 95, 97, 111*f*, 125, 153, 155, 160, 164*ff*, 167–172, 175, 184, 421

transformation/transmutation (*see also* disintegration) **Vol. 8:** 183, 188, 219, 278, 371, 764, 811, 813
Vol. 11: 91, 92, 102, 105, 133, 139, 143, 151, 153, 220, 363, 515, 516, 523, 566, 568, 575–580, 589–597, 599, 613, 614, 616, 622–625, 627, 646, 649, 659, 664
Vol. 12: 42, 51, 52, 59, 61, 63, 67, 79, 81, 84, 211, 215, 216, 256, 257, 260, 261, 266, 277

transition **Vol. 8:** 17, 20*f*, 27*f*, 30*ff*, 34, 117, 198, 200, 215, 276, 285, 381, 396, 400, 411*f*, 501, 505, 586*f*, 598, 605, 657*f*, 662, 675, 683, 685, 687*ff*, 694

between stationary states, *see* stationary state

from micro- to macro-mechanics (*see also* correspondence with classical physics)
Vol. 6: 20, 49, 134, 157, 160, 172–177, 186, 466

probability (*see also* probability)
Vol. 3: 72
Vol. 5: 11*f*, 22, 27, 30, 42*ff*, 58–59, 60, 67*f*, 104*ff*, 124, 146, 150, 256*f*, 259*f*, 263*ff*, 267, 275, 277, 279*f*, 398*f*, 401, 403, 417, 420–421, 490*f*, 496, 506
Vol. 6: 9, 71, 145, 171*f*, 178, 228, 243, 261, 263, 282, 286, 382, 393*f*, 400, 405, 422, 428, 430

Vol. 8: 30, 33f, 208, 275f, 284f, 339, 355f, 379f, 501, 516f, 658, 660f, 675, 687, 694
state for fission **Vol. 9:** 375f, 382f, 387, 470, 663, 666
time **Vol. 5:** 24, 73, 85, 117, 280, 299, 300f, 309, 417, 421, 448, 452
transmutation, artificial **Vol. 2:** 344
transmutation of nucleus, *see* nuclear transmutations
"transuranic elements" **Vol. 9:** 52, 58, 64, 67, 559, 561f, 565, 634, 636
transuranium **Vol. 12:** 84
transverse mass, *see* relativistic mass
transverse temperature difference **Vol. 1:** 389
Traubenberg, H. Rausch von *see* Rausch von Traubenberg, H.
triatomic hydrogen molecule, *see* hydrogen molecule, triatomic
Trier, E. **Vol. 1:** 526, 527
Trinity College (Cambridge)
    Vol. 1: 521, 523, 541
    Vol. 12: 180, 188
    Hall of **Vol. 1:** 525
Trinity College (Oxford) **Vol. 12:** 285
tritium **Vol. 11:** 613, 623
Trowbridge, A. **Vol. 5:** *313*, 333
    Vol. 8: 663
Truman, H.S. **Vol. 11:** 46, 48, 69, 77, 156, 325, 329
Trumpy, B. **Vol. 9:** 42, 537, *653*–657
truth **Vol. 10:** 95, 96, 351, 364, 543, 548
truth, deep/trivial **Vol. 12:** 184, 185, 194, 195
Tube Alloys, *see* atomic bomb project (American–British)
Tube Alloys Consultative Council
    Vol. 11: 17n, 25
tuberculosis screening **Vol. 12:** 327, 330
Tübingen **Vol. 5:** 233, 434, 437
tunnel effect (*see also* barrier penetration)
    Vol. 6: 230–231, 246, 405
Turck, C.J. **Vol. 10:** 590
Turin, J.J. **Vol. 5:** 95
Turner, L.A. **Vol. 8:** 325
    Vol. 9: 71, 385
    Vol. 11: 529
Tuve, M.A. **Vol. 9:** 31, 56, 60f, 64, 256, 350, 353, 383, 550ff
Tuxen, P. **Vol. 11:** 406p

Udden, A.D. **Vol. 2:** 281, 320
    Vol. 3: 21, 241
    Vol. 4: 20, 261, 695
    Vol. 8: *683*, 786
Uehling, E.A. **Vol. 7:** 485
Uexküll, J.J. von **Vol. 10:** 4n
Uhlenbeck, G.E. **Vol. 1:** xxxvi
    Vol. 5: 28, 228ff, 233ff, 237f, 283, 285, 287, 289, 329–333, 339, 371, 373, 406, 461, 463, 484f, 497f, 507f
    Vol. 6: 3, 37, 52, 136, 158, 412, 415, 457
    Vol. 8: 206, 737f
    Vol. 9: 304, 319, 597f
    Vol. 10: 339, 418
Ulfbeck, O. **Vol. 7:** 236n
ultimate subject (*see also* subject/object)
    Vol. 10: xxxix, 95, 160, 185, 286, 292
Umezawa, H. **Vol. 7:** 499
uncertainty **Vol. 8:** 229f, 253, 290ff, 294f, 297, 300, 306, 338, 366, 368, 411f, 438, 441f, 499, 506, 629f, 633ff, 805, 811
of (angular) momentum **Vol. 5:** 475f, 479, 482
of frequency **Vol. 5:** 111
principle/relations (*see also* indeterminacy principle/relations)
    Vol. 5: 61, 94, 95, 215
    Vol. 6: —
    Vol. 7: —
    Vol. 9: 4f, 9, 15, 87, 104f, 111, 139, 305, 320, 518, 523, 548, 618
    Vol. 10: 340, 451, 519, 522
UNESCO **Vol. 11:** 721n
    Vol. 12: 118, 371, 494, 495
uniformization variables **Vol. 3:** 522, 551
Union Académique Internationale
    Vol. 12: 318, 323
United Nations **Vol. 11:** 18, 19, 21n, 24, 28, 47, 65, 66, 68, 70, 78, 79, 81–83, 88, 92, 95, 100, 107, 156, 158, 161, 164, 166, 175, 178, 181, 182, 185, 188, 189, 192, 233, 236n, 238, 239, 287, 288n, 299, 326n, 331, 360, 365, 369, 573, 585, 611, 617, 619, 630, 646, 650
Atomic Energy Commission
    Vol. 11: 65–68, 70, 72, 80, 161, 164, 182, 321, 325n, 326n
    Vol. 12: 116, 359
Charter of **Vol. 11:** 47, 180

International Atomic Energy Agency
  Vol. 11: 83
Moscow Conference, 1943 **Vol. 11:** 21*n*
San Francisco Conference, 1945
  Vol. 11: 47, 288
Secretary General, *see* Hammarskjöld, D.;
  Lie, T.H.
Security Council **Vol. 11:** 238*n*
unity
  of knowledge (*see also* knowledge)
    Vol. 10: xlix, 14, 48, 83, 90, 95, 96, 98,
    141, 143, 152, 157, 160, 287, 293, 559,
    560, 562, 563, 567, 589, 591, 593
  of personality **Vol. 10:** 8, 71, 123, 286, 292
  of science **Vol. 10:** xlix, 15, 16, 53, 64,
    142, 148, 314, 319
  of world picture **Vol. 10:** 85, 103, 158,
    241, 247, 277, 345–347, 350, 351, 358,
    360, 364, 572
unity/concord **Vol. 11:** 134, 137, 140, 144,
    346, 347, 530, 568
    Vol. 12: 3, 93, 95, 183, 193, 221, 243,
    246, 371, 374, 375, 378, 532, 533
Unity of Science movement **Vol. 10:** xli
Unity of Science, Second International
    Congress in Copenhagen, 1936
    Vol. 7: 375
    Vol. 10: 19–21, 36*p*, 39*n*, 538,
    541, 583
UNIVAC **Vol. 11:** 672, 711
universal constancy of angular momentum, *see*
    angular momentum, quantization of
universe, structure of **Vol. 12:** 79, 81, 371
University College, London **Vol. 9:** 19,
    157–158, 641, 643
University of
  Copenhagen, *see* Copenhagen, University of
  London **Vol. 8:** 231, 648, 809
  Manchester, *see* Manchester, University of
  Oslo (Kristiania) **Vol. 8:** 27*f*, 647,
    769–771, 779–782
uranium **Vol. 8:** 237, 239, 244, 313–318,
    320*f*, 325, 331, 333, 376, 385, 541,
    618, 727, 729, 740, 742, 745, 788,
    790, 822
    Vol. 9: 42, 52*f*, 56, 58, 60*ff*, 71*f*, 74*f*,
    285, 295*f*, 317, 329, 343*ff*, 349*ff*, 357,
    361, 365*ff*, 370, 372*ff*, 377*ff*, 382*ff*,
    388*f*, 397*f*, 401*f*, 404, 407*f*, 413,
    431*ff*, 437*ff*, 443, 457*ff*, 462*ff*, 469*f*,
    472*f*, 478, 484, 551*ff*, 558*ff*, 565*f*,

588*f*, 625*ff*, 631, 634, 636, 638, 640,
    658, 662
    Vol. 11: 15, 29, 91, 102, 103, 228, 229,
    359, 360, 363–365, 565, 569, 573,
    579–583, 585, 597–605, 614–616,
    624–629, 646, 649, 650
    Vol. 12: 60, 66, 84, 359
  rods **Vol. 11:** 581, 582, 602, 603
Urbain, G. **Vol. 2:** 429
    Vol. 3: 601
    Vol. 4: 22, 24, 30, 31, 32, 316, 338,
    339, 464, 482, 675, 676, 678, 696
    Vol. 12: 288
Urey, H.C. **Vol. 3:** 659, 660, 661
    Vol. 5: 86, 168, 171, 221, 502
    Vol. 10: 413
    Vol. 11: 531, 569
Urk, T. van **Vol. 4:** 587, 639
Urquhart **Vol. 6:** 195, 278
Ussing, H. **Vol. 10:** 488
Utrecht **Vol. 5:** 230, 240, 258, 264, 279, 329,
    337, 369, 371*ff*, 407, 462*f*, 498
    Vol. 8: 700*f*, 727, 729, 787
Utrecht, University of **Vol. 12:** 116, 348,
    351, 356

vacuum
  fluctuations **Vol. 7:** 4, 445, 482, 483*n*
  polarization **Vol. 7:** 175, 186, 193, 205,
    215, 216, 482, 484
valency **Vol. 2:** 117*f*, 207, 488, 528,
    530, 532
    Vol. 10: 121, 179
    Vol. 12: 45, 46
  electrons **Vol. 2:** 136, 146
Valentiner, H. **Vol. 5:** 137
Valerio, G. **Vol. 11:** 563
Valeton, J. **Vol. 5:** *402*
Van de Graaff generator **Vol. 11:** 568
vaporization/condensation **Vol. 12:** 59, 63, 64
Varder, R.W. **Vol. 2:** 81, 83, 598
    Vol. 8: 151, 153, 720, 725
variation in range and energy loss, *see*
    straggling
Varming, J. **Vol. 11:** 517, 525
Varming, K. **Vol. 11:** 517, 525
Vavilov, S.I. **Vol. 11:** 230
Veblen, O. **Vol. 7:** 279, 458, 459
vector potential **Vol. 7:** 94, 147
vector/tensor analysis **Vol. 11:** 670

[153]

Vegard, L.   **Vol. 4:** 200, 232, 244, 252, 273, 300, 301, 313, 323, 405, 409, 424, 457, 459, 479, 480, 487, 520
   Vol. 5: 251*f*
   Vol. 6: 388
   Vol. 8: 198, 200
   Vol. 12: 139*n*
velocity
   correction, *see* relativistic mass
   curve   **Vol. 2:** 73*ff*
   decrease, *see* absorption of charged particles
   potential   **Vol. 1:** 31, 38
   radial differences   **Vol. 1:** 46, 83, 86
velocity–range relation   **Vol. 8:** 6*f*, 143–148, 236*ff*, 240, 244, 246, 257, 320*f*, 323–333, 375*ff*, 405*f*, 408, 418, 426, 467, 469, 473, 531, 540, 542, 546–565, 668, 750, 752, 822
Venkateswarlu, P.   **Vol. 8:** 519
Verein Ernst Mach   **Vol. 10:** 15
Verlaine   **Vol. 9:** 641*ff*
Verschaffelt, J.E.   **Vol. 3:** 34, 36, 362, 667
   Vol. 6: 35, 36, 37, 100, 318, 478, 481
   Vol. 7: 32*p*
Vesalius, A.   **Vol. 10:** 67, 117
Vibæk, J.   **Vol. 12:** 399, 403
vibrations
   elastic, theory of   **Vol. 12:** 47
   harmonic, *see* harmonic vibrations
vibrator, harmonic, *see* harmonic oscillator model of atom; Planck's theory
Victoria University, *see* Manchester, University of
Vienna circle (Wiener Kreis)   **Vol. 10:** 14
Villars, F.   **Vol. 10:** 558
Vincent, G.E.   **Vol. 11:** 727
Vinding, P.   **Vol. 12:** 152, 162, 163, 175
virtual oscillator/radiation   **Vol. 5:** 7–8, 9, 11, 12, 13, 21–23, 24, 27*ff*, 42*ff*, 47, 52, 55–67, 85, 90, 93, 98, 102, 105–118, 145–147, 149–154, 256, 260, 263, 267, 279, 299, 311, 317, 344, 357, 358*f*, 387*f*, 395, 401, 402–403, 412, 417, 420, 490, 495
   Vol. 8: 31, 208, 330*f*, 396, 400, 498, 500*f*, 505, 511, 636
viscosity   **Vol. 2:** 423
   Vol. 8: 270, 279, 546, 666
   effect of   **Vol. 1:** 30–37, 45, 46
visualization, forms of   **Vol. 12:** 156, 168, 439, 441

vitalism   **Vol. 7:** 376
   Vol. 10: ix, xxviii, 4–6, 10, 15, 19, 20, 33, 45, 46, 61, 68, 70, 71, 77, 117, 130, 199, 279, 538–542, 544, 548
   Vol. 11: 134
   Vol. 12: 157, 170
Vitus Bering Committee   **Vol. 12:** 106, 107, 245*n*
Vitus Bering Medal   **Vol. 12:** 106, 242, 246
Vitus Bering Memorial Foundation
   Vol. 12: 241–243, 245, 246
Vleck, J.H. van   **Vol. 4:** 37, 595, 644, 688
   Vol. 5: 20, 43, 303, 495, 503
   Vol. 6: 318, 349
   Vol. 10: 446
Vlieger, J.   **Vol. 12:** 38*p*
Voigt, W.   **Vol. 1:** xxvii, 245, 276, 278, 356, 381, 383
   Vol. 2: 323, 363
   Vol. 3: 281, 421, 485
   Vol. 4: 367
volition (*see also* free will)   **Vol. 10:** xxv, 62, 71, 93, 94, 105, 143, 152, 160, 184, 185, 199, 200, 286, 291
Volkmann, P.   **Vol. 1:** 62, 64, 65
Volmer, T.   **Vol. 4:** 695
Volta Centenary Congress, *see* Como Lecture
Volta, A.   **Vol. 6:** 109, 113, 147, 148, 264, 287, 319, 374
   Vol. 7: 349
   Vol. 10: 54, 344, 358
   Vol. 11: 530
   Vol. 12: 371
Voorhis, C.C. van   **Vol. 9:** 382, 397, 404

Waddington, C.H.   **Vol. 10:** 553
Waerden, B.L. van der   **Vol. 5:** 7, 20–21, 43, 46*f*, 220, 223, 225, 229, 237, 298, 312, 507–508
   Vol. 8: 677
Wagner, E.   **Vol. 4:** 319
Walcott, C.D.   **Vol. 1:** 115
Wald, G.   **Vol. 7:** 411
Walker, M.   **Vol. 8:** 675
Waller, I.   **Vol. 5:** *240*, 407
   Vol. 6: 191, 304, 432
   Vol. 11: 7*p*
   Vol. 12: 17*p*
Walter, B.   **Vol. 5:** 39, 121, 321*f*, 325
Walter, E.J.   **Vol. 7:** 328

Walton, E.T.S.   **Vol. 7:** 32*p*
  Vol. 8: 652*f*
  Vol. 9: 134, 165, 173, 201, 216, 220
  Vol. 10: 409, 449
  Vol. 11: 568
  Vol. 12: 84
Wambacher, H.   **Vol. 9:** 425, 453, 586*f*
Wang, P.   **Vol. 9:** 357, 383*f*
Wannier, C.H.   **Vol. 8:** 727, 729
Warburg, E.   **Vol. 2:** 199, 320, 321, 323, 326,
  350, 540, 607–608, 616
  Vol. 3: 137, 426
  Vol. 10: 66, 72, 433
  Vol. 12: 121, 534
Warnant, E.   **Vol. 3:** 362
Warner, E.   **Vol. 11:** 310
Warren, R.P.   **Vol. 10:** 82
Warsaw Conference   **Vol. 7:** 261*p*
Warsaw Lecture, Bohr's (*see also* International
  Institute of Intellectual Cooperation)
  Vol. 7: 260, 266, 276, 377, 378
Wartenberg, H. von   **Vol. 2:** 238
Washington conference
  February 1937   **Vol. 9:** 30, 31, 227,
    264, 540
  January 1939   **Vol. 9:** 56, 59, 63, 64, 66,
    550*f*, 555, 564, 622, 624
  April 1939   **Vol. 9:** 68–69, 359–361,
    386, 663
Washington Declaration, 1945   **Vol. 11:** 65,
  66, 305, 325*n*
water molecule, structure of   **Vol. 2:** 138,
  139, 150, 152, 231–232
water surface
  age of   **Vol. 1:** 88
  tension in recently formed   **Vol. 1:** 81*ff*
Watson, E.M.   **Vol. 11:** 284
Watson, J.B.   **Vol. 10:** xli
Watson, J.D.   **Vol. 10:** 24, 168, 474, 475, 485,
  486, 551, 555
wave
  amplitudes, finite   **Vol. 1:** 37–44, 45
  equation   **Vol. 6:** 9, 13, 49, 70*f*, 96*f*, 126,
    128*f*, 154*f*, 172, 264, 287, 423,
    460, 465*f*
    Vol. 7: 244, 272, 332, 347, 495*n*; 208,
    210, 292, 337, 445, 629*f*
    Vol. 8: 208, 210, 292, 337, 445, 629*f*
    Vol. 9: 492, 614
  equation, relativistic, *see* Dirac equation

function (*see also* matter waves; proper
  vibration)   **Vol. 6:** 9, 10, 39, 71, 97,
    101*f*, 104*f*, 127, 140, 142, 154*ff*, 164,
    170, 179, 190, 229*ff*, 245*f*, 270, 295,
    349, 359, 363*ff*, 393*ff*, 398, 416, 431
    Vol. 7: 242, 334, 427–430; 36, 207,
    230, 271, 280, 289, 292, 303, 306,
    411*f*, 456*ff*, 510, 536, 629, 634, 678*f*,
    754, 811
    Vol. 8: 36, 207, 230, 271, 280, 289,
    292, 303, 306, 411*f*, 456*f*, 510, 536,
    629, 634, 678*f*, 754, 811
    Vol. 9: 5, 81, 146, 246, 494*ff*, 501, 509,
    513*ff*, 518*f*, 549, 649*f*, 658, 669
    Vol. 12: 22, 26, 39, 47, 499, 501
  group/packet   **Vol. 6:** 21*ff*, 29, 31, 33, 49,
    59, 61, 71, 76–79, 84–86, 93*f*, 97–98,
    102, 105, 117–119, 129, 132, 134–135,
    142, 149–150, 151, 155, 157, 172–174,
    176, 366, 393, 419, 421, 433, 465*f*
    Vol. 7: 310, 429
    Vol. 8: 271, 280, 291*ff*, 297, 306, 339*f*,
    343, 348, 411*ff*, 439, 510, 630
    Vol. 9: 45*f*, 87, 373, 524, 611
  mechanics   **Vol. 6:** 3*f*, 7*ff*, 14*f*, 39, 50, 52,
    55–56, 58, 96*f*, 126*f*, 130, 137*f*, 140,
    153–155, 169, 184, 230, 245, 265, 288,
    395, 405, 417, 422*ff*, 439*f*, 462*f*, 478
    Vol. 7: 348, 499
    Vol. 8: 30, 207, 210, 226, 228*ff*, 253*f*,
    272, 274, 276, 281, 283, 285, 289*ff*,
    296, 298*ff*, 305, 330, 343, 347*f*, 351,
    355, 372, 385, 388, 390, 437, 447, 450,
    452, 455*f*, 458*f*, 484, 567, 629*ff*, 637,
    663, 805, 811, 813
    Vol. 9: 5, 7, 11, 15, 87, 95, 181, 392,
    407*f*, 508, 510, 592, 649*f*
    Vol. 10: 402, 403, 409, 427, 441, 442,
    448, 449
    Vol. 12: 11, 30, 33, 47
  number   **Vol. 10:** 373–375, 377
  picture   **Vol. 11:** 660, 665, 671
    Vol. 12: 47, 61, 68, 69
  primary/secondary   **Vol. 8:** 271*f*, 280*f*,
    289, 292*f*, 295*ff*, 303, 337, 343, 345*f*,
    348, 350*f*, 379*f*, 395, 399, 445*f*, 449,
    456*f*, 536, 629*ff*, 638
  theory   **Vol. 12:** 10, 47
  wave/particle   **Vol. 10:** 17, 44, 103, 120, 166,
    178, 440, 518, 521

[155]

dualism/duality **Vol. 6:** 8, 9, 27, 59, 61, 63, 108
  Vol. 12: 11, 13, 47, 61, 68
wavelength, de Broglie **Vol. 8:** 253, 270*f*, 276*f*, 279*f*, 285*f*, 292, 298, 303*f*, 339, 343*f*, 348*f*, 379, 440, 446, 452, 485, 536, 630*f*, 634, 638
  Vol. 9: 154, 181, 187, 240, 244*f*, 247*ff*, 253, 255, 277, 311, 325, 376*ff*, 381, 388, 392, 429, 456, 491, 500, 509, 511, 517
wavelength, irregularities **Vol. 1:** 82
  variations **Vol. 1:** 86, 87
wavelengths, determination when different concentric parts have different velocities **Vol. 1:** 83*ff*
waves on deep water, finite amplitudes
  Vol. 1: 67–78
waves, effect of gravity and surface tension
  Vol. 1: 67–78
waves, non-existence of certain
  Vol. 1: 77, 394
Wayne, P. **Vol. 10:** 557*n*
Weber, S.T. Holst-, *see* Holst-Weber, S.T.
Weber, W. **Vol. 1:** 171, 172, 297, 413, 446, 534, 535
Weekes, D.F. **Vol. 9:** 248
Weierstrass, K. **Vol. 3:** 639
Weigert, F. **Vol. 5:** 145, 148
weighing experiment, Einstein's **Vol. 7:** 280, 319, 366, 369
Weis-Fogh, T. **Vol. 10:** 126, 592
Weisgal, M. **Vol. 11:** 375
  Vol. 12: 129, 130, 486*p*, 487, 488
Weiss magneton, *see* magneton
Weiss, P. **Vol. 2:** 128, 129, 233, 254, 256, 261, 262, 265, 626
  Vol. 3: 361, 363, 613
  Vol. 10: 434, 446
Weisskopf, V.F. **Vol. 5:** 23, 220, 238
  Vol. 6: 10, 451
  Vol. 7: 10*n*, 14, 15, 16*p*, 34*n*, 35–37, 220, 257*n*, 474*n*, 477, 480
  Vol. 8: 224, 714, 716, 803
  Vol. 9: 26, 27, 28, 30, 80, 209, 260, 262, 314, 326, 375, 383, 385, 499, 508, 510, 544
  Vol. 10: 418, 452, 489, 531, 532, 555, 556, 558
  Vol. 11: 17, 698*n*

Weizmann Institute
  Vol. 11: 375, 691–693, 696–699
  Vol. 12: 129, 130, 486*p*, 487, 488
Weizmann, C. **Vol. 10:** 397
  Vol. 11: 376, 690, 691, 691*p*, 692, 692*p*, 693, 695, 695*p*, 696, 696*p*, 697, 698, 698*p*
  Vol. 12: 129
Weizmann, V. **Vol. 11:** 697*n*
Weizsäcker, C.F. von **Vol. 6:** xx
  Vol. 7: 435, 445, 446, 474, 477, 513, 520
  Vol. 8: 224, 637, 652, 797*f*, 812
  Vol. 9: 10, 26, 28, 42, 70, 235, 239, 294, 317, 329, 369, 582*ff*, 603*f*
  Vol. 11: 29, 243, 244, 264, 265
  Vol. 12: 17*p*
Welsbach, C.A. von **Vol. 4:** 32
Welsh, E. **Vol. 11:** 29, 264, 265, 270
Welton, T. **Vol. 7:** 34
Wendt, G.L. **Vol. 2:** 344, 476, 485
Wentzel, G. **Vol. 3:** 510, 539
  Vol. 4: 326, 327, 409, 411, 492, 516, 525, 547, 606, 652
  Vol. 5: 3, 251*f*, 465*f*, 468*f*, 472, 474
  Vol. 6: 138, 439*f*
  Vol. 8: 207*f*, 210, 271, 280, 407, 663
  Vol. 9: 543
  Vol. 10: 455*p*
Wereide, T. **Vol. 2:** 619, 620
  Vol. 4: 25, 731
Wergeland, H. **Vol. 8:** 262, 452, 568
  Vol. 12: 38*p*
Werner, S. **Vol. 3:** 9, 182, 524, 553, 657, 658, 660, 662
  Vol. 4: 32, 378, 405, 591, 604, 607, 641, 650, 653
  Vol. 5: 87, *371*, 372, 418, 421, 426, 430
  Vol. 6: 304, 308
  Vol. 8: 457, 568, 687
  Vol. 9: 626, 628
  Vol. 11: 513, 521
  Vol. 12: 14, 118, 210, 215, 368, 386, 387, 518, 519
Wertenstein, L. **Vol. 7:** 302
  Vol. 8: 565, 568
Wertheimer, E. **Vol. 2:** 166
West, J. **Vol. 2:** 597
West, W. **Vol. 5:** 380
Westergaard, M. **Vol. 10:** 487, 488

Weyl, K.H.H. **Vol. 4:** 686
  Vol. 6: 439f, 455
  Vol. 10: 82
Weyssenhof, J. **Vol. 7:** 302
Wheeler, B.I. **Vol. 2:** 618, 619
Wheeler, J.A. **Vol. 1:** xliv
  Vol. 6: 159
  Vol. 7: 7n, 40, 42p, 45, 50, 56, 482n
  Vol. 8: 225, 232f, 250, 301–306, 647,
  727, 729, 790, 795, 797ff, 803f
  Vol. 9: 16–17, 20, 31, 39, 56, 67f,
  69–74, 81–83, 345, 359–389, 398,
  399–404, 413, 443, 469, 484, 538, 551,
  561f, 657–671, 679f
  Vol. 10: 413, 420
  Vol. 11: 12, 327n, 367, 564, 569, 639
Whiddington's law **Vol. 2:** 34, 35, 37, 105,
  125, 132, 137, 148, 211, 410, 629
  Vol. 8: 10f, 66f, 88, 104
Whiddington, R. **Vol. 2:** 18, 28, 34, 36, 37,
  211, 410, 629
  Vol. 6: 386
  Vol. 8: 10, 50, 60, 66, 68f, 76, 88,
  91f, 104, 819
  Vol. 10: 386, 387, 417
White, P. **Vol. 8:** 792
White, S. **Vol. 11:** 327, 328, 330, 331
Whittaker, E.T. **Vol. 3:** 454, 492, 639, 640,
  641, 642
wholeness (*see also* holism) **Vol. 10:** xxxiii,
  xlviii, 4–6, 8, 13, 14, 69, 71, 76, 87, 88,
  91, 96, 103, 104, 111, 112, 117, 120,
  121, 127, 129, 142, 149, 160, 183, 195,
  209, 278, 284, 285, 290, 291, 376, 405,
  428, 432, 544, 548, 559, 560
Wick, G.C. **Vol. 10:** 455p
  Vol. 11: 358p
wide-angle scattering, *see* Rutherford
  scattering; scattering, multiple
width
  of energy level, *see* level width
  of spectral lines (*see also* line width)
  Vol. 6: 13, 133, 138, 143, 156, 173,
  396, 460
Wiedemann-Franz law **Vol. 1:** 144, 339
  deviations from **Vol. 1:** 342
Wien, W. **Vol. 1:** 448
  Vol. 2: 331, 558, 561
  Vol. 3: 73, 691
  Vol. 4: 695
  Vol. 5: 506

  Vol. 6: 10
  Vol. 8: 19f, 209, 719, 722
Wiener, N. **Vol. 6:** 138
Wiesner, J.B. **Vol. 11:** 367, 721, 722
Wightman, A.S. **Vol. 8:** 590
  Vol. 9: 528
  Vol. 10: 455p
Wigner, E.P. (*see also* Breit–Wigner formula)
  Vol. 6: 29, 439f
  Vol. 7: 4, 260, 261p, 302
  Vol. 9: 29, 31, 32, 47, 189, 209, 312,
  325, 375, 377, 491, 507f, 510, 539
  Vol. 10: 403, 455p
Wiley and Sons **Vol. 12:** 15, 510
Wilke, J.K. **Vol. 10:** 347, 361
Wilkins, T.R. **Vol. 8:** 554, 568
Williams, E.J. **Vol. 1:** xxiv
  Vol. 5: 94, 216
  Vol. 8: vi, 15, 209, 212–214, 216,
  221–231, 238f, 249, 255, 274, 283,
  287–300, 330, 355, 371f, 380, 407,
  427, 438, 475, 479, 483ff, 500, 512,
  516, 519, 523, 568, 627–638, 648f,
  664, 678, 698, 709f, 737f, 753,
  791–813, 822
  Vol. 9: 13, 127, 578f
  Vol. 11: 7p
Willows, R.S. **Vol. 1:** 147
Wills Memorial Lecture, Bohr's, 1931
  Vol. 6: 317, 319, 361–370, 481
Willumsen, J.F. **Vol. 12:** 103
Wilson chamber, *see* cloud chamber
Wilson photographs, *see* cloud-chamber tracks
Wilson, A.H. **Vol. 8:** 679
Wilson, C.T.R. **Vol. 2:** 85, 96
  Vol. 5: 5, 81ff, 178, 186, 194, 200,
  336, 382
  Vol. 6: 36, 37, 415
  Vol. 8: 34, 155, 163, 170, 173, 177
  Vol. 10: 383, 384, 435
Wilson, H.A. **Vol. 1:** 102, 105, 174, 189,
  191, 247, 256, 263, 268, 271, 299, 311,
  312, 313, 358, 365, 371, 374, 377, 423,
  427, 474, 528, 529
Wilson, R. **Vol. 8:** 590
  Vol. 9: 528f
Wilson, T.G. **Vol. 8:** 634
Wilson, W. **Vol. 2:** 37, 38, 89, 451, 464, 465
  Vol. 3: 84, 344, 352, 430, 463
  Vol. 4: 196, 270, 567, 624
  Vol. 6: 52, 389

Vol. 8: 5, 69*f*, 76, 91, 159, 725
Vol. 10: 395, 419
Wilson, Woodrow **Vol. 8:** 673
Winge, Ø. **Vol. 11:** 347
Winter War **Vol. 12:** 506*n*
Winther, A. **Vol. 8:** 590
Vol. 9: 528
Winther, C. **Vol. 6:** xxii
Wisniewski, F.J. **Vol. 7:** 302
witchcraft **Vol. 11:** 124
withdrawn paper 1916 **Vol. 3:** 3, 4
Wleugel, P.J. **Vol. 10:** 347, 361
Wolfskehl Foundation **Vol. 4:** 23
Wolfson, I. **Vol. 11:** 698*n*
Wood, A.B. **Vol. 1:** 484
Wood, R.W. **Vol. 2:** 116, 124, 177, 178, 337, 339, 448, 564, 566, 568, 570, 604
Vol. 3: 237, 254, 255, 405, 591, 592
Vol. 4: 695, 712
Vol. 5: 39, 44, 57, 64, 114, 122, 145*ff*, 148, 151*ff*, 335, 398
Vol. 6: 80, 88
Vol. 8: 683
Vol. 10: 401
Woodhead, G.S. **Vol. 1:** 522, 523
Woodward, R.S. **Vol. 1:** 115
world
citizenship **Vol. 11:** 184
organization **Vol. 11:** 21, 234, 236
picture **Vol. 12:** 374, 375, 462, 464
security (*see also* security) **Vol. 11:** 25, 47, 73, 97, 99, 111, 113–115, 119, 124, 127, 176, 181, 236, 237, 288
Vol. 12: 278
view **Vol. 11:** 727
World Bank **Vol. 11:** 68
World Federation of Scientific Workers **Vol. 12:** 111, 277, 529
World War I **Vol. 10:** 228, 233, 396, 437
Vol. 12: 112, 126, 172*n*, 183, 191, 244*n*, 274, 318, 323, 458
World War II **Vol. 10:** 220, 415, 451
Vol. 12: 102, 126, 128, 189*n*, 244*n*, 245*n*, 294*n*, 459*n*, 469–472
Wouthuysen, S.A. **Vol. 12:** 38*p*
Wright, W.H. **Vol. 3:** 164
Wrinch, D. **Vol. 12:** 12, 54
Wu, C.S. **Vol. 10:** 454
Wyckoff, R.W.G. **Vol. 5:** *322*
Wynn-Williams, C.E. **Vol. 10:** 412

xenon **Vol. 11:** 580, 599
X-rays (*see also* spectra) **Vol. 6:** 77, 84, 224, 237, 359, 365, 375, 379, 386*ff*, 390, 394*f*
Vol. 7: 174, 180, 186, 190, 346
Vol. 8: 10*f*, 30, 34, 64, 66*f*, 82, 86, 88*f*, 97, 101, 104*f*, 143, 160, 169*ff*, 173*ff*, 177, 179, 235, 289, 672, 694, 723, 761, 780, 782, 785*f*, 797
Vol. 10: 13, 19, 22, 376, 387, 393, 434, 435, 438, 440, 471, 472
Vol. 11: 375, 565, 574, 587, 658, 663, 726
Vol. 12: 20, 43, 46, 47, 60, 65, 66, 252, 286–288, 348, 351, 503, 504
absorption of **Vol. 2:** 196
characteristic **Vol. 2:** 32, 34–35, 73, 90, 109, 125, 131–134, 137, 148, 162, 178–179, 190, 206–212, 236, 261, 303–306, 311–316, 335–336, 395, 398, 410–413, 419, 420, 424, 429, 469, 494*f*, 516*f*, 530, 544–547, 579, 586, 589, 590, 627*ff*
diffraction of **Vol. 2:** 212, 428*f*, 578, 580
photography **Vol. 12:** 327, 330
reflection of **Vol. 5:** 4, 38–40, 115, 119–124, 318, 319–326, 407, 496, 512
scattering of (*see also* Compton effect/recoil; scattering of X-rays) **Vol. 2:** 10, 189, 290, 500
Vol. 5: 3*ff*, 39, 42, 83, 105, 115*f*, 122*f*, 178, 190, 194, 204, 213, 215, 237, 274*ff*, 280, 323, 336, 382, 400–401
therapy **Vol. 12:** 503, 504
X-ray levels
abnormal **Vol. 4:** 33, 509, 540
and the theory of atomic structure **Vol. 4:** 505*ff*, 537*ff*
complexity of **Vol. 4:** 254, 324
dependence on atomic number **Vol. 4:** 442, 461, 473, 481, 510*ff*, 541*ff*
determination from experimental data **Vol. 4:** 492, 525
diagrams of **Vol. 4:** 325, 326, 417
normal **Vol. 4:** 33, 509, 540
tables of **Vol. 4:** 495*ff*, 527*ff*
theoretical estimation of **Vol. 4:** 500, 533
X-ray spectra (*see also* spectra, high frequency) **Vol. 3:** 199, 379
Vol. 4: 54*ff*, 102*ff*, 194, 195, 247*ff*, 267*ff*, 318*ff*, 335*ff*, 407*ff*, 460*ff*

Vol. 5: 3, 38*f*, 121–124, 244, 248, 251*f*, 272, 278, 282, 284, 288*f*, 336, 483*ff*, 501, 504, 506
absorption and emission of **Vol. 4:** 319*ff*
absorption edges **Vol. 4:** 249, 319, 409
and atomic constitution **Vol. 4:** 321*ff*, 462*ff*, 481*ff*
and periodic system **Vol. 4:** 32*ff*, 486–518, 519–548
and reorganization of atoms **Vol. 4:** 93, 255, 271, 318*ff*
classification of **Vol. 4:** 323*ff*, 490, 523
emission lines **Vol. 4:** 249, 409
relativity doublets **Vol. 4:** 507*ff*, 539*ff*
screening constants **Vol. 4:** 33, 502, 536
screening doublets **Vol. 4:** 507*ff*, 539*ff*
table of **Vol. 4:** 494, 526

Yale **Vol. 5:** 382, 486
Yale University **Vol. 11:** 564
Yamanouchi, T. **Vol. 9:** 76–77
Yang, C.N. **Vol. 10:** 454
Yost, F.L. **Vol. 9:** 384
Young, T. **Vol. 8:** 727, 729
    Vol. 11: 445
Yukawa, H. **Vol. 7:** 227
    Vol. 9: 586*f*
    Vol. 10: 452, 455*p*
    Vol. 11: 569
    Vol. 12: 12, 54

Zacharias, J. **Vol. 12:** 15
Zachariasen, W.H. **Vol. 8:** 759
Zagreb Academy **Vol. 12:** 516
Zahn, H. **Vol. 1:** 284, 389
Zeeman effect **Vol. 1:** xxxvi, 569
    Vol. 2: 123, 124, 130, 255, 257–258, 264, 319–326, 345, 350, 363, 366–367, 368, 378, 398, 445–446, 469–470, 511, 526, 554*f*, 589*ff*, 603, 608, 627, 632
    Vol. 3: —
    Vol. 5: 35, 50, 80, 87, 130, 137, 145, 147, 148, 152*f*, 232, 234, 237*f*, 247, 251*f*, 255, 261, 274, 277*f*, 280, 282, 284, 288*f*, 339, 354*ff*, 360, 362, 391, 393, 406, 411, 423–424, 425, 427–428, 434, 437, 447*f*, 451*f*, 454, 456, 467*ff*, 473, 475*f*, 484, 486, 493, 498, 503
    Vol. 6: 136, 158, 311*f*, 359, 391
    Vol. 7: 346
    Vol. 8: 169, 172, 177, 673, 679, 687

Vol. 11: 566
    Vol. 12: 45, 340, 341, 368
anomalous **Vol. 3:** 42, 183, 198, 280, 527*ff*, 554*ff*
    Vol. 4: 391, 573*ff*, 629*ff*, 647, 661
    Vol. 10: 438
Lorentz' explanation of **Vol. 3:** 231
normal **Vol. 3:** 5, 50, 197
    Vol. 10: 337, 339, 340, 375, 394, 395, 438
of hydrogen **Vol. 3:** 7, 8, 62, 145–164, 197, 264*ff*
Zeeman, P. **Vol. 1:** xxxvi
    Vol. 2: 325, 350, 362, 368
    Vol. 3: 32, 202, 231, 361, 393, 394, 419, 420, 454, 631
    Vol. 4: 34, 44, 214, 216, 434, 469, 470, 658, 702
    Vol. 5: 130, 137, 227, 274
    Vol. 10: 329, 336*p*, 337, 340, 394
    Vol. 12: 42, 100, 118, 356
Zeise, W. **Vol. 10:** 352, 365
Zeleny, J. **Vol. 5:** *486*
zero-point
    energy/motion **Vol. 2:** 338, 341, 456, 459–460, 565, 567
    Vol. 6: 183, 365
    Vol. 7: 30, 74, 133, 201, 222, 226
    Vol. 9: 34, 74, 316, 329, 342, 372–373, 374, 376, 583*f*, 666
    Vol. 12: 54
    field **Vol. 7:** 20, 24, 201
    fluctuations/vibrations **Vol. 7:** 4, 22, 30, 43, 204, 468
Zeuthen, H.G. **Vol. 1:** 7, 15, 95, 504, 505
    Vol. 10: 510, 511
    Vol. 12: 318, 323
Zimmer, K.G. **Vol. 10:** 22, 168, 471, 472
zinc **Vol. 11:** 576, 591
    spectrum of **Vol. 2:** 409
Zinchenko, C.E. **Vol. 11:** 104, 238
Zinn, W.H. **Vol. 9:** 384, 386
zirconium **Vol. 11:** 580, 599
Zlobicki **Vol. 1:** 62, 65
Zuntz, N. **Vol. 1:** 534, 535
Zurek, W.H. **Vol. 6:** 159
    Vol. 7: 7*n*, 56
Zurich **Vol. 6:** 326, 438, 439*f*, 442, 444*ff*, 449*f*, 455, 459, 462, 468, 472
"Zwang" (*see also* dualism/duality)
    Vol. 5: 61, 228, 251*f*, 255, 262, 278, 289, 360*f*, 446*ff*, 451*f*, 454*ff*

[159]

letter from Kramers (for Bohr) to
  Landé, A.
    (28 Feb 21) **Vol. 4:** 723
letter to H.A. Kramers from Ehrenfest, P.
  (25 Mar 1921) **Vol. 3:** 615
letter to Mrs. Niels Bohr from
  Ehrenfest, P.
    (Nov. 22) **Vol. 4:** 683
letters from and to Niels Bohr
    Vol. 5: 294–298,
    Vol. 6: 412–414,
    Vol. 7: 434, 435,
    Vol. 8: 642–648,
    Vol. 9: 534–538,
    Vol. 10: 460–462
letters from Niels Bohr to
  Adam, E.P.
    (12 June 1920) **Vol. 1:** 118
  Barnes, E.L.
    (8 May 1912) **Vol. 1:** 107, 546, 547
  Bohr, Ellen
    (2 Oct 1911) **Vol. 1:** 522, 523
    (31 Oct 1911) **Vol. 1:** 106, 532, 533
    (6 Dec 1911) **Vol. 1:** 534, 535
    (28 Jan 1912) **Vol. 1:** 106, 538, 539
    (5 Feb 1912) **Vol. 1:** 540, 541
  Bohr, Harald
    (12 Mar 1909) **Vol. 1:** 496, 497
    (17 Mar 1909) **Vol. 1:** 496, 498
    (27 Mar 1909) **Vol. 1:** 498, 499
    (20 Apr 1909) **Vol. 1:** 500, 501
    (26 Apr 1909) **Vol. 1:** 94, 502, 503
    (4 May 1909) **Vol. 1:** 502, 503
    (15 May 1909) **Vol. 1:** 504, 505
    (9 June 1909) **Vol. 1:** 504, 505
    (1 July 1909) **Vol. 1:** 94, 506, 507
    (4 July 1909) **Vol. 1:** 506, 507
    (8 Sep 1909) **Vol. 1:** 508, 509
    (12 Dec 1909) **Vol. 1:** 510, 511
    (25 June 1910) **Vol. 1:** 95, 510, 511
    (26 June 1910) **Vol. 1:** 95, 510, 511
    (5 July 1910) **Vol. 1:** 96, 512, 513
    (28 July 1910) **Vol. 1:** 96, 514, 515
    (24 Nov 1910) **Vol. 1:** 96, 516, 517
    (2 Jan 1911) **Vol. 1:** 96, 516, 517
    (22 Apr 1911) **Vol. 1:** 518, 519
    (29 Sep 1911) **Vol. 1:** 104, 518, 519
    (23 Oct 1911) **Vol. 1:** 104, 526, 527
    (28 Jan 1912) **Vol. 1:** 538, 539
    (29 Jan 1912) **Vol. 1:** 538, 539
    (5 Feb 1912) **Vol. 1:** 540, 541

(9 Feb 1912) **Vol. 1:** 542, 543
(19 May 1912) **Vol. 1:** 107, 544, 545
(27 May 1912) **Vol. 1:** 107, 548, 549
(28 May 1912) **Vol. 1:** 108, 552, 553
(12 June 1912) **Vol. 1:** 109, 554, 555
(19 June 1912) **Vol. 1:** 109, 558, 559
(17 July 1912) **Vol. 1:** 560, 561
(22 July 1912) **Vol. 1:** 560, 561
(23 Dec 1912) **Vol. 1:** 562, 563
(30 July 1913) **Vol. 1:** 562, 563
(3 Aug 1913) **Vol. 1:** 564, 565
(13 Dec 1913) **Vol. 1:** 566, 567
(20 Apr 1914) **Vol. 1:** 568, 569
(1 Nov 1914) **Vol. 1:** 568, 569
(2 Mar 1915) **Vol. 1:** 114, 570, 571
(15 Apr 1915) **Vol. 1:** 576, 577
(29 July 1915) **Vol. 1:** 578, 579
(10 Oct 1915) **Vol. 1:** 578, 579
(3 Mar 1916) **Vol. 1:** 585
Debye, P.
  (undated) **Vol. 1:** 402
Francis, W.
  (1 Mar 1912) **Vol. 1:** 112
  (27 Mar 1912) **Vol. 1:** 112
Hoffmann, F.
  (undated) **Vol. 1:** 120
Hume-Rothery, W.
  (28 Feb 1928) **Vol. 1:** 122
Larmor, J.
  (6 Feb 1909) **Vol. 1:** 9
  (4 Apr 1910) **Vol. 1:** 10
Lederer, E.L.
  (7 June 1911) **Vol. 1:** 399
Livens, G.H.
  (undated) **Vol. 1:** 474
  (11 Feb 1914) **Vol. 1:** 478
McLaren, S.B.
  (17 Dec 1911) **Vol. 1:** 432
Oseen, C.W.
  (6 Sep 1911) **Vol. 1:** 103, 126
  (1 Dec 1911) **Vol. 1:** 422
  (4 Dec 1911) **Vol. 1:** 426
  (3 Mar 1914) **Vol. 1:** 114, 127, 128
Reinganum, M.
  (9 Nov 1911) **Vol. 1:** 404
  (11 Dec 1911) **Vol. 1:** 406
Richardson, O.W.
  (12 Feb 1914) **Vol. 1:** 114
  (9 Apr 1914) **Vol. 1:** 115
  (29 Sep 1915) **Vol. 1:** 116, 117, 482

(16 Oct 1915) **Vol. 1:** 117, 489
(12 June 1920) **Vol. 1:** 118
Christiansen, C.
(1 June 15) **Vol. 2:** 493
Fokker, A.D.
(14 Feb 16) **Vol. 2:** 499
Fowler, A.
(15 Apr 14) **Vol. 2:** 504
(28 Apr 14) **Vol. 2:** 506
(23 Feb 15) **Vol. 2:** 508
Hansen, H.M.
(12 May 15) **Vol. 2:** 515
Henriques, V.
(16 Jan 16) **Vol. 2:** 522
Herzfeld, K.F.
(2 Mar 14) **Vol. 2:** 526
Hevesy, G.v.
(7 Feb 13) **Vol. 2:** 529
(13 Oct 13) **Vol. 2:** 533
Langmuir, I.
(17 Dec 13) **Vol. 2:** 539
Moseley, H.G.J.
(21 Nov 13) **Vol. 2:** 546
Oseen, C.W.
(5 Feb 13) **Vol. 2:** 551
(3 Mar 14) **Vol. 2:** 553
(10 Mar 14) **Vol. 2:** 556
(28 Sep 14) **Vol. 2:** 557
(20 Dec 15) **Vol. 2:** 564
(17 Mar 16) **Vol. 2:** 571
(28 Feb[17]) **Vol. 2:** 573
Rutherford, E.
(18 Jan 12) **Vol. 2:** 576
(6 July 12) **Vol. 2:** 577
(4 Nov 12) **Vol. 2:** 577
(31 Jan 13) **Vol. 2:** 579
(6 Mar 13) **Vol. 2:** 581
(21 Mar 13) **Vol. 2:** 584
(26 Mar 13) **Vol. 2:** 585
(10 June 13) **Vol. 2:** 586
(22 June 13) **Vol. 2:** 587
(16 Oct 13) **Vol. 2:** 587
(21 Dec 13) **Vol. 2:** 590
(31 Dec 13) **Vol. 2:** 591
(10 Mar 14) **Vol. 2:** 591
(21 Mar 14) **Vol. 2:** 592
(20 May 14) **Vol. 2:** 592
(19 June 14) **Vol. 2:** 594
(29 Nov 16) **Vol. 2:** 595
Schwarzschild, K.
(23 Feb 14) **Vol. 2:** 600

Sommerfeld, A.
(19 Mar 16) **Vol. 2:** 603
Stark, J.
(14 Dec 13) **Vol. 2:** 606
Warburg, E.
(8 Jan 14) **Vol. 2:** 608
Weber, S.
(31 May 17) **Vol. 2:** 610
Ehrenfest, P.
(May 1918) **Vol. 3:** 11, 12
(22 Nov 1920) **Vol. 3:** 611, 612
(20 Jan 1921) **Vol. 3:** 29, 30
(23 Mar 1921) **Vol. 3:** 614
(28 Mar 1921) **Vol. 3:** 616, 617, 618, 619, 620
(12 Apr 1921) **Vol. 3:** 620
(10 July 1921) **Vol. 3:** 621, 622, 623
(July 1921) **Vol. 3:** 624, 625
(1 Sep 1921) **Vol. 3:** 626
(16 Sep 1921) **Vol. 3:** 626, 627
(10 Oct 1921) **Vol. 3:** 630
(19 May 1922) **Vol. 3:** 630, 631, 632
Einstein, A.
(24 June 1920) **Vol. 3:** 634, 635
Franck, J.
(18 Oct 1920) **Vol. 3:** 644, 645
Haas, A.
(11 Apr 1922) **Vol. 3:** 647
Heisenberg, W.
(31 Jan 1924) **Vol. 3:** 649, 650
(18 June 1924) **Vol. 3:** 650
Kramers, H.A.
(15 July 1922) **Vol. 3:** 656, 657, 658, 659
Landé, A.
(9 May 1922) **Vol. 3:** 664
Lorentz, H.A.
(13 June 1920) **Vol. 3:** 666
(29 May 1921) **Vol. 3:** 667
(1 Sep 1921) **Vol. 3:** 668
Oseen, C.W.
(28 Feb 1917) **Vol. 3:** 670, 671
Richardson, O.W.
(15 Aug 1918) **Vol. 3:** 14, 15
Rubinowicz, A.
(22 Aug 1921) **Vol. 3:** 680
Rutherford, E.
(27 Dec 1917) **Vol. 3:** 682, 683, 684
Sommerfeld, A.
(27 July 1919) **Vol. 3:** 688, 689, 690
(30 Apr 1920) **Vol. 3:** 691, 692

letters to Niels Bohr from
  Adam, E.P.
    (28 Apr 1920) **Vol. 1:** 118
    (8 July 1920) **Vol. 1:** 119
  Baedeker, K.
    (6 May 1911) **Vol. 1:** 398
  Barnes, E.L.
    (7 May 1912) **Vol. 1:** 106, 546, 547
  Bjerrum, Niels
    (27 Apr 1911) **Vol. 1:** 101, 125
  Bohr, Harald
    (23 May 1912) **Vol. 1:** 548, 549
    (fall 1913) **Vol. 1:** 566, 567
    (8 May 1916) **Vol. 1:** 584
  Debye, P.
    (30 May 1911) **Vol. 1:** 400
  Francis, W.
    (26 Feb 1912) **Vol. 1:** 111
  Høffding, Harald
    (25 Jan 1907) **Vol. 1:** 8, 16
  Hoffmann, F.
    (14 Apr 1921) **Vol. 1:** 120
  Hume-Rothery, W.
    (19 Feb 1928) **Vol. 1:** 121
  Larmor, J.
    (28 Jan 1909) **Vol. 1:** 8
    (16 Oct 1911) **Vol. 1:** 104
  Lederer, E.L.
    (26 May 1911) **Vol. 1:** 399
  Livens, G.H.
    (11 Jan 1914) **Vol. 1:** 476
    (8 Feb 1914) **Vol. 1:** 477
    (14 Feb 1915) **Vol. 1:** 479
  Oseen, C.W.
    (5 July 1911) **Vol. 1:** 101, 125
  Owen, E.A.
    (1 July 1912) **Vol. 1:** 110
    (23 Feb 1913) **Vol. 1:** 110
    (28 Nov 1913) **Vol. 1:** 110
    (11 Nov 1914) **Vol. 1:** 110
  Reinganum, M.
    (3 Nov 1911) **Vol. 1:** 403
    (22 Nov 1911) **Vol. 1:** 405
  Richardson, O.W.
    (4 Apr 1914) **Vol. 1:** 115
    (23 Sep 1915) **Vol. 1:** 116
    (9 Oct 1915) **Vol. 1:** 117, 484
    (19 June 1920) **Vol. 1:** 119
  Thomsen, Julius
    (23 Feb 1907) **Vol. 1:** 7, 15

  Zeuthen, H.G.
    (23 Feb 1907) **Vol. 1:** 7, 15
  Christiansen, C.
    (12 May 16) **Vol. 2:** 496
  Fowler, A.
    (27 Oct 13) **Vol. 2:** 503
    (20 Apr 14) **Vol. 2:** 505
    (6 Mar 15) **Vol. 2:** 508
  Garbasso, A.
    (19 Jan 14) **Vol. 2:** 511
  Haas, A.E.
    (6 Jan 14) **Vol. 2:** 513
  Henriques, V.
    (23 Dec 15) **Vol. 2:** 520
  Hevesy, G.v.
    (15 Jan/29 Jan 13) **Vol. 2:** 528
    (6 Aug 13) **Vol. 2:** 531
    (23 Sep 13) **Vol. 2:** 532
  Knudsen, M.
    (12 Apr 16) **Vol. 2:** 535
  Kramers, H.A.
    (25 Aug 16) **Vol. 2:** 537
  Langmuir, I.
    (2 Dec 13) **Vol. 2:** 539
  Miers, H.A.
    (7 June 16) **Vol. 2:** 542
  Moseley, H.G.J.
    (16 Nov 13) **Vol. 2:** 544
  Nagaoka, H.
    (27 Dec 13) **Vol. 2:** 549
  Oseen, C.W.
    (11 Nov 13) **Vol. 2:** 552
    (3 Mar 16) **Vol. 2:** 568
  Rutherford, E.
    (27 Jan 12) **Vol. 2:** 576
    (11 Nov 12) **Vol. 2:** 578
    (24 Feb 13) **Vol. 2:** 580
    (20 Mar 13) **Vol. 2:** 583
    (25 Mar 13) **Vol. 2:** 585
    (11 Dec 13) **Vol. 2:** 589
    (20 May 14) **Vol. 2:** 593
    (13 Dec 16) **Vol. 2:** 596
  Sommerfeld, A.
    (4 Sep 13) **Vol. 2:** 603
  Debye, P.
    (6 June 1918) **Vol. 3:** 607
  Ehrenfest, P.
    (14 May 1920) **Vol. 3:** 609
    (19 Aug 1920) **Vol. 3:** 609, 610
    (17 Oct 1920) **Vol. 3:** 610
    (27 Dec 1920) **Vol. 3:** 612, 613

(25 Mar 1921) **Vol. 3:** 615, 616
(26 Apr 1921) **Vol. 3:** 621
(17 July 1921) **Vol. 3:** 623, 624
(7 Aug 1921) **Vol. 3:** 625, 626
(27 Sep 1921) **Vol. 3:** 627, 628, 629
Einstein, A.
(2 May 1920) **Vol. 3:** 634
Epstein, P.S.
(14 May 1918) **Vol. 3:** 637, 638, 639, 640
(2 Oct 1918) **Vol. 3:** 641
(24 Oct 1918) **Vol. 3:** 641, 642
Kramers, H.A.
(25 Aug 1916) **Vol. 3:** 652
(12 Mar 1917) **Vol. 3:** 652, 653, 654, 655
(New Year 1917/18) **Vol. 3:** 655, 656
(11 Oct 1923) **Vol. 3:** 659, 660, 661, 662
Landé, A.
(26 Oct 1920) **Vol. 3:** 664
Lorentz, H.A.
(20 May 1921) **Vol. 3:** 666, 667
(17 July 1921) **Vol. 3:** 667, 668
Pauli, W.
(16 July 1923) **Vol. 3:** 673, 674
(Aug–Sep 1923) **Vol. 3:** 674, 675
Planck, M.
(7 May 1920) **Vol. 3:** 677
Sommerfeld, A.
(18 May 1918) **Vol. 3:** 686, 687
(5 Feb 1919) **Vol. 3:** 687, 688
(11 Nov 1920) **Vol. 3:** 690, 691
Born, M.
(4 Mar 23) **Vol. 4:** 669, 670
(17 Apr 23) **Vol. 4:** 672

Coster, D.
(16 July 22) **Vol. 4:** 677
(7 Dec 24) **Vol. 4:** 679, 680
Einstein, A.
(11 Jan 23) **Vol. 4:** 686
Epstein, P.S.
(23 Dec 22) **Vol. 4:** 688, 689
Fajans, K.
(25 June 21) **Vol. 4:** 691
Franck, J.
(29 July 22) **Vol. 4:** 696, 697, 698
(23 Dec 22) **Vol. 4:** 698, 699, 700
Konen, H.
(16 Dec 22) **Vol. 4:** 702
Kossel, W.
(15 Aug 21) **Vol. 4:** 708
Ladenburg, R.
(18 June 20) **Vol. 4:** 710
(27 Aug 20) **Vol. 4:** 712, 713
(12 Nov 22) **Vol. 4:** 714, 715
Landé, A.
(25 July 20) **Vol. 4:** 720
(21 Feb 21) **Vol. 4:** 722
Lorentz, H.A.
(10 Dec 22) **Vol. 4:** 29
Rubinowicz, A.
(18 Apr 21) **Vol. 4:** 733, 734
(16 Nov 22) **Vol. 4:** 735
Schrödinger, E.
(7 Feb 21) **Vol. 4:** 737
Sommerfeld, A.
(7 Mar 21) **Vol. 4:** 740
(25 Apr 21) **Vol. 4:** 740, 741
Nature, letters to
of March 1921 **Vol. 4:** 10ff., 72–82
of October 1921 **Vol. 4:** 19ff., 177–180

Printed and bound by CPI Group (UK) Ltd, Croydon, CR0 4YY
03/10/2024
01040333-0015